DEVELOPMENT CENTRE STUDIES

APPROPRIATE TECHNOLOGY

PROBLEMS AND PROMISES

Edited by

Nicolas Jéquier

DEVELOPMENT CENTRE
OF THE ORGANISATION
FOR ECONOMIC CO-OPERATION AND DEVELOPMENT

The Organisation for Economic Co-operation and Development (OECD) was set up under a Convention signed in Paris on 14th December, 1960, which provides that the OECD shall promote policies designed:

— to achieve the highest sustainable economic growth and employment and a rising standard of living in Member countries, while maintaining financial stability, and thus to contribute to the development of the world economy;
— to contribute to sound economic expansion in Member as well as non-member countries in the process of economic development;
— to contribute to the expansion of world trade on a multilateral, non-discriminatory basis in accordance with international obligations.

The Members of OECD are Australia, Austria, Belgium, Canada, Denmark, Finland, France, the Federal Republic of Germany, Greece, Iceland, Ireland, Italy, Japan, Luxembourg, the Netherlands, New Zealand, Norway, Portugal, Spain, Sweden, Switzerland, Turkey, the United Kingdom and the United States.

The Development Centre of the Organisation for Economic Co-operation and Development was established by decision of the OECD Council on 23rd October 1962.

The purpose of the Centre is to bring together the knowledge and experience available in Member countries of both economic development and the formulation and execution of general policies of economic aid; to adapt such knowledge and experience to the actual needs of countries or regions in the process of development and to put the results at the disposal of the countries by appropriate means.

The Centre has a special and autonomous position within the OECD which enables it to enjoy scientific independence in the execution of its task. Nevertheless, the Centre can draw upon the experience and knowledge available in the OECD in the development field.

***

The opinions expressed and arguments employed in this publication are the responsibility of the authors and do not necessarily represent those of the OECD.

TABLE OF CONTENTS

PREFACE

by

Paul-Marc Henry

President of the OECD Development Centre

Nicolas Jequier has approached the problem of appropriate technology with a much needed thoroughness and clarity. In 1975, at long last, appropriate technology has come of age. The need is now felt by many developing countries for a basic reappraisal of the type of technology which responds best to their requirements. High cost technology can be highly productive, it often provides the solutions to a basic problem of development of natural resources taking into account market and time constraints.

But it is in fact the direct product of advanced industrial societies which had to operate in a competitive world in time of war and in time of peace. From the point of view of defence as well as of economic survival, the most industrialised countries in the world could not afford to remain behind in the race for power and survival. This means that the developing countries of today are faced with a well integrated and complex system of production and trade which is essentially related to structures belonging to the temperate zone and representing a certain combination of accumulated knowledge, innovative capacity and skilled manpower.

It has been taken for granted, until very recently, that this formidable system could be extended to the whole planet and serve the needs of the billions of people now living in the tropical and equatorial zones. On the face of evidence, it is now clear that this industrial system has to be re-examined and adapted in order to take into account the basically low income of the developing countries, their fast growing population, their vast needs for essential productions such as food, shelter and health. This should be done within a relatively short period of time, if we want to avoid a fundamental imbalance between human populations and their capacity to survive through development of their natural resources.

Mr. Jéquier points out that "the appropriate technology movement (the term movement is used here for want of another more adequate description) can probably be viewed as a cultural revolution".

I agree with him. Not only it means a revolution for the people directly concerned in the developing countries which have to adapt "the imported technology under very unequal terms of negotiations", but even more for the people of the more advanced countries who feel that their own quality of life is in danger by what many call "aggressive" technology, aggressive by its scale, by its demand on human energy and by the stress it imposes on the whole fabric of society. At the same time, this technology is serving well the hunger for consumer goods, characteristic of an over-consuming society.

This is why this cultural revolution concerns the world as a whole. The problem of adaptation of technology is not only a technical problem, it is also a political one. The present debate goes in depth and has a direct bearing on the future of economic, political and cultural relations between the most industrialised countries of the Northern temperate zone and the less industrialised in the tropical and equatorial zones.

Mr. Jéquier's presentation should be placed in the perspective of the long term work undertaken by the Development Centre of OECD when the first Seminar on Transfer of Technology was held in Paris in November 1972 (Choice and Adaptation of Technology in Developing Countries). One can see from the galaxy of contributors to the present publication that many distinguished economists, technologists and industrialists and development administrators from many countries are already engaged in the new quest.

I am happy that the Development Centre could help this important endeavour.

INTRODUCTION

by

Mikoto Usui
Head of the Technology and Industrialisation Programme
OECD Development Centre

Low-cost technology, intermediate technology, self-help technology, progressive technology, correct technology, appropriate technology - differences in their nuances apart, much has already been said about these concepts and the kind of principles which are generally understood to underly them. The movement to keep these principles alive in various parts of the world and in countries at different levels of development has been spreading rapidly, if still in a fragmentary way. An increasing number of fora on development issues have come to dwell on the question of "progress from within society" and its technological implications. More and more journals and bulletins have come to spare a space for practitioners in appropriate technology, their new missions and visions.

In face of this emerging trend in the development scenery, there is still a certain degree of ambivalence as to the nature of policy actions which are required at this stage on the part of both the national authorities of developing countries and the bilateral and multilateral aid-giving agencies. Indeed, there is a growing number of analytical studies conducted from the point of view of choice of technology - an important phase of the innovation process: they invite our attention to the existence of a hierarchy of alternative technologies, either on the shelf or as agenda for adaptive development possibilities. However, still relatively poorly explored are the ways and means of fostering innovative minds throughout local societies, and especially in the context of rural development.

Orientation of this book

This book, Appropriate Technology: Problems and Promises, is an attempt to improve our understanding of the structural problems facing the appropriate technology movement from the standpoint of a national innovation policy. The phrase, Problems and Promises, hints at the general scope of the writings collected in this book. It covers the vast grey area which lies between a) the generalities debated often with an ideological trait, which point to certain difficult "problems" in adopting definitive, extensive policy actions, on the one hand, and b) the anecdotal specifics of micro-level field missions, which corroborate "promises" and crave for more determined policy supports, on the other.

The reader will see that the state-of-the-arts review presented here refers to the various forms of technical innovation projects and institutions linked for the most part to the people in rural areas. The main thrust of this exercise is not an inventory-taking of on-going relevant activities in the world, nor an appraisal of the internal technical problems of individual innovation projects. Rather, this is an attempt to reflect upon the problématique of the underlying socio-economic mechanisms and policy environment which looms up against the spread of practitioners' actual preoccupations.

Origin of this book

Procedurally, the origin of this book was the meeting of practitioners in low-cost technology organised on 17th-20th September 1974 by the OECD Development Centre. The interim report of this meeting, issued earlier, included only partial proceedings and a preliminary synthesis of discussions(1). The Centre has since continued its communications with various circles of practitioners and researchers in order to sharpen and broaden the synthesis section of the earlier report. Many participants have kindly provided us with their contributions anew, rewriting their earlier papers and taking in account the new questions raised at the meeting. This book has thus been evolved on the whole as a joint project with these many collaborators - a strategy which one would consider the most reasonable in view of the very participatory character of appropriate technology.

Being a collaborative project need not imply forcing heterogeneous particularities into common logics and rhetorics. In Part Two - The Practitioners' Point of View - the treatment of different themes, experiences and propositions are ascribed to individual writers. Not all of the articles in this part may be considered as policy-oriented, but they altogether constitute a set of important underpinnings on which our overall policy discussion is to be structured.

---

1) Low-Cost Technology - An Inquiry into Outstanding Policy Issues (English only), Technology and Industrialisation Programme, OECD Development Centre, CD/TI(75)1, January 1975.

Since these elements of Part Two have been gathered on an ad hoc, voluntary basis, they still leave a substantial degree of freedom as to ways in which the various threads of thought on "problems and promises" could be fitted together in a coherent perspective of policy study. This latter task is pursued in Part One - The Major Policy Issues. In a way, this part serves as a synthesis of the discussion held at the September 1974 meeting, as well as a review of the Part Two material. But essentially this may well be taken as an additional "think-piece", conceived and developed within the framework of the Development Centre's study programme. Naturally, it is not free from the private views of the particular author, Nicolas Jéquier, who has been charged with the responsibility for editing the entire book(1).

Major questions addressed

Without pretending to recapitulate the complex substance of this report, let me stress some of the questions to which the various authors of this book were invited to respond in one way or another. The 1974 meeting on low-cost technology was prompted by our earlier inquiry into the mechanisms of technology transfer for small-scale industries in developing countries. A number of questions had evolved out of that inquiry(2) that needed a closer examination in the light of practitioners' experiences. In the course of the meeting, the list of specific questions became further diversified, rather than simplified, partly because there were many different ways of looking at certain general issues and partly because the meaning of the term "policy" varied, depending on the level of decision problems corresponding to each discussant's daily preoccupations.

Leaving aside rhetorical variations, and in an attempt to make some bold outlines of the themes of this book, let me just select several basic questions which seem particularly important from the standpoint of general strategy for development co-operation.

First, in relation to problems of rural industrialisation,

- How has the low-cost technology movement stimulated the participation of local metal-works, repair shops for farm equipment and implements and other local industrial or semi-industrial operators?

---

1) As is the case with most work by the Development Centre, the particular position taken in this book is that of the author and should not necessarily be regarded as representing the official position of the OECD.

2) See the OECD Development Centre, Transfer of Technology for Small Industries, Paris, 1975; see especially the concluding part of my Review of Discussions in Part I.

- To what extent, and in what forms, should there be an "insulation" policy to protect the small-scale decentralised industrialisation base from any competitive pressure emanating from large-scale technology products?

In relation to the nature of science and technology policy in the context of rural development:

- How does the attribute of low-cost technology (i.e. "continuity from indigenous technology") permit a progressive linkage to formal educational systems and scientific disciplines?
- What is the role of universities in the low-cost technology movement? How is it affecting the local university curricula as well as primary and secondary school curricula?

In relation to the basic options in innovation policy design (general versus specific policy instruments):

- How far have the demonstration effects of specific micro-development projects become visible? Do we know enough about the threshold scale for micro-project assistance below which success would become problematic?

From the standpoint of foreign aid strategy:

- To what extent, and in what special ways, can the assistance from foreign sources play a crucial role in stimulating the local self-help effort?
- What sorts of information services prove particularly helpful to guide the search, choice and application phases of low-cost innovations in developing society?

Outstanding issues inviting further study

Framed in such conceptual terms, the questions stated above do not expect a straightforward, monochromatic answer. But the reader will find in the contributions collected in this volume various practical ways of interpreting the meaning of these questions and the gravity of opinions regarding the desired solutions. Nicolas Jéquier's writing in Part One offers a fairly detailed, but reasonably comprehensive treatment of these questions.

Admittedly, this book is not as much advanced yet in regard to the analysis of a complete arsenal of policy measures and instruments that are relevant for stimulating and directing an indigenous innovation process in developing societies, as in regard to the analysis of basic strategic issues and policy options. A more systematic analysis of concrete policy instruments and techniques of deploying them would be an important task envisaged for the future. This is not the place for drawing up a detailed prescription for such a task. But in concluding this Introduction, let me mention a few messages emerging from this book that I consider particularly worth stressing.

Firstly, there is a considerable scope for more active governmental support. What is important at this moment is perhaps not so much the perfectionist preoccupation concerning the real benefits and costs of individual micro-projects as how to help cross the threshold of change towards a new social attitude or a new prestige concept commensurable to the animation of local innovative minds. The current appropriate technology movements are apparently still a long way from the threshold. A quantum jump in aid resource inputs would be desirable.

Secondly, great care should be exercised in designing direct governmental interventions. Important in an initial phase would be an appropriate mix of "specific" promotion policies for local small-scale initiative-taking on the one hand and of a certain "toleration" policy aimed at the evolution of a competitive trial-and-error process, on the other.

Thirdly, there should be a clear discriminatory policy to obviate premature competition with the larger-scale modern-technology ventures fostered under the earlier import substitution strategy. But there is a burdensome regulatory task which calls for extensive R and D and elaboration of appropriate standards geared to the industrial and semi-industrial undertakings being fostered to cater for the needs of rural villages and towns. And such a task should be set in a dynamic perspective, its primary mission being to facilitate, and not to stifle, gradual, positive interactions between local self-help efforts and the scientific and technological resources from the modern sector.

And finally, reverting to the first point above, further innovative arrangements seem to be needed as to the channelling of any quantum jump in aid resources to spur local innovation projects. To package such resources into large (integrated) rural development projects is apparently an expedient. But the innovation policy, or at least aspects of it which concern a broad social process as such, ought to be stretched beyond the demonstration-oriented greenhouse projects. Especially for multilateral aid authorities, an outstanding challenge is how to respond to the needs of informal, individually small-sized, and mostly risky ventures emerging in a large number.

ACKNOWLEDGEMENTS

As mentioned earlier, our work on the subject was conceived as a collaborative project and was helped by various forms of contribution, formal and informal, from a large number of people. Among others, we owe thanks to all the participants in the September 1974 meeting in Paris, who provided us with a large spectrum of valuable observations and thus laid down major groundwork for our subsequent work. Not all the written contributions - even some of those which their authors kindly took the trouble to revise after the meeting - could be incorporated in this edition due to lack of space. But clearly the editor has benefited from these, too, in substantiating his analysis in Part One. We thus owe both apologies and thanks particularly to Messrs. B. Behari (Ministry of Industrial Development, Government of India), R.J. Congdon (V.C.O.A.D. International Development Centre, London), T. de Wilde (Technische Hogeschool Eindhoven, Netherlands), S. Dutta (SIET Institute, Hyderabad, India), P. Martin (IRFED), V. Martinez (Centro de Desarrollo Industrial del Ecuador, Guayaquil), J. Müller and H. Kristensen (Institute for Development Research, Copenhagen), and N.N. de Silva (Co-operation Wholesale Establishment, Colombo, Sri Lanka) whose interesting contributions we had to part with in view of editorial constraints.

We are also grateful to Messrs. S. Dedijer (University of Lund, Sweden), P. Borel (Institut International de Recherche et de Formation Paris), E. Fallot (Société d'Aide Technique et de Coopération, France) P. Gonod (then with the O.A.S. Pilot Project on Transfer of Technology M. Greene (VITA, U.S.A.), K. N'Gangbet (then at the Industrial Promotion Bureau of Chad), and K. Philip (Chairman of the Industrialisation Fund for Developing Countries, Denmark), for the highly illuminating interventions they gave at the 1974 meeting. Special thanks go to the Commonwealth Foundation whose grant facilitated the participation of several practitioners from Africa and Asia.

Part One

THE MAJOR POLICY ISSUES

by

Nicolas Jéquier*

*OECD Development Centre, Paris

Chapter 1

THE ORIGINS AND MEANING OF APPROPRIATE TECHNOLOGY

A growing number of development experts and national policy-makers are beginning to question the wisdom of massive technology transfers from the industrialised countries to the developing nations. The large-scale capital-intensive technologies developed in Europe, North America or Japan may well be very efficient, but their introduction into poorer, less developed societies often raises more problems that it can solve. They are usually very costly relative to the income of the local populations, they require an educational and industrial infrastructure which takes decades to build up and their disruptive social consequences tend to be much more sudden than in their culture of origin. But perhaps most important of all, their introduction often inhibits the growth of the indigenous innovative capabilities which are necessary if "development" is to take place.

The symbols of modernity, in the form of steel mills, chemical plants, automobile factories or squadrons of military aircraft can be purchased on the international market, but development is a complex social process which rests in large part upon the internal innovative capabilities of a society. Imports of foreign ideas, values and technologies have a major part to play, but few societies in history have developed exclusively on the basis of such imports. One of the major tasks facing the developing countries is to create, nurture and more often than not rehabilitate their internal capacity to invent and innovate. As far as technology is concerned, this implies not only a much greater selectivity in the choice of imported equipment, plants and methods of production, but also - and this is much more important - the invention and diffusion of new types of technology and new forms of organisation which are better suited to local conditions.

I. THE SEMANTICS OF APPROPRIATE TECHNOLOGY

The technologies which meet this requirement are variously described as 'appropriate', 'low-cost' or 'intermediate'. The exact difference between these three types of technology is the subject of lively if somewhat inconclusive theoretical debates, and it is well to recognise that for the time being there are no widely accepted definitions of what constitutes an appropriate, low cost or intermediate technology. For practitioners working in the field, these terms, however, are perfectly clear and require little elaboration. Rather than try here to give a standard set of debatable definitions, it may be more appropriate to illustrate them with a few concrete examples.

The ox-plough, introduced in several tropical African countries by the agricultural extension services, religious organisations and rural development specialists, is a good example of an intermediate technology. It stands, so to speak, half way between the traditional hand-operated hoe and the modern diesel tractor. Intermediateness is of course relative: in the societies of the Middle East and Asia which have known and used the ox-drawn plough for thousands of years, such a technology can be called traditional, and the intermediate level of technology would more adequately be represented by the small two-wheel tractors of the type developed by the International Rice Research Institute in the Philippines(1) or by the industrial co-operatives of Sri Lanka(2). In the tropical African societies which do not have any tradition of livestock breeding and which still use very simple implements, the ox-drawn plough is a major innovation, and from a technological point of view, it represents a big step forward.

Another example of intermediate technology is the gari machine developed in Nigeria. Gari, a dehydrated cassava product, is a staple food in most of West Africa. The traditional manual preparation method for extracting the prussic acid from the cassava root was brought to West Africa in the 18th century by former slaves from Brazil who had borrowed it from the local Indian population(3). With the rapid rate of urbanisation, industrial production methods were required, and a modern large-scale technology was developed in the 1960's by a British firm in Gambia in co-operation with Nigerian technologists. Parallel to this, a smaller scale and somewhat simpler technology was developed at the time of the Nigerian Civil War on the Biafran side. The capacity of this intermediate plant is smaller but the total investment required for the same output is at least four times lower and its profitability substantially higher. Of particular

---

1) See the article by Amir U. Khan, "Mechanisation Technology for Tropical Agriculture", in the second part of this book.

2) D.L.O. Mendis, "The Reorganisation of the Light Engineering Industry in Sri Lanka", ibid.

3) Thomas R. De Gregori, Technology and the Economic Development of the Tropical African Frontier, The Press of Case Western Reserve University, Cleveland 1969.

interest here is the fact that the intermediate technology was an indigenous innovation. The modern plant, by contrast, relied heavily on technological contributions from a highly industrialised country(1)

The definition of what constitutes a low-cost technology is at first sight relatively simple. The ten dollar rural latrine developed by the Planning Research and Action Institute in India is quite clearly immensely less expensive than the modern flush toilet(2) and the water filtration system developed in Thailand, and which uses coconut or rice husks as the filtering media, is so inexpensive that it can be considered for all practical purposes as a zero-cost technology: supplying a family with pure water for one month costs around $0.20(3). However, as soon as one goes beyond the simple home-living 'self-help' technology, cost calculations can become extremely complex, and it is often very difficult to determine whether a new manufacturing technology for instance is cheaper than the one it replaces or supplements.

The small-scale sugar plants developed in India, and which now account for more than 20 per cent of the country's production, are a good case in point. The average investment per ton of output is two and a half times smaller than in the large modern plants, and the investment per worker nine times lower. Differences in production costs however are much smaller (less than 20 per cent), and the present balance in favour of the small-scale technology could easily be tilted(4). This in fact is what happened with a rather similar type of technology in Ghana. Analyses made in 1969 showed that the small-scale low-cost sugar technology was more attractive from an economic point of view, but four years later, with the rise in wages and the improvements in the modern large-scale plants, the situation was completely reversed. Economics can of course also operate in favour of the small-scale lower-cost technology. In India for instance, the sudden rise in the price of imported oil has helped to make cow dung gas cookers much more attractive.

Low cost, like intermediateness, is a relative notion which varies in space and time, and much depends upon the assumptions made

---

1) See the article by P.O. Ngoddy, "Gari Mechanisation in Nigeria: The Competition Between Intermediate and Modern Technology", in the second part of this book.

2) See the article by M.K. Garg, "The Upgrading of Traditional Technologies: Whiteware Manufacturing and the Development of Home Living Technologies", ibid.

3) R.J. Frankel, "The Design and Operation of a Water Filter Using Local Materials in Southeast Asia", ibid.

4) M.K. Garg, "The Scaling Down of Modern Technology: Crystal Sugar Manufacturing in India", ibid.

about the price of inputs. If interest rates are kept artificially low to foster industrialisation, as happens in most developing countries, the modern capital-intensive technology will automatically appear much more profitable in private terms (but not necessarily in social terms) than the local indigenous technology which employs many people but requires little capital. If on the other hand, employment is a real (as opposed to rhetorical) national priority, the profitability of the small-scale technology cannot be measured exclusively on the basis of the effective wages paid to the workers. Employing a person who would otherwise be out of work is a net gain for the economy, and this factor can be taken into account by using a shadow wage rate and measuring the opportunity cost of employment(1).

Measuring the real cost of a new technology brings out very clearly the conflict between micro-economics and macro-economics, and between the national planner and the entrepreneur. The planners who have to choose for instance between a small-scale sugar plant and a large one must take into account such factors as the social cost of unemployment, the effective price of foreign exchange or the import-saving value of a new project. The small entrepreneur who is investing his money and his work in a plant is interested mainly in making a profit, developing his company and securing the cash-flow which will allow him to pay his workers and buy his raw materials. If there is no link between the planner and the entrepreneur - for instance by ensuring that the latter will get some immediate financial compensation to make up for the difference between the wages he effectively pays out and the shadow wage rate upon which the profitability of his project was assessed - the only test of a new technology's cost or value is its acceptability by the market.

When speaking of low-cost technology, one is focusing primarily on the economic dimension of innovation. The concept of intermediate technology on the other hand belongs more specifically to the field of engineering. As for appropriate technology, which tends today to be somewhat more popular than low-cost or intermediate technology, it represents what one might call the social and cultural dimension of innovation. The idea here is that the value of a new technology lies not only in its economic viability and its technical soundness, but in its adaptation to the local social and cultural environment. Assessing the appropriateness of a technology necessarily implies some sort of value judgement both on the part of those who develop it and those who will be using it, and when ideological considerations come into play, as they often do, appropriateness is at best a fluctuating concept.

---

1) For a further discussion of these issues, see the paper by A.J. Bhalla, "Low-Cost Technology, Cost of Labour Management and Industrialisation", ibid.

The solar pump developed by a French firm in co-operation with the University of Dakar, and which is currently being introduced on a large scale in Mexico, is probably a very good example of appropriat technology. It uses a widely available source of energy - the sun - to provide villagers with a scarce but vitally important commodity - water(1). Although it is technically very sophisticated, it blends rather well into the social environment: it requires virtually no maintenance and seems to have a potentially very long working life. In the same way, a number of the technologies developed or popularised by the Brace Research Institute in Canada can be considered as particularly appropriate, be they solar coffee dryers for Colombia, small-scale iron foundries for Afghanistan or solar distillers for the water-less villages of Haiti(2).

The terms low-cost, intermediate and appropriate technology are generally considered to refer to technologies which are used, developed or imported by developing countries. Most of these technologies are, however, equally relevant to the highly industrialised countries, and are in many ways rather similar to the soft or alternative technologies promoted by a rapidly growing number of organisations and individuals in North America and Western Europe. These soft or alternative technology movements all emphasize the need for a much greater attention to the ecological impact of new technology and to the real needs of society.

Among the soft technologies developed in the industrialised countries, one might mention the aquaculture system and the windmills of the New Alchemy Institute in the United States(3), the community technologies (e.g. basement fish-farming, solar kitchens, etc.) suited to an urban environment which are propounded by Karl Hess's Community Technology(4), Robin Clarke's efforts in the United Kingdom to explore new paths in non-polluting agricultural technology and the use of renewable energy resources(5) or the wind generators and methane digesters of an organisation such as Conservation Tools and Technology(6).

---

1) See the paper by J.P. Girardier and M. Vergnet, "The Solar Pump and the Problems of Integrated Rural Development", ibid.

2) See T.A. Lawand et al., "Brace Research Institute's Handbook of Appropriate Technology", ibid. This paper contains an elaborate set of criteria of a technology's appropriateness.

3) "New Alchemy Institute: Search for an Alternative Agriculture", Science, Vol.187, No.4178, 28 February 1975.

4) "Karl Hess : Technology with a Human Face", Science, Vol.187, No.4174, 31 January 1975.

5) See Robin Clarke and Geoffrey Hindley, The Challenge of the Primitives, Jonathan Cape, London 1975.

6) See Andrew McKillop, "Technological Alternatives", New Scientist, 22 November 1973.

The examples of intermediate, low-cost, appropriate or soft technology given here should be viewed simply as illustrations, and not as an attempt to present a global overview of what has been achieved to date. What these examples suggest is that at this stage, the delineation between these various concepts is still in a state of flux. Appropriate technology is very close to, but not entirely identical with, intermediate technology, and a low-cost technology, while often particularly appropriate to the conditions in a developing society, does not necessarily always meet the criterion of appropriateness. In fact, each of these concepts might be viewed as a set of overlapping but nevertheless distinct areas, the frontiers of which are rapidly changing under the impact of recent experiments, new innovations and progressive changes in perspective. For this reason, the terms appropriate, low-cost, intermediate and soft can for the moment be used almost interchangeably, and the choice of one term in preference to another is a reflection of differences in emphasis rather than of fundamental difference in nature.

II. THE HARDWARE AND THE SOFTWARE

The term 'technology' invariably suggests the idea of _hardware_ be it in the form of factories, machines, products or infrastructures (roads, water distribution systems, storage facilities, etc.). Hardware is something visible, and even if it is not understandable, it stands out very conspicuously. Technology however goes much beyond the hardware, and also comprises what can be called, by an analogy taken from the computer industry, the _software_. This includes such immaterial things as knowledge, know-how, experience, education and organisational forms. This distinction between hardware and software is just as important in the case of appropriate technology as in that of modern large scale technology.

The societies which today are highly industrialised owe their development not merely to the invention and widespread application of new types of machinery, from the steam engine of the first industrial revolution to the electronic computer of today, but also to major innovations and gradual improvements in organisational forms and institutional structures. The importance of these non-material innovations often tends to be underrated by historians of technology. One of the major innovations in organisational forms in the first half of last century was a legal invention, that of the limited company. This new form of association allowed potential entrepreneurs to escape from the stifling restrictions of the professional guilds inherited from the Middle Ages. It also consecrated the dismantling of the King's monopoly on industrial and commercial entrepreneurship, which had been institutionalised over the centuries by the system of royal charters. In fact, had it not been for the invention of the limited company, which released the entrepreneurial drive of a society in transition, the industrial revolution may well have aborted despite all the inventiveness in hardware. The history of classical China, Imperial Rome or late medieval Europe suggests that the ability to invent and develop new types of hardware is not alone sufficient to generate the equivalent of an industrial revolution. What is required is an entrepreneurial class and perhaps more important a system of values - cultural, social or religious - which can legitimise and encourage social and economic change.

The problem facing developing countries today is not very different. The range of new hardware which is available to them as a result of the industrial research undertaken in the advanced countries is so wide and increasing so rapidly that it could, in theory if not in practice, meet a large part of their immediate needs. What is really lacking is the software and this is perhaps the area where the appropriate technology movement has the most to contribute. Hardware and the technical ability to produce it in an imitative way can generally be transferred from one country or culture to another. Organisational forms and social values are, by contrast, much more culture-specific and hence generally more difficult to transpose deliberately from one society to another.

The case of the agricultural extension system illustrates this point very clearly. These services, developed in North America and protestant Europe in the 19th century, have proved extremely effective in diffusing new technology and transforming the agricultural system. But when transferred without any substantial modification to the developing countries, they often turn out to be more costly and much less effective. In fact, the problems of adaptation are much more complex than for hardware. Such an adaptation can however be done. One successful example in this respect is provided by the Institut africain pour le développement économique et social (INADES), a mission-based organisation working in most of Black Africa(1) which has developed a low-cost extension service that lays great stress on the inventiveness and entrepreneurship of the farmers to whom it is addressed. In fact, the work done by INADES, or for that matter by the Panafrican Institute for Development(2) provides very vivid examples of the ways in which appropriate software (i.e. education, agricultural extension, knowledge, etc.) can be created and modified to promote development within a poor community.

---

1) See the paper by P. Dubin, "Education as a Low-Cost Technology for Agricultural Development", in the second part of this book.

2) See John W. Pilgrim, "The Role of Non-Governmental Institutions in the Innovation Process", ibid.

The ineffectiveness and high cost of Western-style educational systems in the developing countries is another illustration of the difficulty of transferring software from one country to another. In fact, a number of developing countries have tried to meet some of the criticisms levelled against their educational systems by developing new types of hardware and software which are much better suited to local conditions. One might mention here the educational games developed in some African countries (EcoNiger in the Republic of Niger and the Agricultor in the Central African Republic), the integration of artisanal activities in the curriculum of the primary schools (Chad) or the emphasis placed on the traditional knowledge of the community in which the school is located. The number of experiments currently going on in the educational field is very large, and only just beginning to be documented in a systematic way(1). These examples clearly suggest that appropriate software can be developed and, more important, help to build up the inventive and innovative capability which is necessary to development.

At another level, one might note two small but revealing examples of the way in which specific problems can be solved with a minimum expenditure in money and without using any new hardware. The first is the way in which the bus system in the city of Delhi was reorganised to provide better service and make more efficient use of the existing hardware(2). The other example also comes from India. In the State of Punjab, the agricultural crisis which followed the rise in fertiliser prices in 1974 led to an apparently very simple administrative solution to a problem inherited from the colonial period. The water distribution system and the agricultural extension service which were run separately without any co-ordination were merged under a single authority. As a result of this simple administrative reorganisation, the average grain yields increased in a significant way, which helped to overcome the problems posed by the relative shortage of fertilisers.

Many other similar examples could be found throughout the developing world. Some of them may seem trivial, but development is a process which consists for a large part in thousands of small improvements and modifications in software, rather than in sudden and massive leaps forward in hardware. Software however lacks visibility, and often tends for this reason to be overlooked, not only by national planners and policy makers, but also by many of the small-scale organisations now active in the field of appropriate technology.

---

1) A very interesting summary of current experiments in this area can be found in Développement de méthodes et de techniques adaptées aux conditions propres aux pays en voie de développement (mimeo) UNESCO, Paris, September 1975.

2) See Jon Tinker, "How Delhi Makes the Buses Run on Time", New Scientist, 9 January 1975.

One should not of course overlook the importance of hardware, but an exclusive preoccupation with hardware tends to overshadow the importance of the enormous potential resource represented by software. The ways in which this resource can be mobilised in a systematic way are not entirely clear, but many experiments have pointed to the general direction. One example in this respect is the Comilla Project in what was then East Pakistan(1). This was an attempt to initiate community development at the grassroots level, not so much by introducing new hardware, but rather by reorganising existing resources and promoting innovation from within. The organisation of inexpensive public health systems is another illustration: without any substantial increase in hardware, it is feasible to provide an effective health system at an annual cost of some $10 per person, and a sanitation system for ten times less(2).

A national strategy for appropriate technology (or for that matter the strategy of a small private or public appropriate technology organisation) could and should focus both on hardware and software. In fact, the most appropriate technology for many developing countries is often the one which has a greater software component. Software is much more difficult to develop and diffuse than hardware but strategies for mobilising it in an effective way can be developed(3).

III. THE CULTURAL ORIGINS OF APPROPRIATE TECHNOLOGY

The idea that developing countries should resort to appropriate technology in order to promote development is in the process of becoming 'respectable', both for national policy makers in these countries and for aid-giving organisations in the industrialised countries, and appropriate technology is progressively entering into the mainstream of development aid. This transition from marginality to acceptance is most conspicuous in the United States(4), Canada, the United Kingdom, Sweden and the Netherlands, and there are signs

---

1) See Akhter Hameed Khan, "The Comilla Project - a Personal Account", International Development Review, 1974, No.3.

2) See James S. Pollock McKenzie, "Putting a Price Tag on Health: A Ten Dollar Health Plan", International Development Review, 1974, No.1.

3) See for instance Robin T. Miller, Pattabhi N. Raman and George R. Francis, "Mobilising National Talent for Development", International Development Review, 1974, No.2.

4) One example of this is the bill passed by the U.S. Congress in September 1975: the Agency for International Development has been allocated $20 million to be used for activities in intermediate technology in the 1976-1978 fiscal years.

that this interest in appropriate technology is gradually spreading to other aid-giving countries, notably Germany and France. This shift in public attitudes however is not matched, for the moment at least, by a corresponding shift in the nature of aid programmes: attitudes are several years ahead of concrete realisations, and it will take time before this interest in appropriate technology can be translated into projects and achievements similar in scope to those which have been undertaken under the 'conventional' or 'modern' large-scale aid programmes.

This interest of aid-giving countries and organisations in appropriate technology, important as it may be for the future of the movement, is in fact at the root of a very widespread misconception, namely that appropriate technology is primarily an aspect of development aid. It certainly has a part to play in development aid, but the philosophy which underlies it is precisely the opposite: appropriate technology should first and foremost be an indigenous creation of the developing countries themselves and the central problem they have to face is that of building up an indigenous innovative capability and not that of importing more foreign technology.

The appropriate technology 'movement' (the term movement is used here for want of another more adequate description) can probably be viewed as a cultural revolution. A number of factors have contributed to bringing it about, both in the industrialised and the developing countries, and it might be useful to distinguish here between the immediate origins and the more complex and deeper origins.

The most conspicuous of these immediate origins is the realisation, shared by aid-giving and aid-receiving countries alike, that development aid and a Western style of industrialisation have neither fulfilled the initial hopes which were placed in them nor been fully capable of solving the basic problems of development. This problem has been vividly expressed by Dr. E.F. Schumacher in his influential book, Small is Beautiful, which perhaps more than any other, has contributed to popularise the concept of intermediate technology, both in the developing countries and in the industrialised nations(1).

The disillusion with foreign aid, which several studies have amply documented(2), is not due so much to inefficiency, lack of money, ignorance or the importation of inappropriate technology, as to the fact that while we know quite a lot about the reasons why a

---

1) E.F. Schumacher, Small is Beautiful, Blond and Briggs, London, 1973. The author is the founder of the London-based Intermediate Technology Development Group (ITDG). For further details, see the articles by George McRobie, "The Mobilisation of Knowledge on Low-Cost Technology: Outline of a Strategy" and M.M. Hoda, "India's Experience and the Gandhian Tradition" in the second part of this book.

2) See for instance Tibor Mende, From Aid to Recolonisation - The Lessons of a Failure, Pantheon, New York, 1973.

particular society has developed, we know very much less about the ways in which such a process can be deliberately and successfully engineered.

The idea that all societies can and should 'develop' is by historical standards a very new one. The concept of 'human progress', which is the cornerstone of several great religions, is not of course very recent, but it is primarily a moral notion, and only very recentl did progress come to be equated with a steady rise in material well-being rather than with moral betterment. Development, or for that matter 'human progress', is closely associated with a society's perception of time and with its image of man's place in nature. In many traditional societies, time, as exemplified by the recurrence of the seasons, tends to be perceived as a cyclical phenomenon: the present is a repetition of the past, the future will be the same as the presen and man has no control over time. Progress or for that matter development, implies that the future will be different from or better than the past; time comes to be viewed as a linear, irreversible process, and not a recurrent cycle. These differences in the perception of time are linked for the most part to religious traditions, and take a long time to change.

The second immediate origin of the appropriate technology movements can be found in the industrialised countries themselves. The worldwide student revolts of the 1960's, the debates about "Limits to Growth", the ecology craze and the oil panic, the reactions against the consumer society and the patterns of living imposed by industrial necessity are the most conspicuous symptoms of Western society's growing doubts about its values, its way of life and its long-term future(1). The phenomenon, incidentally, has also spread to Eastern Europe and the Soviet Union(2). Technology, which has probably been the most conspicuous factor in bringing about the social and cultural changes which the majority has now begun to question, has come under attack and this has paved the way to the search for alternative technologies, for a better balance between man and nature and for a greater responsiveness of technology to the 'real' needs of man.

---

1) See for instance Charles Reich, *The Greening of America*, Allen Lane, London, 1971, Theodore Roszak, *The Making of a Counterculture* Doubleday, New York, 1969. Alvin Toffler, *Future Shock*, Random House, New York, 1970. Jean-François Revel, *Without Marx or Jesus*, Doubleday, New York, 1971.

2) See for instance Aleksandr Solzhenitsyn, *Letter to the Soviet Leaders*, Harper and Row, New York, 1973, and Radovan Richta, *Civilisation at the Crossroads* (mimeo), Prague, 1967. Solzhenitsyn's letter, interestingly enough, is probably the first plea made in the Soviet Union for intermediate or appropriate technology. As for Richta's book, which in fact was a collective work, it played a major part in fostering the "Prague Spring" of 1968.

This search for alternative technologies is only just beginning in the industrialised countries, and for the moment is less conspicuous by its practical achievements than by the intensity of the ideological debates which surround it. The reason for this is simply that the development of technology is conditioned not only by the imperatives of engineering but also by the cultural and ideological values of the society which produces the technology. Changes in technological trends must be preceded by changes in culture, which in turn will bring about modifications in the demand for new technology. The modern large-scale technology we have today is basically a result of the cultural demand, or values, of Western society in the last fifty years, and a change in values and in the perception of what technology can achieve is the prerequisite for a reorientation of the research and industrial innovation system.

What is true of technology is equally true of science: a society develops the type of knowledge, or science, which is consonant with its values, and not necessarily that which conforms to scientific truth. Witness for instance the way in which early Christian society gradually "forgot" the heliocentric theory of the universe, only to retain for more than 1,000 years the scientifically incorrect, but theologically more appropriate geocentric view of the world.

The disillusions about development and foreign aid, and the growing doubts of the industrialised societies about their own future are the two immediate and most conspicuous factors accounting for the growing interest in appropriate technology. In fact, the origins of this movement can be traced much further back into history and in particular to the industrial and technological experiences of three major countries, India, China and the United States. In India, this interest in appropriate technology, even if it was not defined in such terms, goes back as we shall see later to the end of last century(1). In China, the philosophy which underlies Mao Tse Tung's ideas about technology could be traced back not only to the civil war of the 1920's and the reaction against the big capitalism of the Kuomintang society(2), but also to the peasant rebellions which have always been one important element in China's history. As for the United States, its industrial history illustrates both the problems of industrialisation in an underdeveloped country and the fact that all the modern large-scale technologies of today were originally small-scale, inexpensive and in certain respects appropriate technologies.

---

1) See M.M. Hoda, op.cit.

2) One very revealing analysis of China's industrialisation problems can be found in the famous novel of Mao Dun, Midnight, (translated into French under the title Minuit, Editions du Seuil, Paris, 1968).

IV. THE UNITED STATES AS A MODEL OF APPROPRIATE TECHNOLOGY

American technology, more than any other, symbolises the large-scale approach to which many developing countries, and many innovators in the industrialised countries, are seeking an alternative. However, in its early years, the United States was in many respects a developing nation, and if what happened in the past does not necessarily prefigure what will happen in other places and other times, there are many lessons to learn from what was, by any standard, one of the outstandingly successful national experiences in technology(1).

Twenty-five years after the Declaration of Independence, the United States still had to import from England such apparently simple things as nails, axes and cloth. Many leaders, Benjamin Franklin and Alexander Hamilton among others, realised the dangers of the situation and tried in their own way to promote what today would be called industrial independence or technological self-determination. Conditions, however, were unfavourable. The manufactured goods imported from Britain were much more competitive than American products, and they were often sold at dumping prices. Protective tariffs were of little avail: the country had little industry, and few of the traditions of craftsmanship which form one of the main bases of industry. And at any rate the small industries which did exist (notably in Connecticut and New York State) were quite unable to meet the needs of a rapidly growing population and the settlers were more interested in opening up new farming land than in building an industrial society on the European model.

Things were to change dramatically with the Napoleonic Wars. For close to two decades, the United States found itself practically cut off from its British suppliers, and had to rely on its own ingenuity to make all the products it needed, from textiles to agricultural tools, from weapons to transport equipment. The country however had very few skilled people capable of manufacturing the wide range of goods required by the market. This situation, coupled with the interruption of trade with Europe, paved the way to the mechanisation of productive processes, for this was the only way to overcome the shortage of skilled craftsmen and meet market needs.

Looking at things in retrospect, the development of a country always appears to be a logical and orderly process. In fact, history is composed for a large part of accidents and unforseeable discontinuities and the development of American technology was immeasurably

1) One of the best short histories of technology in the United States is Roger Burlingame's Machines that Built America, Harcourt, Brace and Company, New York, 1953. For further details, see H.J.Habakukk, American and British Technology in the 19th century, Cambridge University Press, Cambridge, 1962.

less simple and straightforward than this picture would suggest. What is important in this context is not so much the details of the process as some of the conclusions which can be drawn from it. First is the fact that free trade, especially in manufactured goods, is not conducive to the development of industry and technology in the importing country. Second is that a society which for some reason or another is suddenly forced to rely on its own resources can often do so. Third is the crucial importance of demand, or rather of a need for the products and the technologies which were formerly imported. Fourth is that the development of new industries is not necessarily incompatible with the absence of craftsmen and a structural shortage of skilled labour. In fact the production processes of American industry were designed specifically to overcome this drawback. Hence the emphasis on machine tools, interchangeable parts, automatic processes and the parcellisation of work tasks.

Virtually all the industries which grew up in the United States in the nineteenth century started on a very small scale, often as one-man operations. The death rate of new enterprises was also very high: history always remembers those which succeeded, but one should not forget the tens of thousands which failed. In fact, a death rate of some 90 per cent is a normal phenomenon, and in any new industry, seldom more than one per cent of new enterprises turn out to be really successful(1). Technology is a dynamic process, and firms which do not grow in size or sophistication are almost always eliminated from the market. This suggests that one of the crucial factors in development, both at the national level and at the level of the individual firm is the ability to innovate, and to innovate successfully on a continuous basis.

Another lesson from the American experience is that contrary to what happened in most European countries, a high proportion of the inventors and entrepreneurs came from the rural communities. Oliver Evans, the inventor of the automatic milling machine, was brought up in a Delaware farm; Eli Whitney, who was to play a crucial part in the development of the textile industry, and later the machine industry, grew up to manhood in his father's farm in Connecticut; Cyrus McCormick, whose name became the major trademark in agricultural machinery, was also a farmer's son, and Henry Ford himself came from a Michigan farm. Clearly, the American farming community of the nineteenth century was very different from the peasant societies of many

---

1) The automobile industry is a good illustration of this phenomenon: literally hundreds of independent companies have disappeared and today almost all the world production is accounted for by less than twelve large firms. In the computer industry, less than one third of the 70 companies active in this field between 1950 and 1965 had managed to survive until 1970.

other countries: the farmers were free men, and they knew that the future would be what they wanted it to be. These few examples are given here to suggest that development is not necessarily an exclusively urban phenomenon and that inventiveness and entrepreneurship in the rural sector are extremely important. This point must be emphasised since more than 70 per cent of the world population today still lives in rural communities. No society can be considered as truly 'developed' unless it has a healthy agriculture, and the social and economic level of the agricultural sector is generally a good indicator of a country's overall level of development.

Any technology in its early stages is a low-cost technology, in the sense that it requires only small investments and is applied on a limited scale. But if one compares it with technologies which develop subsequently, it is usually both inefficient and expensive. Compare for instance today's jumbo-jet with the old DC3, the hand-crafted European automobile of 1910 with the mass-produced car of today, or the modern combine harvester with the horse-drawn equipment of the 1920's. In fact, the concern of most practitioners in appropriate technology today is not to repeat the experiences of the past and turn back the technological clock but to develop alternative technological and social solutions to problems which for the time being, given the lack of resources and the particular nature of local conditions, cannot be met successfully through large-scale modern technology. Intermediate technology is a complement to, rather than a substitute for, modern technology and might be viewed as an expression of what Ignacy Sachs has called "technological pluralism"(1)

V. IDEOLOGY AND SELF-RELIANCE : INDIA AND CHINA

Historically, low cost and small scale can generally be associated with the early stages of a new industry or a new technology, and the current interest in appropriate technology is linked both with a certain disillusion of industrialised societies with their own way of life, and with the realisation, shared by aid donors and aid receivers alike that technical assistance and development aid as they have operated until now have not really been successful. However, if we look at the two countries which today are probably the most advanced as far as intermediate or low-cost technology is concerned, namely India and China, it will appear at once that this interest in appropriate or intermediate technology is in fact much less recent

---

1) Ignacy Sachs, Daniel Théry and Krystyna Vinaver, Technologies appropriées pour le tiers-monde: vers une gestion du pluralisme technologique (mimeo), Centre International de Recherche sur l'Environment and le Développement, Paris, 1974.

and has a lot to do with their social and political history since the end of the last century.

In India, the rehabilitation and development of traditional village industries, encouraged by such reformers as the Maharaja of Baroda, Rabindranath Tagore and later Mahatma Gandhi was closely linked with the fight against British rule and the attempts to reform Indian society(1). India's pioneering role in the development and application of appropriate technology was part of a much wider movement of national liberation, both from foreign domination and from the country's internal social structures.

India's modernisation efforts, or for that matter those of several other developing countries, clearly show that technology in general, and large-scale modern technology in particular, is neither egalitarian nor socially neutral, and tends to accentuate the social and economic differences between the small minority which can profit - or benefit - from it as consumers or producers and the vast majority of the population living at subsistence levels in the rural areas. In this perspective, appropriate technology might be viewed as the 'survival technology' for the hundreds of millions of farmers who have been completely left out of the development process.

Appropriate technology may well be the only solution to the development, or more modestly the survival, of the rural communities. But its development and diffusion raises a number of political and social problems. One of these is the allocation of resources: even though appropriate technology is comparatively inexpensive, if one measures its cost in terms of investment per workplace, its large-scale diffusion requires important sums of money which might otherwise be used for big projects based on modern technology. The latter have much greater visibility, they are somewhat easier to manage since they deal mainly with hardware, they benefit from the social prestige of modernity and are usually the winners in the political competition for scarce resources. Since investment decisions in the developing countries today are taken for the most part by public authorities (e.g. the planning ministry) and not by private entrepreneurs, the development and growth of industries based on appropriate technology depends very much upon political options taken at the highest level. In the case of India, there are indications that appropriate technology, which was in the political and ideological mainstream in the pre-independence period, lost much of its pre-eminence as a result of the post-1948 large-scale industrialisation efforts.

The political history of intermediate technology in India in the last thirty years is in fact a good illustration of the fight between the modern and the appropriate. This fight is perhaps best illustrated by Gandhi's failure to have the 'charkha' (spinner's

---

1) See M.M. Hoda, op.cit

wheel) retained as the national symbol on India's flag(1). After Independence, the modernist, or rather Leninist, trend - exemplified by the large-scale electrification programme(2) - gained the upper hand over the Gandhian trend. The latter went underground - or rather developed at the local and private level - while the former emerged forcefully at the national policy-making level. In India, as in any other country, political attitudes towards technology change with time, and it is interesting to observe that in the early 1970's, Gandhi's ideas began once more to gain ground(3), as a result of a number of factors, among them the oil crisis, and the shortcomings of industrial Leninism.

The other pioneering country in the field of appropriate technology is China, and one cannot help being struck by the fact that the development and diffusion of such technologies is not only an economic or technical problem, but also an ideological and political issue. China's social revolution, partly by accident and partly by choice, was based upon the farming communities and the rural areas, and incidentally paved the way to what was probably the most significant innovation, or revision, in marxist ideology since Karl Marx - the recognition that the peasantry rather than the urban proletariat was the driving force for social revolution. This accounts to a large extent for the fact that China's development effort, unlike that of many other poor countries, was based to a large extent on the rural areas(4).

If Mao Tse Tung can rightly be seen, along with Gandhi, as one of the main prophets of appropriate technology in Asia - in the same way that Dr. Schumacher is its European prophet - it would be somewhat misleading to interpret his ideology in this exclusive perspective. One should not forget that one of Mao's most vivid experiences of childhood was a peasant rebellion which ended with the public execution of its leaders. Such memories, which were shared by many other people of his age, probably accounts to a large extent for his recognition that the peasantry was indeed the driving force of revolution.

---

1) Dominique Lapierre and Larry Collins, Freedom at Midnight, Simon & Schuster, New York, 1975. On Gandhi's background and psychology, see also Erik Erikson, Gandhi's Truth, Norton C. Norton, New York, 1969.

2) Lenin's famous equation "Communism = Soviets + Electricity" was revised to "Socialism = Panchayat (or 'village councils') + Electricity".

3) Jacques Leslie,"India's Woes Stir Rebirth of Study of Gandhi's Ideas", International Herald Tribune, 21 October 1974.

4) Jon Sigurdson, "Rural Industrialisation in China", in China: A Reassessment of the Economy (A compendium of papers submitted to the Joint Economic Committee, Congress of the United States), U.S. Government Printing Office, Washington, 1975.

From a somewhat narrower ideological point of view, it is also well to note that Mao's views, or for that matter those of other leaders of the Chinese Communist Party were formed for the most part in the 1920's, before the ideological purges against 'left' and 'right'. Among the Soviet ideologists who had the most direct influence upon the Chinese leaders was Nikolai Bukharin who argued unsuccessfully against Stalin that the foundation for industrial growth was a prosperous agriculture(1). The Bukharinian tradition in Maoist ideology did not in fact prevent China from following the Soviet industrialisation model after 1949 and building up large-scale modern urban industries based in large part on imported Soviet technology. In the agricultural field, this commitment to modernity took the form of large-scale, and not always very successful, infrastructural projects, and in the scientific and technological field, in the building up of a strong capability in pure science and advanced military technology.

The focus on appropriate technology with the emphasis it implies on decentralisation, local initiative and self-sufficiency, only reasserted itself after 1960, when the ideological and technological break with the Soviet Union left China with no other option but to 'walk on two legs' and 'rely on its own forces' - to quote two of Mao's most popular slogans. In this perspective, China's commitment to appropriate technology was probably not as deliberate a choice as it is often made to be by outside observers.

The difficulties experienced by China after 1960 and the way in which they were met, clearly suggest that a society's innovativeness, both in the modern sector and in the rural communities, can be greatly stimulated by adverse circumstances, and notably by the sudden interruption of the inward flows of foreign aid and foreign technology. This is not to say that a developing country should, or could, cut itself off from the outside world in order to stimulate its internal inventiveness and innovativeness. The social costs of such isolation, as the Chinese themselves have pointed out, are in the short run very high. In fact, the countries which at one time or another in their history have found themselves in such a situation, did not deliberately choose it, but were forced to accept it because of war or other circumstances(2). Furthermore, isolation as such can

---

1) Bukharin's book Historical Materialism (1925) was translated into Chinese in the late 1920's. His views on science and technology can be found in Science at the Crossroads (1931), reprinted by Frank Cass, London, 1971. Two other very interesting studies on Chinese ideology - and other subjects - are Simon Leys, Ombres Chinoises, Union Générale d'Editions, Paris, 1974, and Jacques Guillermaz, Histoire du parti communiste chinois, Payot, Paris, 1972.

2) One of the few countries to have deliberately cut itself off from the outside world in modern times was Tokugawa Japan, from the beginning of the sixteenth century until the middle of the nineteenth century.

be conducive to invention and innovation only if there already is in the country a pre-existing capacity for development and change. If not, it can only lead to stagnation, regression or the return to a low-level ecological equilibrium. The case of China does however suggest that isolation can have positive effects on the development of technology, and notably on a society's ability to rely on its own inventive forces.

Another example which might be mentioned here is that of Nigeria during the Civil War: the survival technologies developed on the Biafran side in fields like food production, oil refining and weapons were typical examples of appropriate or intermediate technologies based exclusively on local innovative capabilities(1). The innovation potential was obviously present, if in dormant form, before the war, and all that was needed was a powerful motivation to awaken it and exploit it as efficiently as possible. But the big problem is perhaps elsewhere: what happens when the incentives disappear? Mobilising existing resources in time of crisis is often less difficult than sustaining their development over a long period when incentives are more diffuse, and alternatives all too easily available in the form of imported technology.

VI. THE SCEPTICS AND THE DISSENTERS

The growing interest of governments and aid-giving agencies in appropriate technology does not mean that attitudes are universally positive. In some developing countries, notably the oil producers with very ambitious industrialisation programmes, appropriate technology tends to be viewed as a rather pointless diversion from the real issues of large-scale development. At worst, it is considered by some critics of the world's industrial system as an attempt to institutionalise the technological status quo by offering to the developing nations technologies which are inefficient, obsolete and unable to evolve further.

Such criticisms cannot be dismissed lightly, for they touch upon certain very real problems and reflect a number of misconceptions which some of the most ardent supporters of appropriate technology have themselves contributed to generate. One of these misconceptions is that low-cost technology, and more widely the whole range of appropriate technologies, should serve as a substitute for the modern technologies of our industrial system. Modern industry with its large-scale production processes, its sophisticated technology and its

1) See P.O. Ngoddy, op.cit. and E.O.Nwosu, "Scientific Technology and the Future of Africa", paper presented at the West Africa Conference on Science, Technology and Society, University of Ghana Legon, 24-30 March 1972.

elaborate management methods is in a large number of cases by far the most efficient and cheapest way of manufacturing the large quantities of products required by the market. Dismantling it or slowing down its growth for the sake of providing more employment or for the sake of smallness as such is an unrealistic proposition.

For the moment, we do not really know what types of modern manufacturing processes can be economically scaled down to meet the particular conditions of a developing country. The growth of industry in the last hundred years has been closely linked with the steady increase in the size of production units and with continuous innovations in technology. As a result, large scale and high technological sophistication seem to be among the main imperatives of efficiency, and there have been very few incentives to explore other approaches which reconcile efficiency with small size and technical simplicity. In some sectors, such alternatives may well be viable, and in fact an increasing number of large industrial corporations are beginning to look into them. This is the case for instance, of several electronics firms, of the automobile industry and of some food processing companies. The search for industrial processes which are both efficient and small in scale is still in its early stages. Marginal as it is, this trend is nevertheless important, for it suggests that such alternatives are economically and technically feasible, and well adapted to the local conditions in the developing countries.

The dividing line between large-scale modern technology and the small-scale technology is far from clear. In fact, the choice offered to the industrialist, the government planner or the farmer is not one or the other, but rather a whole spectrum of technologies ranging from the small and simple to the large and complex. What the intermediate technology proponents are trying to do is to open up the spectrum and find solutions which are better suited to local conditions. The aim is generally not to replace the existing industrial system - as the term 'alternative technology' somewhat misleadingly suggests - but to promote technological innovation in the areas where it has been, until now, either weak or ineffective.

Another misconception about intermediate technology is that it represents a second-best, or inferior technology. The ox-drawn plough is certainly less complex than a tractor, the solar water heater may not be quite as flexible as the electric boiler and the cow-dung gas cooker makes little sense in a city where there are few cows and a well-organised distribution network for gas and electricity. To the educated and urbanised elite of the developing countries, such technologies may indeed seem inferior. But for the vast majority of the population, notably in the rural areas, which cannot afford the amenities of the consumer society and which has been completely left out of the development process, they represent a big step forward as well as a means of meeting in a more effective way some of the basic needs for food, health, energy and shelter.

The function of low-cost technology, however, is not only to meet these needs in a more effective way, but also to help initiate a process of development by stimulating the innovative forces which exist in any community. One of the side-effects of industrialisation has been to paralyse these forces, either through competition on the market place, or more subtly by bringing to light the growing gap between traditional technologies and those of the large industrial corporation. What the intermediate technology proponents are trying to do is to turn development into an autonomous process of innovation and growth from below. Socially and ideologically, this approach is very different from the idea of 'growth from above' which until now has largely dominated the theory and practice of development. But it may well be more effective in getting as wide a number of people as possible to take an effective part in the development process.

VII. DORMANT TECHNOLOGY AND THE INFORMAL SECTOR

The scepticism expressed by many people about appropriate technology is no less great than the enthusiasm and missionary zeal of its proponents, and it is difficult for an outside observer to assess the effective importance of the appropriate technology movement and evaluate its achievements. The field of appropriate technology is in fact immensely wider than the activities of the various appropriate technology groups would suggest. These groups have in a sense crystal lised the problem and brought it to the public attention, but the pool of appropriate technology which already exists in a number of developing countries goes far beyond the new hardware and software which these groups have helped to develop.

To put things in perspective, one should bear in mind the fact that the total amount of money spent on developing and diffusing appropriate technologies on a world-wide basis by the organisations which view themselves as 'appropriate technology' institutions is currently (1975) of the order of 10 million dollars a year. The yearl budget of VITA for instance (Volunteers in Technical Assistance), a U.S. based organisation which is concerned primarily with the provisi of technical assistance by correspondence to developing countries, and which has been in operation since 1959, was around $450,000 in 1975. The Intermediate Technology Development Group (ITDG), founded in London in 1967 by Dr. E.F. Schumacher, has an annual budget of some £80,000. Leaving aside the Indian appropriate technology organisations (notably the Gandhian Institute of Studies and the Planning and Action Research Institute in Lucknow), ITDG and VITA are among the oldest and largest organisations in this field(1).

Of this expenditure of some 10 million dollars, less than half (i.e. under $5 million) is spent on research and development. Compare this R and D expenditure with the 60 billion dollars or so spent on developing new modern technologies. Given the time needed for an R and D programme to be translated into a viable innovation, and the difficulty of reorienting both men and resources to other priorities, it is obvious that the appropriate technology movement as such cannot make a significant impact in the very immediate future.

Alongside the information-diffusion and research activities of the formally organised appropriate technology groups, there are a number of organisations involved in fact if not in name in this area. These can be divided into three groups. First are the large modern industrial corporations which have developed new products or new technologies which in one way or another can be considered as particularly appropriate to the developing countries. One example here might be the simplified plant for assembling radios and television sets, which was developed by Philips, the Dutch electronics firm. Another is the 'basic vehicle' or the 'developing nations tractor' (DNT) of Ford Motor Company(2). At the same time, a number of firms are developing soft technologies which are of particular interest to the industrialised countries: Boeing for instance is applying its aerospace technology to the development of new types of windmills(3).

The second group includes a large number of public and especially private institutions working in such fields as engineering, consulting, management assistance, information or the provision of services to small industry. Although they are not concerned primarily with appropriate technology, many of them have in fact developed appropriate or intermediate technologies. This is the case for instance of CENDES (the Center for the Industrial Development of Ecuador), which has contributed to the development of a small scale inexpensive production technology for polyurethane flexible foam(4). In Colombia, the Fundacion para el fomento de la investigacion cientifica y tecnologica (FICITEC) has developed an electricity generator based on a modified

---

1) See the articles by M.K. Garg, M.M. Hoda and G. McRobie in the second part of this book.

2) William O. Bourke, "Basic Vehicle for Southeast Asia" in Technology and Economics in International Development, Agency for International Development, Washington, 1972. See also A.I.D's Appropriate Technologies for International Development, Washington, 1972.

3) "Tilting the Windmills", Time, 7 July 1975.

4) Victor Martinez, "Specialised Information Services for Technological Innovations in the Less Developed Countries", paper prepared for the OECD Seminar on Low-Cost Technology, Paris, 1974.

bicycle as well as several small machines for rural and cottage industries(1). In France, the Société d'aide technique et de coopération (SATEC) has developed among other things a small rotary tiller(2) which is now manufactured by a wide number of village artisans in Madagascar. In Senegal, the Institut de recherche et de formation (IRFED) has experimented with the use of plastic-covered roofs to collect rainwater in the villages of semi-arid zones(3).

Hundreds, if not thousands, of similar examples could be mentioned here. The obvious conclusion is that the current innovation effort in appropriate technology is in fact considerably wider than generally suspected. The nature of the organisations involved is equally varied. It ranges from large multinational corporations to charitable organisations like Oxfam or the Christian Relief and Development Association(4), from small companies making agricultural implements and machines to public development corporations like the National Industrial Development Corporation of Swaziland(5). It also includes, and this often tends to be overlooked, the public and private aid agencies of the industrialised countries which have tried, and often been quite successful in developing and introducing technologies which are particularly appropriate to local conditions(6). For the moment, there is no overall inventory of the appropriate technologies developed by these organisations, let alone any global or even piecemeal evaluations of the successes, failures and problems of innovation in this field.

The two groups mentioned here - large industrial corporations and organisations such as SATEC, CENDES or FICITEC - represent what might be called the 'organised but peripheral' sector in appropriate technology in the sense that their primary activities do not lie in the field of technological innovation. When trying to draw an overall - and necessarily very sketchy - picture of appropriate technology one must take into account a third group which might be described as

---

1) Information supplied by FICITEC, Bogota.

2) "Matériel agricole pour les opérations de développement: la houe rotative", Techniques et Développement, November, 1972.

3) P.Martin, "Technical Note on the Collection and Storage of Rainwater" (mimeo), paper prepared for the OECD Seminar on Low-Cost Technology.

4) Oxfam for instance has been involved in a very successful brick-making project in Northeast Brazil. The Christian Relief and Development Association set up in 1975 an Appropriate Technology Advisory Unit in Ethiopia.

5) The Swaziland National Industrial Development Corporation has developed the 'Tinhabi', a simplified tractor with a diesel air-cooled engine.

6) One case in point might be the cheese-making technology brought to Nepal by Swiss foreign technical assistance.

'informal' or 'unorganised'. It includes individual innovators - small industrialists, inventors and tinkerers - and an immensely large number of peasants, artisans, teachers and tradesmen who have developed, transmitted, or who are currently using some form or another of appropriate technology.

This 'informal' technological potential is not of course specific to developing countries. In the industrialised countries, most of the big technological innovations of today stem from the organised R and D efforts undertaken in the laboratories of private industry and government and in the universities. The industrial laboratory however is a relatively recent institutional innovation which goes back to the end of the last century(1). The Industrial Revolution started long before laboratories were set up, and many of the important innovations which contributed to transforming Western society, from the automobile to the aircraft, from the steam engine to the railway, originated from the innovative effort of individual inventors and entrepreneurs working either alone or, when within an industrial firm, without the support of a formally organised laboratory.

The same is still partly true today. In the industrial firm, a relatively important proportion of the innovations still stems from outside the R and D laboratory, and the total research effort of a country as measured by its national expenditures on R and D tends to overshadow the importance of the 'informal' innovation system. The large number of innovations which stem from this informal sector, small or big, include incremental changes in production methods, new forms of organisation, rediscovery of old knowledge, transfers of technology from one sector to another or the better utilisation of existing resources.

In developing countries, the 'formal' innovation system, as institutionalised in the research laboratory, is very small, and contributes proportionately much less to innovation than in the industrialised countries. However, apart from this 'formal' innovation system, which belongs to the modern sector, there is in all developing countries a large 'informal' innovation system represented by thousands of small industrial workshops, individual entrepreneurs, innovative farmers in the rural communities, and institutional entrepreneurs in the service sector (e.g. missionaries, charitable organisations, private associations, money lenders, etc.). This 'informal' sector represents a vast pool of technology, both in hardware and software, which is often relatively simple, but which plays an immensely important part in the economic system.

---

1) An interesting history of the growth of industrial laboratories can be found in W. Rupert MacLaurin, Invention and Innovation in the Radio Industry, Macmillan, New York, 1949. See also The Research System, Vol. I, II and III, OECD, Paris 1972-1974.

The technology of this informal sector is more sophisticated than the traditional technology inherited from the past, but less capital-intensive and generally more simple than the technology used in the modern sector. Its very survival and development, despite the strong competition from modern technology, testifies to its appropriateness and its vitality. Two or three examples can be given here as illustrations. One is the case of a Philipino entrepreneur who built up a starch separation plant using as only equipment 10,000 dollars worth of second-hand washing machines. The efficiency of his plant was so great, and his production costs so competitive, that the 'efficient' one and a half million dollar plant set up in the same region had to be closed down(1). Another case is that of a small Algerian factory making first class razor blades with completely outdated equipment. A modern plant, imported from abroad, would undoubted look much better, but stands little if any chance of outcompeting the appropriate technology of the indigenous plant(2). In Pakistan, there are several small pump manufacturers who have not only survived but also managed to improve their technology without any help from the government planners, without any access to the cheap credit facilities accorded to modern industry and without any opportunity to purchase the inexpensive imported raw materials which are channelled in priorit to the modern sector(3). In Thailand, the sea water which is left to evaporate in the large salterns south of Bangkok is pumped by dozens of small windmills (the technology was apparently brought by the Dutch in the 17th century) which are economically competitive and technologically well-adapted to the local environment.

Thousands, if not millions of similar examples can be found throughout the developing world. This vast pool of appropriate technology has never really been surveyed, and few if any systematic attempts have been made to facilitate its diffusion, even at the local level, improve it where possible and integrate it into a national development effort. This technology, which represents an enormous potential resource, is to a large extent an unexploited, or dormant resource. It plays an immensely important role in the survival of thousands of small and medium-size entreprises and helps provide its customers with products and services, which would otherwise not be available. Yet it is almost totally neglected by national planners and policy-makers.

---

1) M. Perichitch, Transnational Operations of U.S. Engineering Consulting Firms (mimeo), OECD Development Centre, Paris, 1975.

2) N. Jéquier, Vers une politique de la technologie - le cas de l'Algérie (mimeo), Centre d'études industrielles, Geneva, 1973.

3) Edgar Owens and Robert Shaw, Development Reconsidered - Bridging the Gap Between Government and People, D.C. Heath and Co., Lexington, 1972.

This technology, apart from its economic importance, has a psychological and cultural role to play. It represents an indigenous technological creation, and in many cases adaptation, which testifies both to the inventiveness of the local artisans, entrepreneurs and tinkerers and to their capacity for developing appropriate technologies. Its promotion can give developing countries a greater respect for their own creations and show that they also are capable of initiating and mastering the processes of technological innovation.

Technological innovation is not only a question of money or technical knowledge but also and perhaps even more of psychological self-confidence. This applies to individual entrepreneurs as much as it does to nations. What massive imports of the most modern technology have done to developing countries is not so much to disrupt the economic basis of their traditional industries as to undermine the self-confidence of the local innovators and the respect of the national policy-makers for indigenous innovations. What is foreign has come to be viewed as better, and the wider the gap between the existing indigenous technology and the imported modern technology, the greater the loss of self-confidence. This loss of confidence almost inevitably translates into a decline, or even disappearance of the capacity to innovate in an autonomous way.

This problem of self-confidence is among the most complex that policy-makers, and in particular the authorities responsible for industrial and technological innovation, have to face. It touches not merely on technology in the narrow sense of the word, but on such things as values, traditions and a society's image of its place in the world. It is interesting to observe that some of the nations which, in modern times, have been the most innovative from a scientific and technological point of view - France in the eighteenth century, England in the nineteenth century, Germany between 1870 and 1940 or the United States today - are also those which at the time, were at the height of their political power. In these cases, scientific and technological supremacy can be viewed as the cultural manifestation of political supremacy and, in a more subtle way, as the expression of a nation's self-confidence.

But the opposite can also be true. Witness for instance France's tremendous artistic, cultural and scientific vitality in the four decades that followed the disastrous defeat of 1870, England's scientific creativity today, or Japan and Germany's industrial dynamism after 1945. A sudden shock like military defeat or political retreat can in fact have the same stimulating influence on a society as political success. In this sense, innovation seems to be linked not simply to self-confidence as such, important as this may be, but also to the cultural and psychological processes through which a society tries to rebuild the self-confidence which was shattered by defeat, or, as in the case of the developing countries today, by foreign colonisation.

Viewed in this perspective, the appropriate technology movement has a major social function to fulfil. Not simply, as one may somewhat naively believe, merely to develop and diffuse new and more appropriate technologies, but more specifically to build up the self-respect and self-confidence of potential entrepreneurs and innovators in the developing countries, or for that matter of the less favoured communities and ethnic groups in the industrialised countries themselves.

One of the bases for building up this self-confidence is the vast and largely ignored pool of existing appropriate technologies which can be found in every developing country. This technology is seldom glamorous, its social and political visibility is poor, and those who use it have none of the social prestige which attaches even to the most menial work in the modern sector. A large part of it belongs to the world of what the French call 'bricolage', but one must not forget that much of the Industrial Revolution was due to the work of tinkerers, artisans and 'bricoleurs', and not only to modern science and high technology. This technology is for the most part a dormant resource, and one of the basic aims of a national strategy should be to turn this dormant wealth into real wealth(1).

---

1) The ways in which 'dormant' technology could be 'awakened' have been explored among others by P. Gonod in New Challenges for Technology Transfer (Worldtech Report No.1), Control Data Corporation, Minneapolis, 1975.

Chapter 2

THE INNOVATION SYSTEM IN APPROPRIATE TECHNOLOGY

With the wisdom of hindsight, the story of successful innovations, be they industrial or agricultural, would seem fairly straightforward: the need or demand for a new product or service leads to the development of a new technology or a new form of organisation to meet this need. Since the innovation presumably corresponds to a pre-existing demand, it will then diffuse widely throughout the economic and social system, and further stimulate demand. That the process is not quite so simple can be gauged from the fact that the great majority of innovations end up in failure. Success in effect is the exception rather than the rule, and a national strategy for the development of appropriate technology must take this into account.

Two approaches can be envisaged here. The first is to help increase very substantially - by one or two orders of magnitude at least - the total number of innovations in intermediate technology. Even if the proportion of failures remains the same, the global impact will be much greater than it is now. The other approach is to try to reduce the incidence of failure through a more systematic identification of the factors which contribute to the success of innovation. Both approaches are of course much easier to state on paper than to carry out in practice. Before trying to outline some of the measures which might be taken in the framework of an innovation policy for intermediate technology, it may be useful to take a closer look at the innovation system in general, and in particular at that part of the innovation system which deals with intermediate or appropriate technology. Which are the main components of the innovation system? Is innovation necessarily non-egalitarian? Why is the market so receptive to certain types of innovation and not to others? How and by whom are the needs and opportunities for innovation identified?

I. THE COMPONENTS OF THE INNOVATION SYSTEM

One of the basic principles of the various appropriate technology movements is that their new products, technologies or services should be aimed in priority at the poorest and most underprivileged segments of the population. The assumption, which can be verified in a number of cases is that the innovation system is oriented mainly towards the urban population and those people who have the highest income and not those social groups, particularly in the rural areas,which are in greatest need and stand the most to gain from technological innovation What the appropriate technology proponents are trying to do is to change this state of affairs by building up a parallel, or complementary, innovation system aimed in priority at the most underprivileged groups.

The difficulties involved are considerable and their magnitude can be gauged from a comparison between an innovation system which works rather well - that of the industrialised countries of the OECD area - and the innovation system in a typical developing country. One of the most conspicuous components of this system in the industrialised countries is the research and development capability. Today, we know quite a lot about the ways in which to manage R and D both at the corporate and the national level, and Alfred Whitehead observed quite rightly that the greatest invention of the nineteenth century was the invention of the method of invention(1). However, R and D leads to innovation only if it feeds directly or indirectly into the production process through a complex set of managerial, organisational and motivational links. Innovations at the production level - in the form of new products or new processes - must in turn be linked with market demand, and they stand little chance of widespread diffusion if there is not a real need for them.

In order to function effectively, the innovation system requires in fact much more than an R and D capability, a well managed production system and a close link with market demand. It needs an educational infrastructure to supply the qualified manpower, a credit system to finance the risks of innovation and the costs of investment, a transport and distribution network and, even if it is not called by that name, an intelligence or information system. But perhaps most important of all, it requires a reward system: no entrepreneur is willing to launch a new product on the market unless he has some prospect of making a profit, and the same is true of the farmer who buys a new agricultural machine or decides to cultivate an improved variety of wheat.

The reward system, which is institutionalised - somewhat imperfectly - in the price mechanisms, has both a positive and a negative dimension: the incentive to innovate is not always and only the prospect of a profit (positive reward), but also and very often the fear of being eliminated from the market (negative reward). The reward

---

1) Alfred N. Whitehead, Science and the Modern World, London, 1926.

system operates both at the private level - for the individual entrepreneur, farmer or industrial producer - and at the public level. Government policies, and innovations fostered by the action of government, are motivated not merely by economic and social rationality, but also by political needs - to satisfy particular constituencies, to avert social troubles, or even simply to survive. The price mechanism, in this case, is translated essentially into political risk, from industrial strikes to coup d'état, from defeat at the next election to revolution.

The proponents of appropriate technology are addressing themselves to countries, societies and social groups where the innovation system, or parts of it, are weak, ineffective, incomplete or in a state of lethargy. The great difficulty is not so much to develop new technology, complex as this may be, as to build up a more effective system. The problems which have to be solved are extremely diverse.

The first is the absence, or rather the relative weakness, of technical invention. None of the existing appropriate technology groups presently have the equivalent of the industrial firm's R and D laboratory, even though several groups linked with universities have developed, or are beginning to set up, something rather similar in the form of appropriate technology or rural development units(1). All societies have at one time or another in their history displayed a great degree of technical inventiveness, as witnessed by the enormous stock of traditional technology which was built up over the centuries and which is still used today. This capability however lies for the most part unorganised, and the reward system which could encourage its development often tends to operate as a disincentive. One example of these problems is the story of the 5,000 small shoemakers who were suddenly put out of business in one large developing country as a result of the construction of a modern plastic shoe factory. These craftsmen could almost certainly have improved their technology, given a certain amount of technical support and a tangible financial reward. This example incidentally points to one important problem - that of rapidity. Success in innovation depends very much upon timing, yet the people and the social institutions which have the potential ability to develop new technologies often find it impossible to keep up with the pace of change imposed by competition from the industrial sector. Time is a resource, and there is little doubt that if greater attention were given to its exploitation, the innovation system could be made more efficient.

---

1) One good example is the Rural Development Group of the University of the Andes in Bogota, Colombia. See the papers by B.A. Ntim and J. Powell, and by Rufino S. Ignacio in the second part of this book which describe the R and D activities of Kumasi University in Ghana and Mindanao State University in the Philippines.

The second difficulty in building up an innovation system is the absence of an equivalent of the industrial firm, with its institutionalised commitment to growth, its ability to take risks and its mastery of large-scale production processes. The successful village artisan may employ several people, but the transition to an industrial scale of operations is very difficult, and the small firms or workshops which can be found in the urban areas very seldom grow into big enterprises. Government policies to support small industry, while perfectly justifiable in their own right, often contribute indirectly to making this transition to larger scale almost impossible. The fiscal advantages for instance which are made available to companies with less than 50 employees (a typical figure) disappear as soon as the manpower roll exceeds this figure, thus creating a sudden and almost insuperable handicap which serves as a very effective disincentive to further growth.

The third difficulty is the absence of a marketing and distribution system. The farmer can often increase his production very substantially; but it makes little sense for him to do so unless there are roads and trucks to transport it to the consumption centres, as well as storage facilities on the farm or in the neighbourhood which allow him to keep the products for sale at a time when prices are favourable. As for the village artisan or the small industrial enterprise, they can only work for the local market and have no possibility of expanding their activities if there is no effective distribution system(1). The same difficulties arise with the supply system. The raw materials used by the farmer, the village artisan or the small firm are expensive because of transport difficulties, the small size of their orders and the high cost of energy. In the innovation system of the highly industrialised countries and to a lesser extent in the modern sector of the developing countries, the situation is much more favourable: the cost of energy (e.g. electricity or oil) is relatively low, and the same for everyone, the transport system works well and there are well established distribution, marketing and supply networks.

A fourth difficulty is the weakness of the financial infrastructure. Large firms can get low-interest bank loans, but the farmer or the small entrepreneur who seldom has any collateral to offer, must borrow from moneylenders at exorbitant rates (5 to 10 per cent per month is a common figure); such loans are usually short-term in nature, which makes it impossible to run any risk or to engage in any venture which does not have an immediate and very high pay-off. Furthermore, the government's economic and technological policy often

---

1) An analysis of these problems can be found in the paper by M.K. Garg "The Upgrading of Traditional Technologies in India: Whiteware Manufacturing and the Development of Home Living Technologies" in the second part of this book.

works to the detriment of all but the largest and most modern firms. The latter which are well organised, politically visible and economically important can lobby successfully for higher import duties, cheaper raw materials or special allocations of foreign exchange. Since in fact many of them have been set up by the government under the aegis of a national plan, the bureaucracy is politically and psychologically committed to their survival, and therefore prepared to grant privileges and advantages far in excess of what normal political lobbying on the part of their managers can even hope to obtain. Farmers, small firms or village artisans, who are unorganised, geographically dispersed and politically marginal, do not have the same advantages. In one country bordering the Indian Ocean, small innovative entrepreneurs find it almost impossible to import the single pieces of foreign machinery they need for stripping apart in order to build up their own technological capability through imitation. At the same time, large amounts of foreign currency are allocated for raw materials and new equipment to the large-scale modern industrial firms, many of which were built up in co-operation with foreign enterprises(1).

What these few examples and general observations suggest is that innovation is not simply a matter of new technology, even though technology can be vitally important. Innovation is part of a system and it can take place only to the extent that the components, or subsystems, function in an effective way. The great difficulty here is that while we know more or less how to invent, develop and produce new pieces of hardware, we know very much less about the ways in which an innovation system can be built up, and how the different subsystems interact with one another. In many cases, these subsystems and their interactions are more important than the new hardware itself.

The difficulties encountered in the introduction of well tested and apparently simple new technology like the ox-drawn plough in certain tropical African countries illustrates this clearly at the level of the individual farmer or the small community(2). First is the cost of the animals: 200 dollars - the typical price for two oxen - is a very large sum of money, even for a well-to-do farmer. This means that he cannot finance the investment with his own savings, even less with a loan from the traditional moneylender. He must therefore have access to some other form of credit which may not yet exist.

---

1) Such a situation is very frequent in developing countries with a rigid planning system. See the paper by P.D. Malgavkar, "The Role of Techno-Entrepreneurs in the Adoption of New Technology" in the second part of this book.

2) This example is based on a study of the Ombessa Project in the M'bam province of Cameroon and on a number of field reports from other Black African countries prepared by students of the Panafrican Institute for Development in Douala.

Training the oxen to walk in straight lines and respond to simple orders takes time, patience and skills which generally cannot be acquired locally. If the animals are worked for several hours a day, they cannot be left to forage for their own food as all other animals usually are. This means that the farmer has to produce fodder, which requires a lot more work; he must also be in a position to store it, another major problem in tropical climates. The iron plough, the harrow and the seeder, which cost a lot of money, are seldom manufactured locally, and if there is no village blacksmith around, they cannot be repaired when required. Production, despite these problems, can often be increased in a very significant way, but this is irrelevant if there is no way of storing the surplus and the economics of the whole innovation becomes questionable if the entrepreneurial farmer cannot market the additional output at a fair price.

II. INNOVATION AS A DIFFERENTIAL PROCESS

The process of innovation, be it social, cultural or technological, is by nature a differential process. The farmers who are the first to purchase a new type of machine, the industrialist who launches a new product or the family which turns to a new religion represent a very small minority in their community, and it takes years if not decades before the innovation they were the first to adopt or to develop becomes widely accepted. These innovators, who often happen to have close links with the outside world play the role of 'gatekeepers' in the innovation process(1). Their social function is to test the innovation and serve as a first filter in the innovation process. If they are successful, the innovation will acquire the social and economic legitimacy which is essential to its widespread diffusion.

The differential nature of the innovation process means that those who, because of their poverty, their lack of education, their inferior social position or their poor professional qualifications, stand the most to gain from innovation, are in fact those who benefit from it last of all. Innovation first reaches those who are capable of paying for it and those who, for cultural or social reasons, have the greatest ability to absorb it. This phenomenon can be observed in countless instances, from medieval Europe to contemporary Asia,

---

1) See Thomas J. Allen, "Communications in the R and D Laboratory", Technology Review, Vol.70, No.1, October/November 1967 and J.M. Piermeier and F. Cooney "The International Technological Gatekeeper", ibid., Vol.73, No.5, March 1971.

from agriculture to industry, and from religion to ideology(1).

Innumerable efforts have been made by governments, private associations, political groups and other institutions to correct this state of affairs. In the medical field, by building up public health services for those who cannot benefit from the continuous innovations which are taking place in the modern sector of high-technology medicine(2). In the educational field, by investing relatively large sums of money in underprivileged communities. In the industrial field, by supporting small enterprises, subsidising development in the peripheral and poorer regions, and by fostering the redeployment of industry away from the big centres. These objectives have in most countries been institutionalised by government through the taxation system and the social security services.

The proponents of appropriate technology are trying to do the same thing at a somewhat different level, namely, to introduce a greater equality or fairness in the development process by focusing in priority on the social groups which for various reasons have been by-passed by the social and technological innovations which are needed, but which they cannot afford.

The great difficulty is that this objective of equality, or fairness, stands in opposition to the differential and unegalitarian nature of the innovation process. Innovators or inventors are people who are prepared to take certain risks, and one of the social functions of the first recipients of an innovation is to underwrite the risk which is inherent in the use of a new product, a new technology or a new way of doing things. The first innovators pay a high price for the new product they have purchased, but at the same time, they often gain social prestige, quite apart from the convenience of the innovation itself. What this means is that if innovations in intermediate or appropriate technology are to reach the widest number of people, the risk must be minimised or socialised.

A poor farmer who has been persuaded by a well intentioned extension officer to buy a small insecticide sprayer for instance, must be absolutely sure that it will bring in big and immediate benefits; if it breaks down, he will have lost not only a lot of

---

1) See for instance Lynn White, Medieval Technology and Social Change, Oxford University Press, Oxford, 1962. Kalpana Bardhan and Pranab Bardhan, "The Green Revolution and Socio-Economic Tensions: The Case of India", International Social Science Journal, Vol.XXV, No.3, 1973. George M. Foster, Traditional Societies and Technological Change, Harper and Row, New York, 1972. Oriol Pi-Sunyer and Thomas De Gregory, "Cultural Resistance to Technological Change", Technology and Culture, Vol.V, No.2, Spring 1974.

2) One interesting example of this is the training of 'all-purpose' doctors in Cameroon. See "Cameroon's New Kind of Doctor", Development Forum, October 1975. These all-purpose doctors are in fact the equivalent of the Chinese 'barefoot doctor' or the Russian 'feldscher'.

money, but also much of his confidence in new technology and in the wisdom of the extension agent. The small handcarts whose diffusion is actively promoted in the rural areas of tropical Africa, must be sufficiently strong to withstand the excessive loads that are piled onto them, and their wheels must be designed in such a way that the cart does not get bogged down in the forest tracks during the rainy season.

These two small but very typical examples suggest that the design of an intermediate or appropriate technology, contrary to what is generally believed, is often very complex from an engineering point of view. In fact, the weaker the innovation system and its components (e.g. repair services, educational level of the users, credit facilities, transportation network, etc.), the more important the reliability and economic attractiveness of the hardware. The most appropriate technology is the one which entails the least risk for its user: poor people cannot run any risks.

The range of technologies which meet this criterion of low risk and high reliability is not at present very wide and there are strong indications that the appropriate technologies which are currently available in some part or another of the world include a high proportion of products, processes or services which are still largely untested or which, if already tested, are still too unreliable to make an effective impact. The history of the cow dung gas cookers in India is in this respect very instructive. Development work has been going on for almost forty years, major strides have been made in their design but, as M.K. Garg has shown, their cost and reliability are still not what they should be, and their diffusion as a result has been limited(1). In fact, most of the technologies developed by appropriate technology groups are still in the experimental phase, and a much wider and more comprehensive effort is required to make them truly competitive in terms of the environment at which they are aimed. By contrast, the appropriate technologies developed in the informal sector by local entrepreneurs and craftsmen have for the most part already withstood the difficult test of practical application, as the cases of the Pakistani pump manufacturers or the Philippino starch separation plant mentioned earlier clearly show. When evaluating the technological potential of this informal sector, account must be taken of the fact that its products and processes are often a more viable proposition than the experimental and untested technology of a newly-established appropriate technology group.

The differential nature of the innovation process and the fact that the risks of innovation are normally underwritten by the first

---

1) See M.K. Garg, "The Upgrading of Traditional Technologies in India: Whiteware Manufacturing and the Development of Home Living Technologies" in the second part of this book.

and relatively affluent users of a new technology has an important implication as far as national policies for appropriate technology are concerned. If new products (e.g. agricultural machines, durable consumer goods, etc.) are all imported from abroad to meet the needs of the successful farmers or the educated urban consumers, local inventors, innovators and entrepreneurs will as a result find themselves cut off from any access to this initial market represented by the people who have both the ability to pay for a new technology and the aptitude to withstand risk. These imports in effect preempt the market which is so vitally important to the initial success of any new indigenously-designed technology and local innovators and entrepreneurs have to turn to customers who have less money and a lower propensity to take risks. Importing foreign-made tractors for instance may well be economically justifiable, but the market represented by the first farmers who have bought them will as a result be closed to the local manufacturers who desperately need this initial market to build up their technical capability.

III. PUBLIC TECHNOLOGY AND PRIVATE TECHNOLOGY

The problem facing the appropriate technology movement is not only to develop new technology, but also to build up an innovation system. This can be tackled from two complementary angles: production and consumption. Production or supply is represented by the inventors and entrepreneurs who develop a new technology, manufacture a new product or provide a new type of service. Consumption or demand is represented by the people who will be buying or using these new products and services. These two aspects are closely inter-linked and any attempt to build up an effective innovation system must consider them simultaneously. There is little point in developing a new technology, however ingenious or appropriate it may be, if no effort is made at the same time to develop a market for it. In fact, most innovations end up in failure because of this very weakness in marketing.

However, it is somewhat naive to believe that success or failure is determined principally or exclusively by missmatches between demand and supply, or technology and the market. Turning to the developing countries, one can observe that the innovation system, which in some areas appears very ineffective and sometimes totally paralysed, is in some other areas very effective. Witness for instance the very rapid diffusion of products such as transistor radios, watches, metal household goods, mechanical sewing machines or Western styles of clothes, the eagerness with which farmers in tropical areas make the transition from thatched roofs to tin roofs, or the amount of work which goes into saving up enough money for the purchase of a second-hand automobile.

The receptivity of the market to certain types of products or technologies is not simply the result of clever manipulations on the part of advertisers or the symptom of cultural imperialism on the part of the industrialised countries. Rather it is because most products of modern industry correspond to a very real need. They are reliable, easy and convenient to use, relatively inexpensive and meet a need which traditional products and technologies either cannot meet at all (as in the case of the transistor radio or the sewing machine) or cannot meet in as convenient a way. The tin roof lasts for ten, twenty or thirty years, depending on quality, while the thatched roof which often leaks, has to be redone almost every year. As for the motor boat, it is infinitely more convenient than the traditional row boat or sail boat. Even in the most remote regions of the tropical countries, the poor farmers who grow coffee find it much more practical to use instant coffee rather than go through the tedious and rather complex work of preparing a drink from their own crop(1).

The remarkable receptivity of the market to modern products of this type is a very important social and cultural phenomenon. Yet its significance tends to be overlooked both by national policy makers and by the appropriate technology proponents. For the former, these products and technologies appear much less interesting or important than steel mills, dams, petrochemical plants or universities. For the latter, they are all too often seen as the very expression of an industrial society to which they are seeking an alternative.

The rapid diffusion of these types of products and the eagerness with which they are accepted by the market is due only in part to their competitiveness, to the prestige which is associated with their use or to their immediate and measurable usefulness. Another reason is that they belong for the most part to the private economic sphere of man: the purchase of a transistor radio, a sewing machine or a tin roof for instance is a decision which is taken by the individual or his family on the basis of his own resources and ability to pay.

When it comes to something like a village water distribution system, a sewage system or a new type of crop, the decision to innovate is no longer in the hands of the individual, however directly he may be involved. Decisions of this type require some form of social consensus which is much more difficult to achieve than consensus with the family. In the industrialised countries, the farmer can choose, within certain limitations, the type of crop he wants to grow. In traditional societies where land often belongs to the community, or

---

1) This problem of receptivity to certain types of innovations and not others has been noted by many observers. One of the most perceptive analyses of this problem is Peter Drucker in "Modern Technology and Ancient Jobs", Technology and Culture, Vol.IV, No.3, Summer 1963.

where it is attributed on a temporary basis to the individual in accordance with tradition, family ties or social standing, the farmer's decision is in fact dictated to a large extent by the community, the village elders and the extended family. This means that any new technology in this field is subject to a much tighter filtering system. Introducing a new piece of infrastructure is equally complex: not only because the total price is relatively high, but also and especially because it involves changes in the village's patterns of living, and legitimation through some form of social consensus.

When considering the problems of innovation in appropriate technology, it might be in fact useful to make a basic distinction between three types of technology:

- Private technology, which often takes the form of consumer goods and especially durable consumer goods (the tin roof, the radio, etc.) and whose introduction depends almost exclusively on the decision of the individual and his family;
- Community technology which includes not only basic infrastructures like water supplies, education, transport networks or drainage systems, but also most of the production technologies of the individual farmer or craftsman. Infrastructures are by nature a community technology, but so are most of the production technologies since they touch upon the community's collective goods (land, water, economic relations, social structures, etc.);
- Public technology represented by the large industrial firms which produce consumer goods or capital equipment (steel mills, fertiliser plants, etc.) and the national institutions which supply certain basic services (railway transportation, electricity distribution, higher education, credit system, etc.).

Obviously this distinction between the three types of technology is somewhat arbitrary, and the line between them is seldom absolutely clear. The public technology sector for instance produces not only public goods (e.g. roads or education) but also consumer goods whose purchase depends almost entirely on private decisions.

The only function of the distinction at this stage is to suggest that the innovation system does not function with the same efficiency in each of these three sectors. In the private technology sphere it generally operates quite effectively and rather rapidly. The same is largely true in the public technology sector. The decision to invest in large-scale plants or big infrastructural projects are taken by the central government, often with the assistance of foreign technicians, and the effective resistance which faces such innovations is generally limited. In fact, the less democratic the political system, the easier it is to build up the public technology sector, and in this respect central planning, under the aegis of which such innovations usually are introduced, is fundamentally undemocratic since there is no public control over the decisions taken by the technocrats.

In most developing countries, that part of the innovation system which functions least well of all is the one which deals with community technology. There are a number of reasons for this. First is the fact that, unlike what happens in the private technology sphere, there is no effective pricing mechanism: it is virtually impossible to put a price tag on such products as a feeder road, a village-level energy system (e.g. solar pump or methane gas producing unit) or a water distribution system which replaces the individual well. This problem of price is in fact somewhat more complex. Every product, every service and every technology have a price, a cost and a value. In the private technology sphere, there is some sort of equilibrium between the three: the consumer is willing to purchase a product if the price corresponds to his assessment of its value to him, and the producer fixes his prices according to his costs and on the basis of what he thinks the customer is prepared to pay. In the community technology sphere, the balance is much more difficult to achieve. Value is determined not only by economics, but also by social norms, and the consensus which is required in order to take a decision acts as a very effective barrier against innovations which may be very useful, but which are too new or too different from the standard practices. Costs are very difficult to allocate on a collective basis: those who pay the largest share of the cost often expect to receive more than proportional benefits. Finally, the pricing structure often bears little relationship with effective costs: community technology is either underpriced relative to its cost (as in the case of education or agricultural extension services) or overpriced (e.g. health or transport), and it is almost impossible to establish a fair balance between cost, price and value.

The second reason for the weakness of innovation in community technology is that unlike what happens in the parts of the public technology sphere, detailed planning and central decision-making is virtually impossible. Economic analyses can show that a country needs an automobile factory, a big hydroelectric dam or a fertiliser plant. But it is practically impossible to plan in detail the requirements of every village and community in a country and make valid trade-offs between a wide range of possible projects. A third difficulty is that a relatively large part of any community technology is composed of software rather than hardware, which is inherently much more complex and difficult to manage and control.

The appropriate technology movement is concerned primarily but not exclusively with community technology. What is the role of such groups in the innovation system? How do they identify the demand for innovation? What are their motivations? And how do they perceive the need for innovation?

IV. THE ROLE OF THE APPROPRIATE TECHNOLOGY GROUPS

The organisations which are presently playing the most active part in the development and diffusion of new appropriate technologies can be divided into three big categories:

- Higher education institutions, or rather, small groups which originated from and are closely linked with such institutions, like Kumasi University in Ghana, Mindanao State University in the Philippines, the University of the Andes in Colombia or the Technische Hogeschool Eindhoven in the Netherlands;
- Governmental, private or semi-public organisations specialised in intermediate technology and working primarily on a national basis, like the Appropriate Technology Cell of the Ministry of Industry and the Gandhian Institute of Studies in India, or the Appropriate Technology Centre in Pakistan;
- Multinational groups or research centres, like the London-based Intermediate Technology Development Group (ITDG), Volunteers in Technical Assistance (VITA) in the United States or the International Rice Research Institute (IRRI) in the Philippines.

These organisations all play a major part in identifying the needs for appropriate innovations. Most of them are actively involved in the development of new hardware or software. Few of them, however, are directly engaged in manufacturing activities. The International Rice Research Institute for instance has developed many new types of agricultural equipment specifically suited to the small holdings of low-income farmers in Southeast Asia. Once developed, the technology is transferred to local manufacturers and entrepreneurs who, with the assistance of the Institute, turn it into a viable industrial and commercial venture(1). In the same way, the home living technologies developed by the State Planning Institute in Lucknow, India, are brought via a very effective extension service to the local craftsmen who play the role of small-scale manufacturers and salesmen(2).

The appropriate technology groups play in effect the role of a mediator in the innovation process. In the innovation system of the highly industrialised countries, there is no need for a mediator: the

1) See the paper by Amir U. Khan, "Mechanisation Technology for Tropical Agriculture", in the second part of this book.

2) M.K. Garg, "The Upgrading of Traditional Technologies in India: Whiteware Manufacturing and the Development of Home Living Technologies", ibid.

industrial firm, which is one of the main agents of innovation, performs within the same organisational framework the three functions of identifying a demand for innovation, developing the technology to meet this need and bringing the technology to the market place. In rural societies living at subsistence levels, there is no equivalent institution, and the task of the appropriate technology groups, or for that matter of all the organisations which in one way or another are concerned with development, is to try to identify the real needs of the local communities, develop or introduce technologies and organisational means which can somehow meet these needs and, what is probably the most difficult of all, contribute to initiating a process of development based on the internal innovative forces of the local community.

Most of the appropriate technology groups currently working in the developing countries are too young and their experiments too new to allow for any meaningful generalisations about their achievements and their mode of operation. Some general observations can nevertheless be made, notably as far as the identification of needs for innovation is concerned(1). There are in fact three general patterns in the process of identifying needs for innovation.

The first is the major importance of medium to long-term economic assessments of specific local situations. The economic forecasts which underlie the innovations of appropriate technology groups generally go into considerable detail, and the sophistication of the cost-benefit analyses is comparable with that of large-scale industrial projects. Although they are dealing mainly with small-scale technologies intended primarily for the rural populations, it may be interesting to observe that the assessment methods belong culturally and psychologically to the modern, Westernised and highly educated segment of society.

A second pattern which emerges is that the need for appropriate technology tends to be evaluated in macro-economic rather than micro-economic terms. Employment generation, import-substitution, public health, rural development or social equality are concepts which belong to the system of reference of the economic planner or the political reformer, not to that of the individual entrepreneur or the small farmer, who is more concerned with profit, convenience and the practical means of survival. This points to one of the basic problems of innovation in appropriate technology - namely the conflict between the appropriate technology group's emphasis on general economic and social priorities, and the emphasis given by the individual to immediate economic benefits.

---

1) This analysis is based for the most part on the original twenty-five case studies presented at the OECD Seminar on Low-Cost Technology, Paris, September 1974. Most of these studies can be found in revised form in the second part of this book.

The third pattern is what one might call the trans-economic dimension of innovation: in most cases, the identification of a need for innovation goes far beyond pure economics, and touches upon ideology and a long term vision of a more egalitarian society.

V. THE NEED AND THE DEMAND FOR INNOVATION

Innovation does not occur simply as a result of a need, but rather in response to a demand, expressed in terms of an ability to pay a certain price for a product or a service. Need and demand are thus two very different concepts, even though they do overlap to a certain extent. Most of the innovations in low-cost technology promoted by appropriate technology groups attempt to meet a certain need - for energy, for housing, for education, for managerial assistance, for water or for employment - mainly in the rural areas of the developing countries. What is often not very clear is whether these needs can be translated into an effective demand for the products and services resulting from a given innovation.

Two examples can be given here to illustrate the point. The first is that of the Indian methane gas plant, and the second is that of the Sahelian solar pump. Technologically, the Indian gas equipment is a very ingenious piece of hardware(1). From the viewpoint of the national economy, it is well adapted to local conditions, since it uses a local material available in relatively large quantities, namely cow dung. Furthermore it fits in with the traditional patterns of energy consumption. Its price however is relatively high: a household unit costs the equivalent of several months of income of the average Indian family. And some five cows are needed to supply the plant with its raw material. Socially, economically and politically, there certainly is a need for equipment of this type. But the only demand, at this stage at least, is that coming from the relatively rich farmers.

The solar pump developed by SOFRETES is a somewhat similar case(2). It is technologically well adapted to local conditions - virtually no maintenance is required - but its price will presumably remain very high relative to the income of the people for whom it is intended. According to recent estimates, large-scale industrial production could reduce its price, or rather cost, from $40,000 to less than $8,000. This is still a very large sum of money for a

---

1) See M.K. Garg, op.cit.

2) See the study by J.P. Girardier and M. Vergnet, "The Solar Pump and the Problems of Integrated Rural Development", in the second part of this book.

Sahelian village community, and its application will have to be considered in the framework of a much broader public programme for infrastructural development. Furthermore, the social problems posed by its introduction are considerable. In the Mauritanian village where the first installation was made, the two richest people occupied the pipeline and are now selling the water to the villagers. The permanent availability of water has relieved the children of their traditional task of drawing water from the well and carrying it back home, which means that some other way must be found to keep them out of mischief and get them to contribute to the economic life of the community. Finally, the increase in water supplies might well upset the delicate ecological equilibrium which for centuries has allowed such arid zone communities to survive.

That need and demand are two very different concepts is obvious. What is perhaps less obvious is that the notion of need is rather subjective, and depends a lot on one's culture and values. Most appropriate technology projects are motivated by a need, but the need is not necessarily perceived in the same way by those who develop the technology and those to whom it is addressed. The introduction of piped water supplies in one African community brings this difference out very clearly. To those who brought this new appropriate technology, it was a major contribution to public health and the eradication of water-borne diseases, and it corresponded to what they perceived according to their standards as a basic need. However, it was soon found out that the villagers were continuing to use water from the river for their drinking purposes because it tasted so much better! Their perception of the need was in effect completely different from that of the outsiders trying to introduce a new technology.

This example, small as it may be, suggests that one of the important factors in the diffusion and adoption of intermediate technology is the compatibility between two systems of perception of needs: that of the technological innovators, and that of the people to whom the new technology is addressed. The difficulty is that the two systems seldom coincide. Furthermore, the fight between them is somewhat unequal. The innovator is almost always an outsider, he tends to be highly educated and familiar with what is going on elsewhere and he often has at his disposal some form of support - financial, technological or even only moral - which although usually very small, is nevertheless important relative to the resources of the community in which he is working. As a result, his own perception of the needs for innovation tend to gain the upper hand over what the farmers perceive as their most important needs. The innovation may well take place, but once its promoter has left, it runs the risk of being neutralised or even rejected by the community.

One of the most interesting efforts to match these two sets of perceptions is that undertaken in the field of technology extension by INADES, the Institut africain pour le développement économique et social(1). What distinguishes this extension service from similar efforts undertaken elsewhere is the active participation of the farmers in the definition of their educational needs. Interestingly enough, the educational programme suggested by the farmers in co-operation with the extension workers corresponds almost exactly to the package programme of any classical extension service. The difference here is that the programme is not an alien import, but a creation of those who will be using it. Another important characteristic is the emphasis on the 'know-why' rather than on the 'know-how' or 'know-what': the farmer does not only want to know what should be done or how it should be done, but also and perhaps even more why a new type of fertiliser is more effective, or why a new way of tilling the soil will help his plants to grow better.

VI. PARTICIPATION AND COERCION

The success of an organisation like INADES in building up an effective agricultural extension service can be attributed in large part to the active involvement of the farmers in the innovation process. They have come to feel directly concerned by the new technologies which are offered to them, they know that their previous knowledge and experience is not useless, they are learning that progress depends upon their own decisions and choices, and they are finding out for themselves, with the help of the extension agent, why new ways of doing things are better. Many of the government-sponsored extension services in developing countries fail to take this need for participation into account and their impact in the long run is, as a result, somewhat limited. The problem here is not that of technology or knowledge as such - all extension services provide the farmer with the same type of knowledge about crops, implements, fertilisers, land use or management methods - but rather the way in which technology is transferred to and assimilated by the user.

The same problems arise with the diffusion of appropriate technology to small firms, village communities or private entrepreneurs. The individuals or the groups which are trying to promote the adoption of more appropriate technologies find themselves, like the extension agent, in a position of authority vis-à-vis the community they are trying to help, and the temptation is great to use this authority in order to promote change from above. Innovation, or for that matter

---

1) See the study of P. Dubin, "Education as a Low-Cost Technology", in the second part of this book.

modernisation, always involves an element of coercion and it would be naive to believe that it can take place in a balanced and orderly way through participation and initiative from below. This is true of nations and industrial firms, as it is of farming communities or individuals. Peter the Great's modernisation of Russia or Kemal Atatürk's modernisation of Turkey were neither democratic nor participatory, but they were nonetheless quite effective, even if the social cost was very high. And Governor Huey Long's modernisation of Louisiana in the 1930's was anything if not dictatorial.

Innovation in fact is seldom a free choice, and very often it is imposed by necessity. The industrial firms which launch a new product on the market generally do so because they are forced to under pressures from their competitors or as a result of government intervention and regulation. The farmer who introduces a new crop or decides to buy a new machine does it not only because he thinks it might be profitable, but very often because he is forced to if he wants to survive.

The same is often true of nations. In a country like Denmark, for instance, the modernisation of agriculture which took place in the last decades of the nineteenth century was not initiated by careful planning, but imposed by economic disaster. The unsuccessful war against Prussia in 1864 had cost Denmark 40 per cent of its territory and most of the export earnings which until then had come from the sale of wheat from Schleswig-Holstein.

Innovation always involves a certain balance between coercion from above or from outside, and participation from below or from within. An exclusive reliance on coercion may be very effective in the short run, but as the examples of Turkey after Atatürk or Russia after Peter the Great show, the innovation system returns to lethargy as soon as coercion ceases or becomes less effective. In Denmark, where social structures were very different, and farmers much more individualistic, yet bound by long traditions of co-operation and mutual help, the modernisation which was imposed by outside circumstances stimulated rather than stifled the development of invention and innovation, and paved the way to the development of an efficient and well organised agriculture.

Farmers who have always been told what they should do, either by government officials, religious authorities or village leaders, and who remember that their parents and their grandparents, like themselves, never had any control over their own destiny, may continue to accept, or rather adapt to, innovations imposed by outsiders or by their traditional leaders. Innovation in this case is a purely passive process of adaptation.

Participation, which can be deliberately fostered by outside innovators like missionaries, extension agents or appropriate technology specialists, is not merely a means of legitimising innovation or making it psychologically more acceptable by directly involving those who will be affected by it. Rather, it is a tool for building up the self-confidence and sense of independence and control over one's future which is one of the preconditions for invention and for an active involvement in the innovation process.

Fostering participation is a difficult, delicate and slow process. In fact, in the short run, helping people to help themselves by encouraging them to become active rather than passive innovators is often a much less efficient and usually slower way of modernising a society than coercion. Most of the appropriate technology groups and many development agents, both local and foreign, are very conscious of this. Yet there are strong indications that a substantial proportion of the development projects based on appropriate technology are for the time being of a non-participatory nature: the beneficiaries of these innovations are not directly involved in the definition of their major needs, and they do not take any direct part in the development, testing and improvement of the technology which is offered to them.

It may be interesting to observe here that participation and the control of the individual over his own destiny and that of his community are at the basis of the social system of two countries whose agricultural and industrial development conforms most closely to the idea of appropriate technology - namely the United States and contemporary China. In the United States, they result both from the Protestant ethic and from the fact that American society was a society of immigrants, who as newcomers were culturally attuned to the idea that the future was theirs. In China, the Maoist idea that the most important resource of a nation is man, and that every peasant and every worker can and should be an inventor is not essentially very different, and the aim of the 1949 revolution was to destroy the social structures which through their patterns of authority and coercion prevented the individual from playing this very role.

Chapter 3

THE INFORMATION NETWORKS

Most of the existing appropriate technology groups devote a considerable part of their activities to the collection, processing and dissemination of information, and their role as 'knowledge centres' tends to be more important than their activities in the field of research and technical development. This emphasis on information stems largely from a recognition of the fact that one of the first requirements of an effective innovation system is the development of an information network.

Knowledge about modern technology tends to circulate very rapidly. Even in the most remote regions of the world farmers have heard about such things as the moon landing, the jumbo jet or the achievements of modern medicine, and their awareness both of modern technology and of the way of life that goes with it is one of the many reasons for emigration to the cities. This knowledge, which is usually of a rather general nature, is propagated by hearsay, by the stories of town dwellers or foreign emigrants who return to their home village, by transistor radios and by illustrated magazines, whose carefully cut-out pages are pasted on the inside walls of the farmers' huts.

By contrast, knowledge about appropriate technology can take years to travel a hundred miles, and farmers in one village often do not know about the technological innovations which have been developed or introduced by a neighbouring community. The same is true at the national level: the experiences in appropriate technology undertaken in one developing country are seldom well known the other side of the border, and the lines of communication from one developing country to another are rarely direct, and very often have to transit through an industrialised country. There are of course many exceptions to this general pattern, but the fact remains that the awareness of modern technology tends to be much greater than that of the less sophisticated and less glamorous intermediate or appropriate technology.

As a result of this weakness of the information networks, the diffusion of knowledge about newly developed appropriate technologies is usually slow and often ineffective. A more serious problem is that the vast pool of appropriate technology which can be found in the informal sector (as opposed to the formally organised appropriate technology groups) lies for the most part unexploited, except at the local level where it is used by the small enterprises and individual innovators who have developed or improved it.

One of the functions of the information network is to create and stimulate demand. As George McRobie has very rightly noted, the demand or market for appropriate technology depends to a large extent upon the knowledge that such a technology is available(1). A farmer will not think of buying or making a solar heater or a methane gas cooker - to take two cases of relatively well-known appropriate technologies - unless he knows that such equipment can be built or eventually purchased. Nor can he be expected to think of using anything else but mud bricks and a thatched or a tin roof to build his house if he is not aware that better bricks can be made from soil-cement, or that there are better ways of covering his dwelling. The role of an information network is to create an awareness of alternative economic and technological options. This awareness will not necessarily be translated into a demand - poor people seldom have the money for anything but the most basic items - but it is a precondition for stimulating demand and for giving the local inventor or artisan the idea that he can develop something more appropriate.

I. THE HANDBOOK OR CATALOGUE APPROACH

One approach to this problem of information which has been tried out by several organisations is that of the handbook or catalogue. Two generations ago, the Sears and Roebuck mail order catalogue played a major role in bringing to the isolated American farmers and the small towns both the products of the emerging consumer society and the awareness that these products existed. The same role can be played by a handbook of appropriate technology. One example is the Village Technology Handbook prepared by VITA (Volunteers in Technical Assistance) which explains in simple terms and with the help of numerous illustrations how to make such things as a water pump, a better house, a local irrigation system or food storage facilities(2).

Two somewhat similar catalogues are Brace Research Institute's Handbook of Appropriate Technology(3) and the Directory of Appropriate

---

1) See the paper by G. McRobie,"The Mobilisation of Knowledge on Low-Cost Technology: Outline of a Strategy", in the second part of this book.

2) Village Technology Handbook, VITA, Mount Rainier (Maryland) 1975.

3) Brace Research Institute, Handbook of Appropriate Technology, Quebec, 1975. The story of the development of this handbook is told in the paper by T.A. Lawand et al. in the second part of this book.

Technology prepared by the Gandhian Institute of Studies in India(1). These three handbooks which are a good example of appropriate technology - their production costs were small, and the final product is available in many cases free of charge - all contain a wide range of technical information covering the whole spectrum of activities in a typical community of a developing country. Their breadth obviously means that only a selected number of technological alternatives can be presented.

Alongside these general catalogues or manuals, there are a number of more specialised handbooks dealing with a specific topic. One example is Christopher Cone's handbook on automotive operation and maintenance, which gives hundreds of very appropriate pieces of information on the most effective way to operate a car or a truck in a country with bad roads, a weak servicing infrastructure and a difficult climate(2). It deals with problems which are seldom encountered in an industrialised country, and which a driver can otherwise only learn about at great cost and through bitter experience - for instance how to extricate a vehicle from sand or mud or to make repairs when no spare parts are available.

Rather similar in scope if not in subject but aimed at a somewhat more specialised audience is the manual prepared for the Agency for International Development by a group of technologists from the Monsanto Company and which presents a wide range of low-cost roofing materials that are available to, or have been developed in developing countries(3).

These various handbooks, unlike the commercially-oriented catalogue of a typical mail order firm, do not offer products, but rather knowledge and technology. For the user, this knowledge is available in most cases free of charge, but it is up to him to put it to use. In fact, making any one of the devices described in these handbooks is not always very easy and involves not only great care, but also a certain amount of creativity. By presenting a number of practical possibilities and outlining the basic technical steps, such handbooks can contribute significantly to the development of invention and innovation at the local level.

Although it is difficult to evaluate the total cost of preparing such handbooks - the experience of the specialists which took part in their elaboration cannot really be quantified - there are indications

---

1) Gandhian Institute of Studies, Directory of Appropriate Technology (mimeo), Varanasi, 1973.

2) E. Christopher Cone, Automotive Operation and Maintenance, VITA, Mount Rainier, 1973.

3) I.O. Salyer, G.L. Ball, R.A. Cass et al., Development of Low-Cost Roofing from Indigenous Materials in Developing Nations, Monsanto Research Corp., Dayton, 1974.

that it is quite low, and at any rate very inexpensive relative to the benefits it can bring to innovative users: the order of magnitude of the direct expenses involved is around $100,000.

Along with these general or specialised directories and catalogues, there is a rapidly growing number of periodicals concerned in whole or in part with appropriate technology. For instance, the Intermediate Technology Development Group's journal Appropriate Technology, the Quarterly Newsletter of the Engineering Experimentation Station of Georgia Institute of Technology, the Documentation Bulletin of SENDOC (the Indian Small Enterprise Documentation Centre) or the regular publications of CENDES (the Center for the Industrial Development of Ecuador).

These examples show that the process of building up an information network in appropriate technology is already well engaged. At this stage, three types of markets, or target groups, for this information seem to be emerging. The first is that of national policy makers, aid agencies and large industrial firms. The second is that of the extension officers, rural development agents or educators. The third is that of the individual farmers, craftsmen and rural dwellers. Although there is a certain amount of overlapping between these groups, each one forms a distinct market, which calls not only for a different product, but also for major differences in presentation and packaging.

II. THE ROLE OF DOCUMENTATION SERVICES

Handbooks of the type prepared by Brace Research Institute, VITA or the Gandhian Institute of Studies are playing a major role in bringing about an awareness that alternative technological options are available and in stimulating the inventiveness of their users. These manuals which are aimed primarily at the developing countries, are also of relevance to industrialised nations, and it is interesting to note that, especially in the United States, several similar handbooks have been published by individuals or organisations active in the field of soft or alternative technology. One example is Robert de Ropp's guide to the alternate society, which shows for instance how to make use of unconventional energy sources or how to grow food without harming the environment(1).

For the farmer, the city dweller or the development agent, the information provided by such handbooks, while very helpful, is not always sufficient. Making the equipment presented in these manuals sometimes calls for materials which are not available locally, or which could more economically be purchased from an outside supplier,

---

1) Robert S. de Ropp, Eco-Tech, The Whole Earther's Guide to the Alternate Society, Dell Publishing Co,, New York, 1975.

and there is no point in trying to reinvent everything. Because of their rather general nature and the fact that they are not for the most part aimed at any specific country or region, these handbooks cannot give any information about sources of supplies or the purchase price of materials and components. The appropriate technology movement does not have anything equivalent to the highly popular Whole Earth Catalog in the United States, which gives its ecologically-minded reader a vast amount of information as to where and how to obtain the increasingly wide range of products based on soft technology(1).

Another problem is that as soon as one goes beyond the relatively simple, yet important, self-help technologies, information of a much more detailed and technical nature is required. The village blacksmith for instance can find out from such a handbook how to improve some of his tools, but when it comes to production technology this is no longer sufficient, and the knowledge he needs is very different. And the farmer whose crops are suddenly attacked by a pest he has never seen before cannot rely on the rather general and wide-ranging type of information which can be found in such handbooks The two most commonly used methods to solve problems of this nature are the documentation service and the technical assistance service (management consulting, technical extension and agricultural extension

Most of the existing documentation services in developing countries do not deal specifically with appropriate technology as such, but collect, process and distribute a much wider range of information dealing with technical, economic, industrial and agricultural matters. The effectiveness of such centres varies considerably from country to country but the problems they encounter seem to indicate that a traditional documentation and information service is not the most appropriate way to stimulate the diffusion and application of appropriate technology in a developing country.

The first problem is that such a service is generally a rather expensive proposition. Three elements come into play here: the collection of information, its storage, and its diffusion. In developing countries, collection costs are comparatively much higher than in the industrialised countries. This is particularly true if one considers not only written information but also the oral information which is gathered mainly through travels and personal contacts. Storage costs are equally high. According to some estimates which seem reasonably representative, storing one document or piece of information costs around one dollar and a half per year and account must be taken of the obsolescence of information.

---

1) The Whole Earth Catalog, Ballantine, New York, 1971; The Last Whole Earth Catalogue, Penguin, Harmondsworth, no date.

In order to be of some effective use, an information service must have a certain critical mass and be able to collect and store a substantial proportion (maybe 50 to 80 per cent) of all the information concerning a particular technology. In most developing countries, the number of potential users of this information is small: there may well be no more than one or two entrepreneurs or government agencies interested in any one piece of technology. The smaller the country or the market, the greater the relative amount of information which has to be stored for any one application. This of course increases costs and therefore it may be much more economic for developing countries to use the existing information systems of the multinational intermediate technology groups rather than build up their own systems. In the case of solar energy, to take one case, it makes more sense for a developing country to address itself to Brace Research Institute, one of the world's main centres in this field, than to build up at high cost its own documentation service.

Because of these high collection and storage costs, providing technical information to the user is expensive. Rough estimates based on the Indian and Ecuadorian experience show that the cost of one technical answer or piece of documentation about a specific technical problem is somewhere around $150. This is equivalent to the average per capita annual income in a typical developing country. To put things in perspective, one can imagine the problems which would face a documentation service in an industrialised country if each technical answer were to cost the same relative amount, namely between $3,000 and $5,000(1).

No farmer and few if any small industrial firms can afford to pay $150 for a piece of information which may turn out to be irrelevant, and one of the problems which has to be faced is that of pricing. Should the customer get the information free of charge on the assumption that this is a public service which must be financed by the community? Should he pay a price equivalent to the real cost? Or should he contribute merely a token fee? The first and last solutions are the two most frequent. However, it does not seem that the subsidisation of information services makes them more attractive to the potential customer. The real problem lies elsewhere, namely in the ability of the customers to absorb and use new information.

This is clearly illustrated in the very typical case of the techno-entrepreneurs of the Poona region in India(2). Here are young

---

1) In fact, the unit cost of documentation services in industrialised countries is generally small. In the case of the Irish Institute of Industrial Research and Standards, the cost of a typical inquiry is around $30 (this figure comprises only the direct operating costs of a technical information unit).

2) See the article by P.D. Malgavkar, "The Role of Techno-Entrepreneurs in the Adoption of New Technology", in the second part of this book.

industrial entrepreneurs with little practical experience but a good idea of the type of information they require in order to build up their first and very modest production facilities. As it turned out - and this seems to be a fairly general pattern - none of the existing information and documentation services in the region (and this is one of the most developed parts of India) could provide them with the type of information they required. They had in fact to build up themselves a tailor-made information network. First by using the knowledge available from their university textbooks and former teachers, second from the technical catalogues of hundreds of foreign firms whose addresses had been noted down in the foreign consulates of Bombay, and third from the careful study of equipment imported into India.

This study of the Poona techno-entrepreneurs shows very clearly that the search for information is an integral part of the inventive and innovative process. The only person who knows exactly what type of information is required is the one who will be using it, and a documentation centre, however well run and however complete its stock of knowledge, cannot be half as effective. Interestingly enough, the somewhat haphazard and artisanal way of gathering information of these techno-entrepreneurs turns out to be rather inexpensive - a few days or weeks of work at the beginning, plus the cost of letters and stamps - and the value of this information is very high relative to the cost of its collection.

One of the difficulties with the traditional documentation centres is that they tend to accumulate vast amounts of information which is often not used at all or which, when supplied to a customer, tends to be excessively voluminous, insufficiently selective and difficult to translate into usable knowledge. Such information services undoubtedly have a role to play, for instance as transfer agents for appropriate technology developed abroad, but other ways of dealing with knowledge and technology can be envisaged. One of them is for instance the "technology exchange" service which has recently been set up by Control Data, the big American computer firm. The principle is simple: any firm or individual who has a particular technology to sell can, for a relatively small fee, list this technology in a world-wide computer file. Conversely, those who are looking for a particular technology or piece of knowledge write down a request, also for a fee, and can find out within a very short time if the technology they need is available. With the rapidly falling price of computer time, this approach can be potentially very effective, but its main attraction is that it allows for an almost exact matching between a demand for technology (or information) and a supply of technology. Such a system, designed primarily for the industrialised countries, is equally relevant to developing nations, for it can just as well be used for appropriate technology.

There are strong indications that too much attention is currently being given to the building-up of traditional information and documentation services for appropriate technology, and not enough to the requirements of the market for information. The type of knowledge which the poor farmer requires is very different from that which is needed by an industrial firm, small or large. The delivery system furthermore has to be different in each case. The poor farmer, who may be illiterate, will in most cases not take the initiative to go to an official information centre, and the information has to be conveyed to him through an extension service. The large industrial firm by contrast tends to be a much more active customer, with a precise knowledge of what it needs, while the small family firm falls somewhere between these two extremes. Information services must obviously take these differences into account if they are to serve a useful purpose.

III. MANAGEMENT ASSISTANCE AND TECHNICAL CONSULTING

Information can to some extent be organised in a rational way and this is as true of information about sophisticated high-cost technology as it is of information about intermediate technology. However, the emphasis on rationality and organisation must not overshadow the fact that in the innovation process, some of the most creative ideas come precisely from disorder, luck and the use of apparently irrelevant information. Rationality is needed, but there are also virtues in duplications, ignorance and serendipity.

In the same way that the economies of the developing countries are characterised by the presence of a modern 'formal' sector and an 'informal' sector, the information networks in appropriate technology have a set of informal, or somewhat less well organised channels through which knowledge is conveyed to potential users. This informal information network has a number of components. Two of the most important are the various technical assistance services provided by appropriate technology groups, and the managerial assistance given by specialised organisations to small local firms.

One example of the former is the service provided by Volunteers in Technical Assistance. As its name does not suggest, VITA does not send volunteers abroad, but provides the equivalent of a specialised technical assistance service by linking the developing country customer with a special technical problem to a specialist in an industrialised country, who supplies this knowledge free of charge on a voluntary basis. Although no comprehensive evaluations of this service have yet been carried out, it appears to be very effective. One of the main reasons for this is that the customer has already clearly identified his problem and is anxious to use the information or the technology which is supplied by his adviser.

Technical assistance by correspondence may not be quite as effective as bringing a foreign consultant on the spot, but it is immensely less costly. According to VITA, the average direct cost for a piece of technical advice is around $100, and if the total costs are summed up, to include among others the time of the voluntary expert, the real amount is two to three times higher. Somewhat similar services have been organised elsewhere. One is the Development Reference Service of the Society for International Development in Paris, which deals not specifically with technology but with more general problems of development, and the other is the Industrial Inquiry Service of the United Nations Industrial Development Organisation in Vienna which is concerned with industrial problems. (Both services originated in the Question and Answer Service of the OECD Development Centre)(1).

Another approach to the diffusion of information and knowledge on appropriate technology is through technical and managerial assistance to small firms and local entrepreneurs. Practically all developing countries today have some sort of national organisation aimed at promoting the development of small enterprises. One of the usual tools is to provide firms with some sort of managerial assistance, for instance in the form of short training courses or consultancy services. With the exception of some countries like India, where services of this kind have been operating for many years, most of these organisations are too new to allow for a meaningful evaluation(2). There are, however, a number of problems which seem to be very general. The small firms which are being helped today should have the potential to become the big corporations of tomorrow. It is however extremely difficult to identify those which are likely to be the winners in the long run, and even the most promising small entrepreneurs may fail.

What is more, there seems to be little, if any, selectivity in the choice of small firms to which technical or managerial assistance will be extended. In fact, assistance is granted primarily in response to requests coming from the small firms themselves and most, if not all, requests are eagerly seized upon by the assistance-granting organisations. If one considers the problem in terms of demand and

---

1) See Maurice Domergue, "Technical Assistance by Correspondence", International Development Review/Forum, No.3, 1974. An evaluation of UNIDO's technical information service can be found in An Evaluative Review of the Industrial Inquiry Service (1966-1973) (mimeo), UNIDO, Vienna, 1974.

2) See Bepin Behari, "Low-Cost Technologies and Rural Industrialisation" (mimeo), paper prepared for the OECD Seminar on Low-Cost Technology, Paris, 1974. This study describes the consulting activities of the Small Industries Development Organisation in India.

supply, there is some evidence that in many developing countries, the supply of managerial and technical consulting services to small industry greatly exceeds the effective demand for such services. The most conspicuous case in this respect is probably that of Kenya where the Rural Industrial Development Programme has been supporting small firms whose effective technical, economic and managerial viability is questionable to say the least(1).

Another problem with such organisations is that the services they provide tend to be rather expensive. One way to measure this is to compare the number of firms which have received some sort of assistance during a given year with the number of people employed by these organisations. Another indicator is the ratio between the turnover of the firms which have received some form of assistance and the administrative costs of running the consulting organisation. A very rough estimate, based on some of the case-studies, indicates that such an organisation with ten staff members can provide meaningful assistance to a group of small firms which, taken together, employ some one hundred people. This is clearly not the most efficient way to foster industrial development on a large scale, nor is it the best way to use the highly qualified people who are on the staff of these organisations.

When the small firms to which technical or managerial staff is supplied are of a very low level, the outside consultants or advisers are very often tempted to go much beyond the supply of advice, and thus end up by virtually running the firm they are trying to help. This turns out to be very costly and time-consuming. On the other hand, small firms are often unable to absorb in an effective way the knowledge which is brought to them by a consultant in the course of a typical one-day visit, and it is very difficult to strike a balance between the one-shot approach and the virtual take-over of the assisted company.

In fact, there seem to be very great differences in the quality and effectiveness of national organisations which supply technical and managerial assistance to local enterprises. In one country, the organisation which was set up for this purpose under the auspices of a well-known international organisation ended up after five years of operation with a payroll of over 600 people, mostly in administrative and secretarial positions and, according to the widely-shared feeling of local entrepreneurs, is almost totally ineffective. In other countries such organisations can work very effectively and rather cheaply. This is the case of several institutions in India for

1) See for instance H. Kristensen, "The Technology Problem in Rural Small Scale Industries: A Case Study from Kenya" (mimeo), paper prepared for the OECD Seminar on Low-Cost Technology, Paris, 1974.

instance, and of at least one programme in Korea which was set up in co-operation with the Engineering Experimentation Station of the Georgia Institute of Technology(1).

Supplying managerial knowledge or new technology through such organisations cannot really be dissociated from the much wider problem of entrepreneurship. Information is a commodity, and its diffusion responds to market needs. Supplying information will not alone create the needs for information and new technology. This need can only arise in the much wider context of a favourable entrepreneurial environment which can be built up by legislation, the organisation of credit, or the development of education.

1) See the study by R. Hammond, "The Impact of Micro-Development Projects", in the second part of this book.

Chapter 4

THE ROLE OF THE UNIVERSITIES

The number of universities, both in the industrialised and the developing countries, which are involved in some way or another with the development of new appropriate technologies, is growing rapidly, and there are indications that this interest is more than a passing fashion. It is in fact a reflection both of the university's own questioning of its role in society and of the fact that its dual function of creating and transmitting knowledge has a direct bearing on the development and application of more appropriate technologies in the local innovation system.

Most of the appropriate technology groups which originated from universities, like the Technology Consultancy Centre in Ghana(1), the Regional Adaptive Technology Centre in the Philippines(2), the Division of Micro-projects in Eindhoven, Netherlands(3) or the Grupo de Desarollo Rural of the University of the Andes in Bogota, are too recent to allow for any meaningful generalisations about the role of higher education institutions in the promotion of appropriate technology. These experiments do however point to a number of problems which must be taken into account if appropriate technology is to play an effective part in the process of economic and social development. Are universities the most adequate institutions for fostering innovation in appropriate technology? Can their activities in this field be reconciled with their educational responsibilities? Can their curricula be reoriented to include more appropriate types of technology? These are some of the questions which have to be considered.

---

1) See the study by B.A. Ntim and J.W. Powell,"Appropriate Technology in Ghana: The Experience of Kumasi University's Technology Consultancy Centre", in the second part of this book.

2) Rufino S. Ignacio, "Intermediate Technology and Regional Development in the Philippines", ibid.

3) T. de Wilde, J. Janssen, B. van Wulfften-Palthe and S.P. Bertram, "Remarks on a Policy for Appropriate Technology and Some Cases of Micro-projects" (mimeo), paper prepared for the OECD Seminar on Low-Cost Technology, Paris, 1974.

I. THE UNIVERSITY'S COMMITMENT TO MODERN TECHNOLOGY

In the last fifteen years, higher education has been one of the fastest growing sectors of the economy in the developing countries, as can clearly be seen from the number of new universities which were set up or the dramatic increases in the number of students. One of the main motivations behind this explosive growth of higher education was to bring underdeveloped societies into the modern world by building up a massive technological and educational basis. The somewhat optimistic ideology which legitimised the creation of hundreds of new universities and which was actively promoted by international organisations, aid agencies and private foundations has today lost much of its veneer and given way to a more sober assessment of what higher education can effectively achieve. In some spheres, the pendulum has swung even further, and it is interesting to observe that the far-reaching criticisms of the university made by someone like Ivan Illich are now being frequently quoted, and possibly even put into practice by government authorities which, directly or indirectly, have the responsibility for higher education(1).

Whatever the failings of the university - and here the disappoint ment is the inevitable counterpart of exaggerated expectations - it is still the most effective, or least ineffective, means for providing a wide number of young people with the skills and training necessary in a modern industrialised, or industrialising society. Universities have a basic commitment, institutionalised in their curricula, their entrance requirements, their communications system and their quality control procedures, to modern technology.

This commitment stands in almost complete opposition to the guiding philosophy of the intermediate technology movement. University education is expensive, and for many developing countries prohibitivel expensive (one year of higher education for one student often costs more than the total income of hundreds of farming families), it is concerned with sophisticated modern technology, it is to a large extent a foreign cultural import, and despite the claims to the contrary, it is elitist and non-egalitarian. Can the university as such play a useful role in the field of appropriate technology for which it was never designed and which culturally and politically is so far away from it?

The increasing number of universities which are involved in developing some form or another of appropriate technology, shows at any rate that something can be done and that universities, and especially technical universities, do have a role to play. For the

---

1) Ivan D. Illich, Deschooling Society, Harper and Row, New York, 1971.

moment, this involvement is in most cases marginal, in the sense that it is not yet an integral part of the traditional educational curricula or research programmes. There are as yet no 'universities for appropriate technology' and India is the first country where the possibility of granting academic degrees in intermediate technology has been seriously considered.

The involvement of universities in intermediate technology is also in most cases politically marginal, in the sense that it does not generally result from a deliberate policy of the university as such, but rather in response to pressures on the part of activist student movements and as a result of the entrepreneurial activities of individual faculty members. In this perspective, involvement in appropriate technology might be viewed as a means of letting off steam and neutralising in a constructive way the internal political tensions of the university. What must also be considered here is that in several developing countries, the internal problems of higher education institutions are compounded by their potentially destabilising political role in society. Universities, or rather the student body, are often violently opposed to the ruling political elite of the country and are in effect a revolutionary force. The examples of Thailand, South Korea, Sri Lanka and several Latin American countries show this very clearly.

Appropriate technology, whether its proponents want it or not, has a political dimension, since it is oriented primarily towards the most under-privileged groups in society and assumes, probably rightly, that the development strategies followed until now have benefited primarily the ruling elites. The ideology and values of the appropriate technology movement could, if combined with the political activism of the universities in the developing countries, have truly revolutionary political consequences, and this cannot simply be brushed aside when one is trying to assess what part the higher education system might play in appropriate technology.

The involvement of universities in appropriate technology can of course be interpreted, somewhat cynically, as an attempt to neutralise their outside critics and buy off their student dissenters. This view, which some very specific cases may justify, is nevertheless rather unfair, in that it grossly understates the very genuine concern both of universities and educational authorities about the contribution of higher education to economic and social development. The universities also have a number of assets - intellectual, technological and organisational - which can usefully be applied to the development and promotion of more appropriate technologies.

II. THE UNIVERSITY'S LIMITATIONS

Before trying to outline what the universities could do, a few words of caution are necessary. The first is that the university is primarily an institution of higher education and higher learning. Its function is neither to act as an industrial entrepreneur or management consultant, nor to play the part which belongs institutionally to government agencies, industrial firms or non-profit associations. Faculty members may of course in their individual capacity take on a number of activities outside the university - primarily as advisers and consultants - but the institution to which they belon is geared essentially to education and learning in the strict sense of the term.

The second point is that even if universities did want, or were allowed to play a much wider role than that for which they were originally designed, they probably could not. For one thing the task of educating thousands of students is in itself a formidable affair. The problem is particularly acute in the developing countries: one need only to look at the teacher/student ratios or their financial situation to realise that education alone is more than a full-time preoccupation. There is an obvious temptation to say that the university can do everything. Does it not after all harbour some of the country's best minds? The temptation however is at best illusory, and at worst potentially damaging to its educational mission.

In several highly industrialised countries, universities have played in some cases an important part in industrial and technologica innovation, and one can infer, by analogy, that the higher education institutions in developing countries also have a role to play in thei local appropriate technology innovation system. By historical standards, this involvement of West European and American universities with industry is a recent phenomenon: it started in Germany at the end of the last century, with the development of the chemical industry, but it was only with World War Two that these links became institutionalised in a large way(1). At the same time that the universities were developing the basic knowledge used at a later stage by industry and establishing closer contacts with the business world, the large firms were building up their own R and D laboratories and trying thereby to systematise the innovation process.

University research was never a substitute for industrial research; it was a complement, and while important in certain fields, was nevertheless, and still is, a very small part of the innovation system. What should not be overlooked is that this involvement of

---

1) For a detailed discussion of these problems, see The Research System, Vol.I, II & III, OECD, Paris, 1972-74.

European and American universities in the processes of industrial and technological innovation was to some extent at least an accidental phenomenon rather than the result of a deliberate policy. What has worked in the past will not necessarily work as well in the future. In fact, it is quite possible that these links, built up in the last decades, will progressively deteriorate as a result both of the educational burdens placed upon the universities and of the increasing diversity and pace of change of industrial technology. In the developing countries, the establishment of close links between industry and the universities has until now proven to be rather difficult, and a national strategy for appropriate technology must take into account the fact that the building-up of such a relationship is not necessarily the most effective way of promoting innovation.

Another point of caution about the role of universities is that the development of new intermediate or appropriate technology does not for the most part embody any radically new scientific or technological principles. There are of course certain exceptions: the solar pump developed at the University of Dakar and further improved by a French industrial firm belongs to the world of high technology(1) and the gari machine developed in Nigeria did call for the development of new fundamental knowledge in chemistry(2). By and large, however, appropriate technology belongs to the realm of the craftsman and village artisan rather than to the world of the academic teacher or scientist, and an exclusive preoccupation with what the university might do can lead to a neglect of the crucial role played in the innovation process by craftsmen and small entrepreneurs. As far as education is concerned, this preoccupation carries with it the risk of overlooking the importance of secondary education, and notably of secondary technical education, which produces the middle-level technical specialists required by industry. These specialists, which are in very short supply in most developing countries, have a vital role to play, and could be one of the most effective links between the modern and the traditional sector, and could also form a basis for the innovation system in appropriate technology.

Developing even simple hardware can be very difficult for the local inventor - witness for instance the problem of increasing the reliability of his products and minimising the technical risk to the user. The technical capability of the university can be scaled down to meet the specific problems faced by local craftsmen and entrepreneurs. One successful example in this respect is the way in which

---

1) See the study by J.P. Girardier and M. Vergnet, "The Solar Pump and the Problems of Integrated Rural Development", in the second part of this book.

2) P.O. Ngoddy, "Gari Mechanisation in Nigeria: The Competition Between Intermediate and Modern Technology", ibid.

university teachers in Sri Lanka have been encouraged to act as technical consultants to local blacksmiths(1). Another similar case is that of Soon Jun University in Korea, whose mechanical engineering department has contributed to solving many technical problems of the local metalworking firms and, in the process, to diffusing this improved technology throughout the industry(2).

III. APPROPRIATE TECHNOLOGY IN THE HIGHER EDUCATION SYSTEM

The students of today are the engineers, technologists, industrial entrepreneurs and political leaders of tomorrow, and the type of education they are now getting will determine to a large extent the type of society that will exist twenty or thirty years hence. This time-lag or production cycle of the educational system suggests that if appropriate technology is to play an important part in the development process, the students of today need not only to be familiar with it, but also to have what N. de Silva called, rather poetically but nevertheless realistically, a certain 'sympathy' for it(3). The shift in values and attitudes which this requires cannot be achieved in a simple and straightforward way, but a number of steps have been suggested. One for instance is the Indian idea, already mentioned earlier, of integrating intermediate technology in the university curricula and eventually having degrees in this field. This obviously poses a number of problems, as the experience of Mindanao State University in the Philippines clearly shows(4). Academic curricula, notably in the engineering field, are designed in such a way that it is difficult to add on new courses without either suppressing some existing courses which are essential, or lengthening by several semesters the total amount of time spent by students at the university. And for the time being at least, it does not seem possible to design a full curriculum focusing exclusively on appropriate technology.

Another approach is to accept intermediate technology as a legitimate subject for an engineering student's diploma work. Designing a windmill, a palm-nut crusher or a water distribution system

---

1) See the study by D.L.O. Mendis, "The Reorganisation of the Light Engineering Industry in Sri Lanka", in the second part of this book.
2) Ross Hammond, "The Impact of Micro-Development Projects", ibid.
3) N.N. de Silva, "Rural Industrialisation in Sri Lanka" (mimeo), paper prepared for the OECD Seminar on Low-Cost Technology, Paris, 1974.
4) See the study by Rufino S. Ignacio, "Intermediate Technology and Regional Development in the Philippines", in the second part of this book.

which uses only local resources and which can be manufactured by people with little technical education is just as challenging, difficult and instructive of the student's abilities as the design of a diesel engine, a steel bridge or an electronic circuit. Several universities have already explored this approach. One instructive example in this respect is that of the Asian Institute of Technology in Bangkok. Part of the research and development work for the low-cost water filtration system, which is now being diffused to several Southeast Asian countries, was carried out by graduate students and served as a basis for several master's theses(1).

Encouraging universities to develop intermediate technology requires a modification of their reward and promotion system. In most institutions, the evaluation of a faculty member's performance - which determines his salary level and promotion prospects - is based to a large extent on publications. This yardstick is perfectly legitimate if research, preferably of a high quality, is accepted as one of the main missions of the university. It is, however, strongly biased against those who are very good teachers but poor researchers, and it does little justice to the other missions of a university, and notably to its educational role and its contribution to economic and social development. One consequence of the student unrest which swept many universities of the world a few years ago has been a rehabilitation of 'consumer satisfaction' (e.g. through performance ratings given by the students to their professors) as one of the bases for evaluating a faculty member's salary and promotion prospects.

Changes in the evaluation methods can and do take place, and there is no reason why in the universities of the developing countries, contributions to the development of intermediate technology might not also be taken into account. In fact, from a social point of view, such activities are probably much more useful and rewarding than a publication in a recognised foreign scientific journal, and the knowledge that they would also serve as a basis for the professional evaluation at the end of the academic year would provide a powerful motivation to do more.

The various steps outlined above amount in effect to an institutionalisation of intermediate technology. That they could be useful is more than likely. However, even if they were carried out, and this is beginning to be done in some places, many problems would remain. One of the most important is that the university as such belongs essentially to the modern urbanised segment of society. Students and teachers, even if they come originally from a rural community, have for the most part lost contact with the farmers and craftsmen in the

---

1) R.J. Frankel, "The Design and Operation of a Water Filter Using Local Materials in Southeast Asia", ibid.

villages. If they are to help develop new technologies which are truly appropriate to the daily conditions of the farmers and poor people, they need some much closer contacts with those to whom these new technologies are addressed.

Without going as far and as forcefully as China has done in compelling students to spend a year of real work in the farms and factories and eliminating academic performance as a criterion for granting degrees, there is a lot to say for encouraging much closer contacts between the university and the rural communities. One purpose of such contacts is to make students realise what are the real problems of the farmers and how difficult their daily life can be. It can also serve to give them greater respect and more understanding for their values, their culture, their skills and their technology. One should not forget that the technologies developed by generations of farmers in the developing countries are in many respects very sophisticated and ingenious, primitive as they may seem when compared to what has been achieved in some industrialised countries. This technological basis is in the process of disintegration, and vast amounts of knowledge accumulated over centuries are now being lost within a few years. Much of this knowledge could serve as the basis for developing new and more appropriate technologies. The idea here is not of course to preserve the past at all costs and turn remote villages into museums, but to get farmers and craftsmen to participate in the process of development of new technologies with the help of the university, and to encourage the university to identify, evaluate and analyse the vast pool of dormant technology which can be found in the informal sector.

Sending students off for a few months to the country is not the most popular measure, and to do so on a large scale and with little preparation is to court disaster. One successful example is that of the Pan-African Institute for Development(1). This institution, which is not strictly speaking a university but rather a post-experience management school for specialists in rural development, has a two-year education and training programme, a substantial part of which is composed of field missions in various African countries. The purpose of these missions is not only to study a specific problem, but to achieve something useful and practical. The reports prepared by the students - many of whom already have a number of years of experience behind them - are of very high professional quality and show a rather remarkable perception of the social processes of innovation and development. This work is in fact a very good example of the type

---

1) See the study by John W. Pilgrim, "The Role of Non-Governmental Institutions in the Innovation Process", in the second part of this book.

of new social knowledge which higher education institutions might try to develop and which is crucially important in the diffusion of appropriate technology.

IV. LOCAL RESEARCH AND INFANT TECHNOLOGY

Among the many questions which for the time still remain open, one of the most important touches upon the research and development activities of the university, both in the industrialised and the developing countries. The complexity of the question can perhaps best be expressed by summarising two apparently contradictory views on this problem. Given the obvious need for much more R and D in the field of appropriate technology and the technical capability of the highly industrialised countries, the first view is that the universities in these countries should increase their R and D effort and provide the developing nations with the intermediate technology they need(1). The opposing view is that technical assistance, foreign aid and the introduction of foreign technology tend to paralyse the local inventive and innovative capability. Bearing this in mind, R and D in intermediate technology should be done exclusively in the developing countries themselves by local scientists, engineers and technologists(2).

Presented in this way, the dilemma may seem rather theoretical. In fact, it is a very concrete problem, as the following recent example clearly shows. In one East African country, a local appropriate technology group tried to develop a new type of irrigation pipe based on indigenous materials to replace the imported plastic pipes which had become prohibitively expensive as a result of the rise in oil prices. The problem was urgent and technically complex. Rather than address itself to the local university laboratory, whose technical capability in this particular area was not very strong, the appropriate technology group chose to call upon a Dutch university with a much wider experience. As far as immediate results are concerned, this was undoubtedly the better solution. But in doing this, a major opportunity was missed to help build up a local technological capability.

The greater the R and D effort undertaken by universities of the industrialised countries on appropriate technology, the greater the temptation for developing countries to use this technology rather

---

1) This view, shared by most universities in the industrialised countries, is also supported by several groups in the developing countries. See for instance B.A. Ntim and J.W. Powell, "Appropriate Technology in Ghana: The Experience of Kumasi University's Technology Centre", in the second part of this book.

2) See M.J. Hoda, "India's Experience and the Gandhian Tradition", _ibid_.

than develop their own. In fact, there are strong reasons to believe that the patterns of technological dependence which characterise the relations between industrialised and developed countries are being replicated in the field of appropriate or intermediate technology. It is a well known fact that more than 90 per cent of the world's potential in science and technology is located in less than 25 highly industrialised countries, and that the developing countries taken as a whole do not account for much more than one per cent of the world's output in science, technology and industrial innovation. For the time being it is difficult to determine if the same imbalance prevails in the field of R and D in intermediate or appropriate technology. For one thing, no serious attempt has yet been made to measure on a world-wide basis the total amount of money spent on research in this area. Furthermore, as noted earlier, a large part of the innovations in appropriate technology stem not from a formally organised - and therefore quantifiable - R and D effort, but from the empirical development work done by artisans, small entrepreneurs or innovative farmers who belong to the informal sector.

However, if one takes into account the rapidly growing R and D effort in appropriate technology undertaken in the industrialised countries and the fact that many appropriate technology groups tend, like the one from East Africa mentioned above, to resort to appropriat technologies developed in Western Europe and North America, there are reasons to believe that the same patterns of dependence upon technolog transfers from the industrialised countries could develop. What furthe contributes to reinforcing this trend is the fact that appropriate technology is a non-proprietary technology, available in most cases free of charge. Few if any of the products and processes developed by appropriate technology groups in the industrialised countries are covered by patents or licensing restrictions, and those who have developed them are, for very good reasons, eager to put them at the disposal of developing countries(1).

Such transfers can make an important contribution to development but the problem of technological self-determination remains unsolved. In the same way that government policies for industrialisation have tried to build up local industries by protecting them through tariffs or other means from excessive competition - this is the well known 'infant-industry' principle(2) - a policy for appropriate technology

---

1) One instructive view of the role played by universities of the industrialised countries in developing new appropriate technologies for the less developed nations can be found in R.J. Congdon, "Appropriate Technology in British Universities" (mimeo), paper prepared for the OECD Seminar on Low-Cost Technology, Paris, 1974.

2) This principle was first propounded by the German economist Friedrich List at the end of the last century. See his National System of Political Economy (1885) reprinted by Kelley, New York, 1966.

in the developing countries should consider the arguments in favour of 'infant technology', and in particular of infant appropriate technology. Protecting an infant industry is always expensive in the short run: locally-made products are most costly and often less reliable than those imported from abroad. The same is true of infant technology, and one must accept the fact that the purpose of R and D on appropriate technology in the developing countries and notably in their universities, is not merely to develop new technologies or new products, but also to build up a capability for technological innovation. Mistakes are inevitable and inefficiency very likely, but this is normal phenomenon. As Victor Martinez has noted, before reading Shakespeare, one must know the alphabet, and learning how to master technology calls first for a mastery of the alphabet of technology(1).

V. THE NEED FOR APPROPRIATE MODERN TECHNOLOGY

Universities in the developing countries may well be able to make an important contribution to the development of new intermediate or low-cost technologies, despite their primary commitment to education and their cultural bias towards modern technology. However, if one considers the ways in which their R and D activities could be mobilised for development, it is probably a mistake to look exclusively at innovations in low-cost or intermediate technology. Technologies of this type are no doubt important, but they cannot solve everything. There are a number of complex problems - in agriculture, natural resources, transportation or public health - which can only be solved through the most sophisticated modern technology. One of these for instance is the eradication of bilharzia, a parasitic disease which affects hundreds of millions of people in the poorer parts of the world(2). Another is the mapping of underground mineral resources by space satellites. In the transportation field, there are a number of alternatives to the traditional methods like ox-carts or the modern methods like trucks, but these parallel technologies (e.g. the airship) still require a certain amount of development work before they can become economically attractive(3).

---

1) Victor Martinez, "Specialised Information Services for Technological Innovation in the Less Developed Countries" (mimeo), paper prepared for the OECD Seminar on Low-Cost Technology, Paris, 1974.

2) Recent developments of basic research on bilharzia illustrate very well the role that can be played by universities in the developing countries. An Ethiopian scientist working in his country has discovered that the fruit of the local soapberry plant was a very effective, and natural, means of eliminating the water snails which transmit the parasite.

3) See for instance Alternative Transport Technologies (mimeo), OECD Development Centre, Paris, 1974.

Many of the problems which are still unsolved could probably be dealt with successfully in the next ten or twenty years through a large scale R and D effort involving the universities of the developing countries. The technologies required to provide new types of crop more efficient transportation systems, or a better utilisation of natural resources are in many cases neither low-cost nor intermediate but highly sophisticated and very modern. Yet they are particularly appropriate, in the sense that they try to solve problems which are crucially important to the developing countries but which have been neglected by the research community of the industrialised countries.

The new appropriate technologies required by the developing countries can either be low-cost and intermediate or highly sophisticated and very advanced. In both cases, the universities of the developing countries have an important part to play, both as educational centres and as R and D centres. Their educational mission should not only be to train scientists and engineers and give them the basic qualifications for a professional activity in the modern world, but also to familiarise them with the social and technical problems of the millions of people living in the rural areas. Their R and D activities should probably focus more on the development of technology, but this should not be at the expense of highly sophisticated research in the scientific and technological problems which are specific to the developing countries. Their 'modern' R and D activities might well in fact serve as the basis for the creation of new science-based industries focusing on the problems of development.

Chapter 5

BUILDING-UP NEW INDUSTRIES

The work done by the various appropriate technology groups shows that it is technically possible to develop new products, production processes and services which are better adapted to the requirements of the rural populations in the developing countries than those offered by the capital-intensive high technology industries. Perhaps the most important contribution of these groups is to show that there are alternative approaches and that the technical and human resources of the most underprivileged segments of the population can be mobilised more effectively than they are now. Developing new technologies to meet what seem to be very real needs is however but one aspect of the innovation mechanism. Technology is a process, not an end in itself, and it must somehow be translated into new products and new services if it is to be of any use. Products must be manufactured and sold, and this can only be done by an industrial firm or, if the market is very small, by individual craftsmen. In the same way, a service must be brought to the customer, either by individuals or by an organisation designed for this purpose, and self-help technologies must be taught to the user.

I. INDIVIDUAL ENTREPRENEURS AND COLLECTIVE ORGANISATIONS

Two rather striking conclusions emerge from a comparative evaluation of the work done until now by the various appropriate technology groups working in the developing countries. The first is the enormous importance of individual entrepreneurship in promoting innovation, and a rather general scepticism about the role of government agencies and co-operative forms of organisations. The second is that appropriate or intermediate technology, if it is to make any impact at all, must be economically, socially and technically competitive.

Stressing the role of the individual entrepreneur may appear like a return to Adam Smith and Joseph Schumpeter, and an implicit criticism of the organisational forms and ideologies which have governed the development effort of many countries in the last twenty years. Or it might be viewed simply as a reflection of the pragmatism and realism of the intermediate technology proponents. There is no doubt a certain cultural and social affinity between the philosophy of intermediate technology and the attitudes of the small entrepreneur and this may account, at least in part and somewhat subconsciously, for the major emphasis given by many low-cost technology practitioners to the role of the individual entrepreneur.

The experience of the various appropriate technology groups clearly seems to suggest that individual entrepreneurship is one of the most effective channels of innovation in appropriate technology. This is even more true if one does not consider simply the work done by formally organised appropriate technology groups, but also the much wider informal sector. In that sector, represented by individual craftsmen and small enterprises, individual entrepreneurship is perhaps the single most important factor of success in innovation.

Despite the apparent superiority of individual entrepreneurship, an increasingly large number of developing countries, for reasons which are both ideological and political, have tried to base at least part of their development effort on co-operative forms of organisation and public entrepreneurship is often viewed as the mainstay of modernisation. In Tanzania for instance, co-operatives are the main form of organisation of the country's economic and social life.

The difficulties encountered by the co-operative movement in many developing countries seem to be due to two types of reasons. The first is the high degree of administrative and technical skills required to operate such an organisation. As far as administration and management is concerned, the co-operative is in fact a 'high-technology' enterprise, in the same way that the small family firm or workshop is a 'low-technology' organisation. If all the members of the co-operative are to take an effective part in its activities - this after all is one of the basic principles of the co-operative movement - they need to have a relatively high educational level, and the central group of people, or the individual, who serve as its managers must be skilled administrators and, equally important, they need to be very highly motivated.

The second problem touches upon the very ways in which co-operatives are formed. If we look at the co-operatives which, judging from several decades of experience, have been the most successful - for instance in the Scandinavian countries, in Switzerland or in Israel - one of the main reasons for success is that they were created by the people whom they benefit. One should not of course take an overly idealised view of the social processes of organisational innovation: at the origin of a successful co-operative, there is not only a community of people sharing the same values and interests, but also an institutional entrepreneur and social missionary. Yet despite the crucially important part played by such innovators, the success of a co-operative depends very much upon the involvement of the whole community.

The gradual building up of a co-operative is a long process and quite understandably, many developing countries have tried to accelerate the process by training administrative staff and establishing new co-operatives by government decree. These co-operatives, created by the government rather than by the communities they are intended to serve, are usually faced with tremendous administrative difficulties, and as a result of their political links with the government, tend to be viewed with great distrust by the population, especially in the rural areas. Large financial subsidies do not make them socially more acceptable. In fact, when they do have a lot of public money at their disposal, their members will almost inevitably interpret this as a private grant for their own personal use, or as a fair compensation for the exactions of the government tax collector.

In practice, it is often very difficult to compare the relative merits of co-operative forms of organisation and private entrepreneurship. Co-operatives are usually organised in the service sector (e.g. agricultural marketing, distribution of food supplies or basic consumer goods, etc.) while private entrepreneurs can more generally be found in the production sector, and in each case the conditions are very different. There is, however, at least one practical case where a direct comparison can be made. In India, the Planning Research and Action Institute in Lucknow developed a very efficient small-scale technology for producing crystal sugar(1). This technology was diffused simultaneously to private entrepreneurs and to farmer co-operatives set up for this purpose. A few years later, the results were striking: four of the eight co-operative plants set up between 1958 and 1962 had ended up in bankruptcy and despite considerable encouragement, less than half a dozen new units were created between 1962 and 1974. By contrast, the individual entrepreneurs were outstandingly successful: they had created over 1,200 new production units in less than fifteen years, and currently account for close to 20 per cent of India's crystal sugar production.

This example does not of course mean that co-operatives cannot work as a matter of principle. In fact, there are a number of successful co-operatives throughout the developing world. However, account must be taken of the fact that co-operatives are inherently more difficult to run than a private enterprise. One interesting example in this respect is that of the light engineering industry in Sri Lanka, which was reorganised on a co-operative basis, while at the same time preserving individual ownership of the means of production(2).

---

1) See the study by M.K. Garg, "The Scaling-Down of Modern Technology: Crystal Sugar Manufacturing in India", in the second part of this book.

2) D.L.O. Mendis, "The Reorganisation of the Light Engineering Industry in Sri Lanka", ibid.

This approach, which tried to combine the advantages of private entrepreneurship with the benefits of the co-operative (for instance in the supply of raw materials, the marketing of the blacksmith's products and the development of new technology) appears to be one of the most promising in the diffusion and application of appropriate technology.

II. THE ROLE OF GOVERNMENT

Large public or governmental organisations tend to be beset by bureaucratic inefficiencies and administrative weaknesses and there are some justified doubts as to their ability to promote in a direct way the development of appropriate technology. Government agencies do nevertheless have a very important role to play. Perhaps not as industrial entrepreneurs but rather as the indirect supporters of low-cost technology. In every society there are entrepreneurs and innovators. But what happens is that the initiative of such people is often systematically stifled by the government bureaucracy, the banking and credit system, the marketing and distribution organisations and last but not least the social pressures against innovation and change.

The guarantees required by the credit institutions for instance are such that a new entrepreneur has practically no chance of obtaining any financial aid. Import regulations and foreign exchange restrictions make it very difficult for him to obtain from abroad the equipment he needs either to manufacture goods or to develop his own technology. When he does happen to have money, the priorities of the national economic plan do not allow him to invest it as he sees fit and he must spend vast amounts of time and energy trying to get approval for his projects. From the entrepreneur's point of view, it therefore makes much more sense not to engage in industrial activities and to devote his energies to socially less productive and personally more rewarding activities like trading or speculation in real estate.

One of the tragedies of this situation is that while governments on the one side are trying very hard and often at great social and financial expense to build up new industries, the institutional obstacles to entrepreneurship on the other, coupled with the competition from these highly subsidised industries, is stifling the creation of new small firms and progressively destroying the economic and social bases of the local handicrafts industry. Many developing countries are spending vast sums of money to import foreign technology, but at the same time these imports are contributing to the destruction of the existing technological system in a large number of sectors. Thus for instance the carpet manufacturing machines imported from

Switzerland by one North African country: they have not only ruined hundreds of families who had made their town famous for its carpets, but have also made sure that in five or ten years time, the knowledge of how to make them will also have disappeared completely.

Clearly the process of creating and diffusing new technology is inseparable from the decay and disappearance of older technologies and obsolete knowledge, and a complete preservation of the past can only be achieved at the cost of complete stagnation. But in many countries, this process of destruction of traditional technology is artificially accelerated by large-scale industrialisation and, what is of much greater consequence in the long run, it is destroying both socially and morally the very class of people who have the technical and inventive ability which is so important in the process of development.

This is not to say that the action of government is necessarily harmful or that large-scale industrialisation as such is an evil. Governments can in fact do a lot, for instance by helping to create a more favourable climate for the development of entrepreneurship. Before even considering what positive measures might be taken, it would be useful to start by dismantling at least some of the institutional obstacles to entrepreneurship. These obstacles are innumerable, but in most cases relatively easy to identify: it is only necessary to ask those innovators who have succeeded in overcoming them. Such surveys have seldom been conducted, and if followed up where possible by specific remedies would probably be much more useful and effective than the somewhat grandiose schemes for technical education, management training or information services established by government authorities, often on the advice and with the help of foreign aid agencies.

What such surveys would probably show is that the disincentives or obstacles to entrepreneurship are rather similar from one developing country to another. Among the most conspicuous, one could mention for instance the taxation system, which is often very favourable to large enterprises - including foreign firms - and overly harsh on the small company. Import regulations are another case in point. To give a typical example: in one of the large tropical African countries, a small entrepreneur had decided to start manufacturing aspirins. He had managed to buy the necessary machines and to import in bulk the raw materials he needed. But the refusal of the Ministry of Trade to reduce the enormous duties on small glass bottles, coupled with the duty-free imports of aspirin in tubes from Germany brought the new enterprise to close its doors and ruined the man who had started it. One may of course argue that he should have known about the structure of import duties and import regulations before starting. In this case however - and this is not untypical - regulations are constantly changing, and the entrepreneurs cannot count on the minimum of stability which is required for building up new industries.

III. THE SOCIAL PATTERNS OF ENTREPRENEURSHIP

Creating a more favourable climate for entrepreneurship can be greatly facilitated by a number of relatively simple measures. There are however a number of social factors which cannot easily be changed. Among the most sensitive from a political point of view is the fact that certain types of entrepreneurship are often associated with certain ethnic and linguistic groups. In Cameroon for instance, some 80 per cent of the trade in basic products like oil and soap is in the hands of the Bamileke. In Kenya, the data show that 93 per cent of new companies registered between 1946 and 1963 belonged either to European or Indian settlers, while the total capital of the African-owned companies amounted to less than 1 per cent of that of all new firms(1). In Thailand, a large proportion of the entrepreneurs who have gone into the tractor rental business for the green revolution happen to be Chinese. In Zaire, the Balubas represent some 10 per cent of the population, but reportedly some 70 per cent of the university graduates belong to that ethnic group.

Hundreds of similar examples can be found elsewhere. The social tensions which arise from the ethnic or religious affiliations of a country's entrepreneurs are not of course specific to the developing nations, as the history of the Jews in Europe or of the Protestant minorities in Catholic countries clearly shows. The promotion of entrepreneurship, and in particular of entrepreneurship in intermediate technology, cannot leave out the cultural, ethnic and religious factors(2).

If ethnic or religious affiliations are an important determinant in the patterns of entrepreneurship in a country, the same is also true of a society's system of values or its social traditions, and such things take time to change. That entrepreneurship is good for the economy is one thing, but its social acceptability is another. The motivations of the man who starts up a new enterprise are very complex, and he will tend to go into activities which are approved by the majority and which give him both financial rewards and social prestige. These activities are not necessarily those which contribute

---

1) See H. Kristensen, "The Technology Problem in Rural Small Scale Industries: A Case Study from Kenya" (mimeo), paper prepared for the OECD Seminar on Low-Cost Technology, Paris, 1974.

2) One interesting way to deal with this problem is that of INADES in the Ivory Coast; in that country, most of the trandesmen are foreigners - Lebanese, European or Malian - and INADES has attempted through its training programmes in the commercial field to build up a class of Ivorian merchants and shopkeepers.

most to increasing the gross national product or to developing the country's industrial basis. As P.D. Malgavkar has observed in his study of the techno-entrepreneurs of the Poona area, one of the most positive results of their work had been to bring about a gradual change in values and to make industrial activities socially more acceptable and more prestigious than they were a few years ago. This in turn is motivating other potential entrepreneurs to start up new firms(1).

IV. THE POLITICAL LEGITIMATION OF APPROPRIATE TECHNOLOGY

The growing interest of many governments and aid-giving agencies in appropriate technology can contribute not only to its development and widespread diffusion but also, and this is in some ways more important, to giving it the social and cultural prestige which it now lacks. There are clear indications that intermediate technology is now moving from a somewhat marginal position in development policy to a much more central role. This process of institutionalisation means that more money will be available for new experiments. Needless to say, there are certain risks, and one of the main ones is that of bureaucratisation. However one should not overlook the major social and psychological importance of this new trend.

As long as governments and aid-giving agencies, both national and international, were concerned almost exclusively with large-scale projects and modern industries, it required an enormous amount of independence of mind on the part of the intermediate technology proponents and the few local entrepreneurs in the developing countries to stand up against what everyone else, including the experts, thought was right. Marginality is not of course a deterrent to the true missionary, quite the contrary, but in this case it meant that intermediate technology had little chance of involving the millions of people who form the silent majority, and who stand most to benefit from it.

The active interest of national governments and foreign aid donors in low-cost technology could reverse this situation by legitimising the efforts of local entrepreneurs and innovators in this field and by providing powerful incentives to those who might otherwise have gone into socially less productive activities. In a sense, the social function of this institutionalisation of the appropriate technology movement is not only to develop new technologies, important as this may be, but to create the cultural and psychological conditions for the development of entrepreneurship and innovation.

---

1) P.D. Malgavkar, "The Role of Techno-Entrepreneurs in the Adoption of New Technology", in the second part of this book.

In this respect, it is interesting to note the role which is being played by the United States in the legitimation of appropriate technology. For several decades, the industrial and technological experiences of that country have played the role of model or yardstick for other industrialised countries, both East and West. Witness for instance the way in which other countries have tried to match, in relative terms at least, the American investments in research and development, their fascination with American management methods or their attempt to build up industrial firms which were as large as American companies. At the same time that these countries are trying to replicate some of the United States' industrial and technological experiences, the model country is itself undergoing very deep changes which, as J.F. Revel has noted, are symptomatic of what may well be the only true social and political revolution of the last half of this century(1).

One manifestation of this revolution is the movement of population away from the very big cities and, more recently, the return to the rural areas. The latter trend is now sufficiently important to have become statistically significant. The seventeen mainly rural U.S. states (e.g. Maine, Minnesota, Iowa, the Dakotas, Nebraska, etc.) which between 1940 and 1970 had experienced very high rates of emigration, have witnessed a complete turnaround, and since 1970 have become areas of net immigration(2). This ruralisation trend, which is associated to a large extent with the ecology movements and with American society's critique of its industrial way of life is having a direct effect upon the development of technology. Hundreds of new industrial firms have sprung up in the last few years to meet the needs of this new rural market, and now manufacture such products as horse-drawn ploughs, windmills, solar heaters and low-cost food storage equipment.

Significant changes are also taking place in the political sphere both at the federal and state level. California for instance is considering the ways and means to develop appropriate technology to help solve the problems of its immigrant workers, and several rural states (e.g. Montana) have established, or are planning to set up appropriate technology centres. At the federal level, and notably within Congress, there is a widespread interest in alternative technologies and the creation of a National Appropriate Technology Centre is under consideration. This is likely to have a direct impact in the international field; one illustration is the bill passed by Congress in September

---

1) Jean Francois Revel, Without Marx or Jesus, Doubleday, New York, 1971.

2) William N. Ellis, "The New Ruralism: The Post-Industrial Age is Upon Us", The Futurist, August 1975.

1975 which allocated 20 million dollars to the Agency for International Development to spend on the promotion of intermediate technology for the developing countries between 1976 and 1978.

The fact that this interest in intermediate technology is shared not merely by marginal groups, as it was a few years ago, but also by national policy-makers, is enormously important. In the same way that American experiences in high technology served as legitimation and justification for industrial policies in Western Europe(1) and in many developing nations, the coming of age of intermediate technology, which is rapidly being translated into institutional and technological innovations, could serve as a powerful motivation for similar efforts elsewhere, and notably in the developing nations. If the United States is taking intermediate technology very seriously, this is a strong indication that it is a viable proposition, and in a sense this interest is the political and psychological breakthrough which is necessary to turn intermediate technology into a serious alternative for the development of low-income countries.

One should not of course underestimate the problems which this legitimation of intermediate technology is likely to cause in several developing countries. Socially and politically, intermediate technology is not neutral. One of the basic assumptions of its proponents is that it should be aimed primarily at the poorest segments of the population, most of whom live in the rural areas. How this can be done most effectively is not entirely clear. One of the means is to try gradually, through education, extension services, financial assistance and social rehabilitation to identify and stimulate the entrepreneurial and innovative forces which exist in all communities but which have been weakened and repressed by the existing economic and social system.

What must not be overlooked however is that these entrepreneurial forces, once released, will not focus exclusively on building up small industries and developing new technologies, but will also in many cases address themselves to social reforms and political change. The social background and the psychological dispositions which make a man a good potential entrepreneur are no guarantee that he will become an industrial entrepreneur and thereby contribute to increasing the gross national product. He may just as well become a political entrepreneur and a social revolutionary(2). This is not necessarily

---

1) This phenomenon has been analysed in Raymond Vernon (ed.). Big. Business and the State - Changing Relations in Western Europe, Harvard University Press, Cambridge, 1974.

2) It may be interesting to note here that the psychological dispositions which make a man a good entrepreneur (inventiveness, creativity, need for independence, etc.) are also characteristic, as several recent studies have shown, of the successful criminal and law-breaker.

negative: it may in fact be one of the most effective means of stimulating a real development process in the most underprivileged segments of the population.

V. THE LIMITATIONS OF PLANNING

Another political and social issue raised by the development of appropriate technology touches upon the limitations of economic and industrial planning. A large number of developing countries today have a national Plan and a large bureaucracy to implement it. Aside from ideological choices and political fashions, one of the motivations for planning is that market forces, both national and international, are not conducive to real development. Since human and financial resources are limited, they should be channelled in priority into those areas which are likely to have the largest multiplier effect from an economic and social point of view. If the motivations for planning are perfectly legitimate, the results are often disappointing. For one thing, national economic or industrial planning is from an administrative point of view a 'high-technology' activity which requires not only an efficient information system, but also a high degree of administrative competence. This means not only a large number of highly-qualified people, but also what might be called an administrative or public service tradition. In many developing countries, none of these requirements are fulfilled.

Planning inevitably breeds the idea that it is possible to control and direct all of a country's economic and industrial activities. It may indeed be relatively easy to decide upon the construction of a steel mill, a big dam or an electricity distribution system. Such projects can be analysed in terms of costs and return on investment, and the technical expertise needed to carry them out is usually available from abroad. However they represent only a small fraction of the economic and industrial activities in a country. What makes an economy work are the hundreds and thousands of farmers, craftsmen, tradesmen, repairmen and industrial workers who provide their fellowmen, often on a very small scale, with the vast range of products and services they need.

To the economic planners, the way in which these activities are organised may seem irrational and inefficient, and there is an almost inevitable temptation to reorganise them in a more coherent way. For instance by replacing the thousands of small trucking companies by one large transport co-operative, by setting up price control mechanisms for basic products like bread, rice or flour, or by bringing all the small tradesmen to form a large marketing organisation. What usually ensues from such reorganisation efforts is at best another type of inefficiency, and at worst complete chaos. Even in the most underdeveloped and undifferentiated society, the economic system is probably too complex and diverse to be organised in a rational way.

What appropriate technology has to offer here is not a means of controlling and operating the economic system in a coherent way, but rather a series of technical tools to reduce the inefficiencies in specific areas. A very simple example can illustrate this. In most developing countries, the small farmer sells part of his surplus production on the open market. Since all farmers are doing the same thing at the same time, prices collapse, and the farmer has little incentive to produce more than he needs. To the economic planner, one solution might be to impose some minimum price for agricultural products. To the intermediate technology specialist, the solution lies elsewhere, namely in developing a number of simple storage methods which will allow the farmer to stock his surpluses for a few weeks or a few months, and thereby to obtain a better price from the consumer or the tradesman.

VI. THE COMPETITIVENESS OF APPROPRIATE TECHNOLOGY

One of the most important yet often overlooked factors in the diffusion of intermediate technology is its competitiveness. A technology will not be adopted by the producers or the consumer for the only reason that it is new, quite the contrary. Novelty as such is often an obstacle rather than an incentive to its diffusion, especially in the rural areas. What is important is that it should be technically and economically more advantageous than both the existing traditional technologies and the modern technologies embodied in the products of large-scale industry. It must also be commercially viable and capable of producing a surplus for those people who will be using it.

Two big difficulties arise here. The first is that the range of intermediate technologies which can be developed is considerably wider than the number of those which meet these basic conditions of competitiveness. One can wonder for instance whether the cow dung gas cooker, which has been in the process of development for some forty years in India, is really competitive with the traditional wood or coal cookers, or with the modern butane gas stoves manufactured on a large scale by industrial firms. The same observation might be made about the rainwater collection system promoted by a French organisation in West Africa(1). The plastic sheets covering

1) See P. Martin, "Technical Note on the Collection and Storage of Rainwater" (mimeo), paper prepared for the OECD Seminar on Low-Cost Technology, Paris, 1974.

the traditional roof may well be technically appropriate, but one of the alternative solutions, namely the tin roof, is not only technically simpler, but socially much more prestigious: the tin roof, quite apart from its practical convenience, is one of the most conspicuous signs of affluence and a top priority in a family's investments.

The second problem is that it is generally very difficult to evaluate the competitiveness of intermediate technology. Part of the problem stems from the fact that the concept of 'competitiveness' or 'efficiency' covers three very different notions. The first is what might be called the technical or engineering efficiency of a new product or a new way of doing things. The second is its economic viability, and the third its social and cultural acceptability. On the basis of these three criteria, intermediate technology must then be compared with the traditional technologies and with the large-scale technologies of modern industry. This gives six criteria of competitiveness which can be presented in the form of a simple matrix as follows:

CRITERIA OF COMPETITIVENESS OF APPROPRIATE TECHNOLOGY

	Competitiveness vis-à-vis traditional technologies	Competitiveness vis-à-vis modern technologies
Engineering efficiency		
Economic viability		
Social acceptability		

In theory, the intermediate technology which stands the greatest chance of success on the market is the one which, on the basis of these criteria, happens to be the most competitive or the most efficient. In reality things are considerably more complex and the successful innovator often has to rely more on his own experience and his personal hunches than on rational economic and technical analyses. This set of criteria does however suggest three important points. The first is that social or cultural acceptability is often a major factor in the success or failure of an innovation. Because it is so subjective and ill-defined, there is an obvious temptation to favour the more rational criteria of engineering efficiency or economic viability. The second point is that intermediate technology has to compete not only against the existing traditional technologies, but also against the whole technological system of modern industry which in many cases is remarkably efficient.

The third point is the dynamic nature of the technological system. The criteria indicated here can give some idea of the competitiveness of intermediate technology today. As time goes by, however, they can change very substantially. What is socially acceptable today may no longer be so tomorrow. For instance, in one cotton-growing country of Africa, workers are no longer prepared to do the picking by hand, and large-scale picking machines had to be introduced to meet this problem. In the case of India, one may wonder if the plans of some appropriate technology groups to make a much wider use of cow dung as a source of energy are realistic, considering the inconvenience of collecting the raw material and the increasing reluctance even of unemployed people to do this work(1).

Technology, and in particular the modern technology against which appropriate technology has to compete, is in a process of constant change. And the basic economic assumptions upon which an innovation is based - cost of labour, price of raw materials, import duties, etc. - can change radically within a period of a few years. Many entrepreneurs sense this very clearly, and it is interesting to observe that in a number of cases their technological choices anticipate by a few years what is likely to happen in their country. Industrial wages for instance may be very low today, but most private entrepreneurs expect them to rise, and this, coupled with the development of trade-unionism, almost inevitably brings them to choose more capital-intensive types of technology. This is the case for instance in the Bolivian cigarette industry and the Cameroonian soap industry.

What the shifting nature of these criteria suggests is that appropriate technology, if it is to succeed, must not only be competitive today, economically, technically and culturally, with existing technologies: it must also have what might be called an evolutionary capacity. The problem is not merely to develop new technologies to meet an immediate need, but also to build up an innovative capability, or innovation system. What international technology transfers can do is to introduce new ideas, new forms of knowledge and new ways of doing things. But what they cannot do is to help build up within the importing country the entrepreneurial and innovative basis which in the long run will ensure the widespread diffusion of appropriate technology.

---

1) M.M. Hoda, "The Energy Situation in India" (mimeo), paper prepared for the OECD Seminar on Low-Cost Technology, Paris, 1974.

Chapter 6

POLICIES FOR APPROPRIATE TECHNOLOGY

The design of national policies for appropriate technology is still very much in its beginnings, and there are no real models to serve as reference points. A number of countries, both industrialised and developing, have made some very interesting experiments and the activities of the various appropriate technology groups have shown what were some of the main problems of innovation, and pointed to the directions in which a policy for appropriate technology might move. These experiences however are probably not sufficient to provide a clear-cut answer to the question which national policy makers, aid-giving agencies and private industry are beginning to ask, namely, what can be done in practice to foster the development and diffusion of appropriate technology?

I. THE GENERAL PRINCIPLES

A certain number of general principles are nevertheless beginning to emerge. The first is the need to recognise that appropriate technology does not consist solely of what has been developed by the existing appropriate technology centres. These organisations are playing a vitally important role as pioneers, knowledge centres, political lobbyists and social missionaries, and will probably continue to do so for a long time to come. But a national policy must take a wider view and try to enhance the vast stock of appropriate technology developed by small local firms, individual craftsmen and inventors, farming communities, educational institutions, charitable organisations and institutional entrepreneurs. Along with this pool of technology which belongs for the most part to the 'informal' sector, there is also a vast amount of technology from the industrialised countries which in many cases is equally appropriate, and which could be used in a much more effective way. The appropriate technology groups represent in a sense the tip of the technological iceberg, and one of the aims of a national policy is to explore and exploit that part of the iceberg which lies below the surface.

A second general principle is that centralisation and co-ordination from above is not the most effective approach to innovation in appropriate technology. What is needed is not a monopolisation of the inventive process by government but a much greater degree of initiative throughout the economic and political system or, to put things somewhat differently, a form of technological federalism, with initiatives and innovations coming from all levels. This diversity necessarily implies a certain amount of disorder, duplication and apparent inefficiency. But innovation, like biological processes, is inherently wasteful, and order grows out of disorder.

What this means in practical terms is that a policy for appropriate technology should involve as wide a number of institutions and people as possible. For instance the banks, which have a major part to play in financing new enterprises and industry, the school system which shapes the children's attitudes towards technology, and the big industrial corporations which have a strong technological capability, and whose growth is dependent upon their ability to meet the changing needs of the market. It should also involve government agencies which, like the Finance Ministry through its fiscal policy, can have a major impact on innovation, or which, like the judicial and legislative system, have an important role to play in social regulation. It might equally well involve the armed forces, which in many countries are actively engaged in technical education as well as in the transfer of sophisticated foreign technology.

The third general principle is that a policy for appropriate technology, whether it is initiated by government, private industry, non-profit institutions or any other organisation, cannot and should not focus exclusively on the development of new types of hardware. Hardware is undoubtedly very important but so are new forms of organisation, more efficient uses of existing resources, and faster transfers of knowledge between sectors and between regions.

Until now, the pioneering role in appropriate technology has been played for the most part by private organisations - like the Intermediate Technology Development Group in England, the Gandhian Institute of Studies in India - or by semi-public institutions like the universities. With the progressive institutionalisation of appropriate technology, the time has come to start thinking of the ways in which national governments (including local authorities) could become more effectively involved. Governments may not necessarily be the most appropriate institution to promote intermediate technology, but their impact on innovation is considerable, and it can in many cases be made more effective.

One of the first areas of government activity which can be considered here is the national science and technology policy. The research system represents a substantial part of a country's innovative capability, and the future of intermediate technology depends to a large extent on reorientations within the research system, which in most countries is heavily if not totally supported by public funds.

II. SCIENCE POLICY AND THE CONSTRAINTS OF PENURY

Like any other political constituency, the research community will try to obtain more money from the government, and national science and technology policies have had little difficulty in adapting in the 1950's and 1960's to rapidly growing budgets. The next ten or fifteen years, both in the industrialised and the developing countries, show every sign of being much less easy. Financial restrictions, coupled with a somewhat more sober assessment of what science and technology can really achieve, have brought about not only an effective halt in the growth of most national science budgets, but also in many cases an actual decrease. The same phenomenon can be observed in a number of large multinational corporations. This new era of penury is not of course affecting the whole system in the same way: changing national priorities have led to enormous increases in research and development expenditures in a few selected areas - energy is the most conspicuous case today - matched by equivalent decreases in others.

These changes have a number of implications for science and technology policies in the developing countries. The first is that, with the exception of some oil producing nations, financial restrictions to the growth of R and D activities have become, or are rapidly becoming, even tighter than in the industrialised countries. There are two reasons for this. One is that R and D usually is much more expensive to undertake in a developing nation than in a country with a well developed technological infrastructure, and its productivity is generally much lower. The other is that the social cost, or opportunity cost of research, is one or two orders of magnitude higher than in the industrialised countries. A rough estimate of this cost can be obtained by comparing for instance the cost of a researcher (including his training abroad and the equipment which has to be imported for his work) with the income or production of an average worker, or with the total amount of taxes paid by the latter to the government. The figures show that in a typical developing country, a researcher 'consumes' the economic surplus of between 1,000 to 50,000 farmers, while in an industrialised country, he consumes the surplus of ten to a hundred industrial workers.

The very high social cost of R and D in the developing countries and its relative inefficiency means that investment will increasingly be allocated to immediately productive activities rather than to research.

In the newly-rich developing countries, the same slow-down of R and D activities is likely to occur, but for the opposite reasons. The large sums of money which have suddenly become available have led to enormous increases in industrial investments, which in turn are absorbing most if not all the scientists and engineers who, under other circumstances might have gone into research. The science and technology system of the developing countries, which is still very small, is thus threatened on two sides: in the poor nations by the shortage of money, in the rich nations by the rapidly growing shortage of qualified manpower.

In almost all the developing countries, population is increasing rapidly, consumer demand is moving ahead, expectations are rising and social tensions, real or potential, show little signs of abating. What this means is that the needs for innovation, and particularly for industrial and technological innovation in the modern sector and for agricultural innovation in the traditional sector, will continue to grow rapidly in the years to come. This demand for new technology can be satisfied only in part by imports: foreign technology is expensive and it is often unsuited to the economic and social environment of the importing countries. At any rate the range of technologies required by the developing nations is so wide that an exclusive reliance on imports is in the long run untenable from a financial standpoint, even for the richest of the developing countries(1).

In other terms, an increasingly large share of the demand for technological innovation will have to be met by the local science and technology system. How this can be done in view of the very limited financial means available to the poorer countries, and the dramatic shortage of highly qualified people in the richer countries, is one of the difficult questions policy makers will have to face in the years to come. What is required in effect is a redesign of the somewhat ineffective national policies for science and technology which have been developed in the last ten or twenty years, and the development of what might be called appropriate or intermediate science policies and technology policies.

The science and technology policies which now exist in most developing nations are practically a carbon copy of those which can be found in the industrialised countries. Their institutional set-up is the same, the methods of intervention are similar, and the R and D objectives are not very different. Furthermore, as a result of the communications networks of modern technology, the research system of the developing countries tends to be much more closely integrated into the world research system, which is dominated by a few industrialised countries, than into the local economy. Finally, part of the research performed in the developing countries is aimed at

---

1) This problem is clearly illustrated by one oil-producing country which is currently spending 40 per cent of its oil revenues on the import of food, despite the fact that some 70 per cent of the population is employed (or partly unemployed) in the agricultural sector.

solving the problems of the industrialised countries, and consequentl does not contribute much to meeting local needs.

It would of course be completely unrealistic to suggest that the developing countries should dismantle in a deliberate way the small modern science and technology system they have managed to build up, often at considerable expense, over the last twenty or thirty years. There are however a number of ways in which national science and technology policies could help to make it more effective. The idea here is that national policies for science and technology should take at least part of their inspiration from the principles and the philosophy of the appropriate technology movement.

III. REGIONALISATION AND DECENTRALISATION

In most developing countries as well as in a number of industrialised countries, the science and technology system is highly centralised, both geographically and administratively. This form of organisation, which usually results from history or from political and social traditions, is perhaps justified in the case of certain large-scale projects, but it is generally recognised today that the less centralised systems are more efficient. One of the policy innovations of the late 1950's and early 1960's was the idea of a nationa science plan or technology plan, and the creation of central ministries dealing with science and technology. One of the next innovations, now in the process of development in several countries (United States, Canada, West Germany, Brazil, etc.) is the concept of a regional or provincial science and technology plan(1).

The idea here is not simply to decentralise the research system government decree, but to stimulate, by both public and private means the creation of local systems which are financed, operated and indepe dently controlled by local government authorities and by the local community. This regionalisation of science and technology policies has two advantages. One is the much closer convergence between the real needs of the local community and the R and D activities of the scientific establishment. The other is the possibility for the regional system and its sponsors to initiate research in areas which are unfashionable, or which have been neglected by the central govern ment, but which might nevertheless be of considerable importance to a particular region or community. One illustration here is the research on sailships financed by the Hamburg City Council in West Germany.

---

1) Cf. The Research System, Vol.III, OECD, Paris, 1974. The links between regionalism, competition and creativity were discussed in the first half of the last century by the French economist J.C.L. Simonde de Sismondi in his Histoire des républiques italiennes du moyen âge, Paris, 1840. For an excerpt from this book, see Futuribles, Summer 1975.

Another is the research on oil sands and on coal conversion sponsored by the Alberta Research Council and the Alberta Government in Canada. This type of decentralised or regionalised science policy seems to be particularly well indicated for the development of intermediate technology, at least in those countries which have more than a few hundred thousand inhabitants.

Partly as a result of the absence of a significant infrastructure, most if not all the scientific and technological research activities in the developing countries are controlled and financed by the central government. The growing interest of many governments in appropriate technology, coupled with the tradition of government monopoly in the field of science and technology, carries with it the risk of a governmental take-over of all R and D activities in intermediate technology or, in other terms, of a destruction of the decentralised or 'federal' type of organisation now gradually being established by private and non-governmental organisations. One of the basic aims of national science policies should be to encourage, or at least tolerate, this federalism rather than try to suppress it for the sake of co-ordination and rationalisation. The research system is much too complex and delicate to be organised in an orderly way, and a large part of its internal dynamism stems from duplication, competition and disorder.

It is well to note that decentralisation and local initiative, while necessary to the large-scale diffusion of appropriate technology, may stand in direct opposition to political priorities. One of the major tasks facing developing countries is to build up a viable nation-state. The odds are often enormous: time is short, the economy underdeveloped, ethnic conflicts close to the surface, and communications difficult. Extreme centralisation is usually the answer, and regional initiatives, whatever their origin or their object, often tend to be viewed as a threat to a faltering national unity, or at best, as an obstacle to the smooth working of the central bureaucracy(1). In this perspective, modern technology and the creation of large-scale industries are usually seen as a unifying political tool.

The intermediate technology movement with its emphasis on local initiative, diversity and self-reliance, its distrust of governmental action and its scepticism about modernisation at all costs, clearly represents a potentially destabilising political force. Poor farmers who are gradually taught how to make new tools and run their holding in a more rational way, very quickly learn that some of the big

---

1) The study by D.L.O. Mendis,"The Reorganisation of the Light Engineering Industry in Sri Lanka",in the second part of this book, suggests that the promotion of appropriate technology requires working against the 'system'. In India, at least one appropriate technology group has, at times, been subject to considerable harassment on the part of the bureaucracy.

problems they face are in fact due, indirectly or directly, to the central government's modernisation drive. And the ethnic groups or the regions which, for cultural or social reasons, are more apt than others to seize the advantages offered by intermediate technology, can easily come to represent a political and economic threat to other groups, and notably to the ruling elites. One of the big problems which many developing countries have to face is that of preserving a certain balance between the centrifugal forces of diversity and the centripetal forces of modernity, and any national or regional policy for appropriate technology must take this into account.

IV. THE PROMOTION OF LOCAL TECHNOLOGICAL TRADITIONS

One of the fascinating questions in the social history of technology is the association of certain ethnic, religious or social groups with particular types of technology. The case of the watch-making industry in the Swiss and French Jura and the German Black Forest is a classic example. The same phenomenon can be observed in almost every country: witness the Thai metal working tradition, the Jewish Yemenite silversmiths, the Afghan or North Cameroonian leather works, the Iranian carpets, or the firearms manufacturing tradition in Northern Pakistan. In most cases, we do not know why a particular town or social group came to develop such a tradition, and little is known about the cultural and social processes through which it is transmitted and improved from one generation to another. Most of these technological traditions are disappearing rapidly although they have in some cases been successfully revived: for instance, the Indonesian batik industry. Along with the technology, what also disappears is its underlying culture, its values, its system of learning and such important things as pride in one's work and social consideration.

One of the aims of a national policy for intermediate technology might be to try to identify these local traditions, and attempt to use them as a basis for the development of new but somewhat related technologies. The idea here is not to return to the past and to rebuild an industry with little future, but to help those who have generations of experience behind them make the transition and adapt in an innovative way to the new needs of the community. The craft of the silversmith can be immensely important in making or repairing small machines, the art of the potter can be useful in building up a water distribution system or an irrigation network, and a traditional gun-maker can learn how to make more complex metal-working equipment.

There are no simple ways of building-up such local technological traditions, and time is often the determining factor in success or failure. Several appropriate technology groups have been quite successful in upgrading and developing a local technological basis. One very good illustration is the work done by the Planning and Research Institute in Uttar Pradesh: the traditional skills of the local potters have been systematically upgraded, new products have been developed with the help of extension agents, and this 'new' industry has become sufficiently competitive to be able to meet on equal terms the competition from large-scale industrial firms(1).

What is often overlooked is the fact that these technological traditions carry with them a certain psychological image, which from a marketing point of view can be extremely important. This image of a product or a technology often survives long after the technology has disappeared. Most people in Europe have heard for instance of the steel blades which made the town of Toledo in Spain so famous, yet the technology was lost several generations ago. Many of the traditional technologies which have survived in developing countries today are well known outside their area of origin, and the reputation, or image, which some of them still have, is a major asset which could be exploited in a much more effective way.

It is quite clear that all the existing traditional technologies cannot be systematically upgraded and improved. Furthermore, the disappearance of technology is a normal social phenomenon which is closely linked with the overall processes of technological change. Yet it is surprising to see that virtually no systematic attempts have been made to draw up inventories of these existing stocks of technology. Without such inventories, it is difficult to know what exists, let alone what could be improved and developed. New nations are always anxious to discover, or rediscover, their past(2), yet few of them have really tried to take stock of what they had in the field of technology. Drawing up such inventories is one of the many tasks which a national policy for appropriate technology might try to tackle.

V. THE SCHOOL SYSTEM AND THE UNEDUCATED INNOVATOR

Modern science and technology, and national science policies are closely associated with the universities which supply the system with the highly qualified manpower it needs and perform the greatest

1) See M.K. Garg, "The Upgrading of Traditional Technologies in India: Whiteware Manufacturing and the Development of Home Living Technologies", in the second part of this book.
2) It may be interesting to note in this connection that one of the fastest growing 'industries' in the developing countries today is archaeology.

part of a country's fundamental research. In the same way that the university is the cornerstone of modern science and technology, the primary and especially the secondary school is a basic element in intermediate technology. One of the functions, unfortunately rarely fulfilled, of the primary and secondary school, should be to give children a certain feeling for technology in general, and for the appropriate technologies useful in everyday life in particular. Without trying here to outline how this might be done - there are in fact a number of very interesting experiments now going on in many developing countries(1) - one can suggest that the schools and the teacher training colleges should be one of the main centres of attention of a national policy for intermediate technology.

One of the most widespread assumptions about the science and technology system is that innovators are primarily highly educated people and that most innovations come from large research laboratories As a result, national science policies are almost unanimous in stressing the need for training more university graduates and, until quite recently, in emphasising the vital importance of greater expenditures in research. Industrial history however clearly shows that a large number of important innovations are made by individuals with little if any university education, and even today the correlation in most industrialised countries between inventiveness and educational level is far from being conclusive.

What this suggests for national policies vis-à-vis intermediate technology is that the promotion of invention and innovation should focus in priority on this vast pool of formally uneducated entrepreneurs and innovators which exists in every society, but which is being either neglected or repressed as a result of the unduly optimistic belief in the virtues of higher education. The idea that science policy or technology policy should be based upon people with no training in science and only a very empirical knowledge of modern technology may be anathema to science policy makers who, for the most part, are themselves eminent scientists or engineers in the modern sense of the word. But it may well be more appropriate.

VI. DUALISTIC POLICIES FOR TECHNOLOGY

In the same way that the promotion of appropriate technology should avoid bringing about a dismantlement of the modern technology system which exists in a number of developing countries, an intermediate technology policy along the lines suggested above is not a

---

1) See for instance, Développement de méthodes et de techniques adaptées aux conditions propres aux pays en voie de développement (mimeo), UNESCO, Paris, 1975.

substitute for modern science policies, but rather a complement. The idea here is to have two parallel policies, and not to replace one by the other. One observer has suggested that a country like India needed a Cow Dung Commission rather than an Atomic Energy Commission. In fact, the two are probably needed, and each has a different function to fulfil.

The dualistic policy suggested here is but a reflection of the social and technological dualism which characterises the economic system in the majority of the developing countries. The question which has to be considered here is whether the existing science policy establishment should be given the responsibility for developing a national policy with regard to intermediate technology. To put things in a somewhat caricatural form, is the Atomic Energy Commission, or another similar organisation, the one most suited for establishing and operating a Cow Dung Commission? At first sight, the obvious answer seems to be negative.

The science and technology policy organisations which were created in the last twenty years in the developing countries were geared, like the universities, to the development of modern science and technology. To accuse them of having almost totally neglected intermediate technology is somewhat unfair, since this was never their aim. Their activities were in fact but a reflection of national priorities at the time, and dissenters notwithstanding, corresponded to the ideology and dominant values of society. Times have changed, so have values, and in the more advanced of the developing countries, even the strongest proponents of modern science and technology have come to realise its limitations. The growing understanding of policy makers for low-cost technology does not mean however that they are the people most suited to be entrusted with its promotion, nor does it mean that the institutions they have helped to create are the most adequate for developing new appropriate technologies.

Innovation in general, and institutional innovation in particular, does not generally take place through a transformation of existing organisations, but by the creation of new ones which by-pass them. Intermediate technology, in its institutional and policy aspects at least, has developed and will probably continue to develop along parallel paths rather than in the mainstream of existing organisations. If innovation and originality are to be encouraged, this parallel approach should be emphasised. Although there are at present no clear models of interaction between modern technology policies and intermediate technology policies, one can envisage for instance that central government authorities would continue to be in charge of the modern science and technology system, and that local or provincial authorities, supplemented by private organisations, would have the main responsibility for intermediate technology.

There are some good reasons to believe that national policies for the promotion of intermediate technology can best be carried out by institutions which are not directly concerned with large-scale industrial development or with highly sophisticated modern technology. This is not to say that the modern sector should be cut off from the new developments taking place in appropriate technology, quite the contrary: large industrial firms, university research centres and government departments stand much to gain by becoming more familiar with the work of the intermediate technology proponents, and there must be some sort of cross-fertilisation between the innovators in the modern sector and those in the intermediate sector.

Among the many problems which will have to be tackled by a national policy for appropriate technology, one of the most frequently overlooked is that of the risks of innovation. Ever since the beginnings of the industrial revolution, governments have taken measures to safeguard the community against the risks and dangers of innovation National legislation on new pharmaceutical products or safety regulations for industrial machinery are typical examples. What should not however be overlooked is that innovation almost always implies a certain risk, and that a society must be prepared to accept certain risks if innovation is to take place. This is true not only of modern technology, but also of intermediate technology. What seems to be happening in several developing countries, however, is that new technologies are increasingly being subject to quality standards and safety regulations which are so strict as to make innovation, if not impossible, at least extremely difficult. One objective of an innovation policy might be to relax some of these standards, if only temporarily, in order to stimulate the development of new indigenous technologies.

A national policy for appropriate technology cannot be carried out exclusively by government. It should be a collective effort involving as wide a number of institutions as possible, both public and private and within the public sector, local and regional authorities have as important a part to play as the central government. With the possible exception of China, no country has yet developed a comprehensive national policy for appropriate technology, and it is far from clear what the most effective approach might be. A number of interesting national experiments have been carried out, and the most successful among these point very clearly to the need for taking a rather wide view of the problem of appropriate technology: the real issue is not so much to develop new hardware, but to build up an innovation system.

One very important element in this respect is the credit system. At present, there are relatively few intermediate credit institutions, i.e. organisations or groupings which are neither traditional (like the local moneylenders or the private person-to-person credit mechanisms) nor oriented primarily towards large-scale investment in the modern sector, as are most of the banks. One intermediate organisation is the credit co-operative. These co-operatives, which played a major part in mobilising the existing financial resources for the development of agriculture and small industry in nineteenth century Europe, could be one of the most effective tools in the development of intermediate technology. Several very successful co-operatives of this type have been established in Latin America and in several Asian countries(1).

As the example of the light engineering industry in Sri Lanka clearly shows, one of the key factors in the successful upgrading of the village blacksmith's production methods has been the availability of inexpensive credit, mobilised through a co-operative organisation(2). This example suggests by the way that credit is one of the areas where co-operative forms of organisation are the most appropriate. Co-operatives are seldom good industrial entrepreneurs, as the story of the small-scale sugar plants in India shows, but in the credit field, where the problems are very different, they can be remarkably effective(3).

The mobilisation of a community's resources can also be approached in other ways. One interesting example is the way in which one of the Indian appropriate technology organisations has undertaken to educate the local managers of both public and private banks to the problems of risk capital, industrial entrepreneurship and agricultural development(4).

A second area of interest for an appropriate technology policy is that of testing and evaluating new technology. A large part of the equipment developed by appropriate technology groups and which is currently available on the world market, is untested. Its reliability is still too often open to question, and the economic risk to the user as a result excessively high. Testing and evaluation require time and money. No country has yet set up such testing and evaluation facilities, and this is one of the problems which could be solved on an international or inter-regional basis.

One of the basic principles underlying the appropriate technology movement is to try to make a much more effective use of existing

---

1) See C.L. Kendall, "Accumulating Capital at the Grassroots Level", Co-operation Canada, No.20, May-June, 1975.

2) See D.L.O. Mendis, "The Reorganisation of the Light Engineering Industry in Sri Lanka", in the second part of this book.

3) P. Dubin, "Education as a Low-Cost Technology", ibid.

4) P.D. Malgavkar, "The Role of Techno-Entrepreneurs in the Adoption of New Technology", ibid.

resources,be they human or material. One of the aims of a national policy should therefore be to investigate in a comprehensive way how these resources could be better used. For instance by identifying, and hopefully removing, the vast array of institutional and organisational obstacles to entrepreneurship and to the diffusion of innovation.

VII. THE ROLE OF FOREIGN AID

A growing number of aid granting agencies, both international and national and public as well as private are coming to the conclusion that appropriate technology is, if not the wave of the future, at least a potentially very effective tool in the development process. This evolution coupled with the growth of the intermediate technology movement in the developing countries raises a number of questions, both for donors and for receivers. What for instance should be the relationship between the suppliers of appropriate technology in the industrialised countries, and the users of this technology in the developing countries?

An excessive reliance on intermediate technologies developed abroad, and in particular in the industrialised countries, could in the long run be self-defeating. The function of foreign assistance should not be merely to introduce a new piece of technology, attractive as it may be, but to help initiate, within the receiving community, a process of innovation and self-sustaining growth. Such processes are very complex, and the social and cultural mechanisms which underlie them are often very fragile. In many cases, it needs very little to throw them completely off balance, and the sudden introduction even of a seemingly unimportant piece of technology is sometimes enough to do just this. Needless to say, this is not always and necessarily the case. In many communities the ecosystem of innovation is amazingly resilient, and the introduction of a new technolog can serve as the starting point of a self-sustaining process of innovation. The difficulty of course is that it is very difficult to know beforehand whether the existing innovation system is particularly weak or potentially strong.

In the case of intermediate technology, it may well be that the problem of foreign aid and technical assistance is stated in terms which are far too crude and which reflect to an excessive extent the experiences, often rather disappointing, which have been made in the field of modern technology and large-scale industrial development. This can perhaps best be gauged by looking at the ways in which the various appropriate technology groups working at the international level have been operating in the developing countries. Their work does not consist simply of bringing in a new piece of technology from an industrialised country or another developing nation, but rather in developing new technologies on a multinational and transcultural basis. Intermediate technology today is probably one of the few fields where the concept of international co-operation is more than an empty catchword or a convenient verbal substitute for total dependence.

The growing institutionalisation of intermediate technology will inevitably reflect upon the patterns of foreign aid. Already today, many aid-giving countries and agencies are trying to focus more of their assistance programmes on appropriate technology. This is undoubtedly a positive change and it should allow for a considerable widening of the technology spectrum. In fact, in a few years time, the 'consumer' of appropriate technology will probably find himself in the equivalent of a supermarket, with dozens of different tools or technologies to meet every single one of his needs. This will be very different from the somewhat spartan shop of today where many technologies are missing, unavailable, un-invented or mislaid by history.

Technical assistance programmes as they now stand consist by and large in helping developing countries to acquire and use as effectively as possible new or obsolete technologies originating in the industrialised nations. As far as intermediate technology is concerned, there are some serious doubts as to the validity of this pattern, and in fact the ways in which the various groups active in appropriate technology are now interacting with one another seem to indicate that the concept of foreign aid or technical assistance is, if not obsolete, at least very largely inadequate. There are several reasons for this.

In the field of appropriate technology, which is addressed mainly to the most under-privileged segments of the population and notably to the rural poor, the delivery system of technology is comparatively more important that the technology itself. Developing new construction materials or more adequate agricultural tools for instance is far from simple, but the most difficult problems usually lie elsewhere, and in particular in the social processes which lead to a community's adoption or rejection of an innovation. The industrialised aid-giving countries are in many cases better equipped for developing the hardware - the case of SOFRETES's solar pump or the windmills now being designed by some large aircraft corporations are good cases in point. This comparative advantage does not however attach to all types of hardware. In fact, it probably disappears almost entirely when it comes to very simple low-level technologies. The Directory of Appropriate Technology published by the Gandhian Institute of Studies is a good example of the areas in which a country like India probably has a comparative technological advantage over a European technical university or a large American industrial firm.

When it comes to the software or the delivery system, there are no reasons to believe that the industrialised countries and their aid-giving agencies, public or private, are better equipped or more effective in initiating innovation processes. Some foreign aid-giving organisations have been very successful, but their effectiveness has resulted mainly from their intimate knowledge of local conditions. Insofar as the knowledge of local conditions is concerned, there is little doubt that the insiders - local extension officers, social workers, etc. - enjoy a large potential advantage over the outsiders, be it only that of language and culture.

What this suggests is that the delivery system of appropriate technology is probably the area where the developing countries must do most of the work themselves. There is also another justification for their active participation in the diffusion of low-cost technology and that is the relative weakness of the various international groups which are now trying to promote intermediate technology. The history of foreign assistance shows that aid-donors can exert a considerable influence upon the national policies of the receiving country. Withholding a hundred million dollars for a big dam can bring about substantial changes in the economic policy of a developing country, and it is a well known fact that the granting of aid by some multilateral agencies is often subject to basic economic and social reforms Intermediate technology groups obviously do not have the same leverage withholding the introduction of a new technology for crushing peanuts or for collecting rainwater is at best an unpractical means for getting the government of a developing country to initiate new policies for the promotion of low-cost technology. In fact intermediate technology, contrary to large-scale modern technology, is probably one of the fields in which a country can assert its sovereignty in the most effective and positive way. This means that national governments in the developing countries will have a great impact, positive or negative, on the diffusion of low-cost technology.

Part Two

THE PRACTITIONERS' POINT OF VIEW

I. THE MOBILISATION OF KNOWLEDGE ON LOW-COST TECHNOLOGY: OUTLINE OF A STRATEGY

by

George McRobie*

Mobilising knowledge on low-cost technology and communicating it to potential users in a practical form can be envisaged as a three stage process. First is the task of diagnosis and understanding: what is the need for technology, at what level can this need be identified, what is the kind of technology we should be concerned with? Second is the problem of securing the knowledge itself (sources, methods, organisation); and third is the question of communication: how do we get the knowledge into the minds of planners and administrators, and into the hands of those who can turn it into action? Each of these three aspects may be given different emphasis by industrialised and developing countries, but these are in essence the questions to which all practitioners are addressing themselves.

DIAGNOSIS AND UNDERSTANDING OF THE BASIC NEEDS FOR TECHNOLOGY

There is today a widespread recognition of the need to develop and introduce low-cost technologies. This understanding however is not matched by anything like correspondingly widespread action on the part of those who control the mainstream of aid and development programmes. Most if not all practitioners in intermediate technology share the view that the choice of technology is probably the most critical problem confronting developing countries. With the onset of the energy crisis, this problem of choice has become even more critical, and is now also confronting the rich countries. There is a growing recognition that our present economic system is based on the ruthless exploitation of non-renewable resources and that economists have committed the cardinal sin of confusing capital with income. Fossil fuels are a capital, but we have been using them as an income. Agriculture has been operating as an increasingly energy-intensive

* The author is Director of Communications of the Intermediate Technology Development Group (ITDG) in London.

system, and the over-exploitation of our income resources may well turn them into non-renewable resources.

As far as the developing countries are concerned, there seems to be a fairly general acknowledgement of the following facts:

a) the source and centre of world poverty lie primarily in the rural areas of poor countries, which are largely by-passed by aid and development as currently practised;
b) the rural areas will continue to be by-passed, and unemployment will continue to grow, unless self-help technologies are made available to the poor countries and assistance given in their application;
c) the donor countries and donor agencies do not at present possess the necessary organised knowledge about appropriate technology and do not have the communications system which would enable them to contribute to rural development on the scale which is required;
d) unless the disease of poverty is tackled at its source in the rural areas, outside the big cities, it will continue to manifest itself in three ways - mass migration into cities, mass unemployment, and the persistent threat, or even actuality, of mass starvation.

Within the framework of conventional aid and development programmes, there is in practice no major political or commercial impetus to offer to the poor countries any real choice of technologies. This is especially true when it comes to small-scale equipment that can be wholly made locally and which uses indigenous materials and serves local needs. In the Intermediate Technology Development Group (ITDG) we are thinking along the lines of Mansur Hoda's three 'levels' of technology that are required to fill this gap: the technology that suits the family, the village or community, and the market town or small regional centre, in ascending levels of cost, sophistication and volume of output(1).

Within these categories it is possible to enumerate a very large, even daunting number of possible technologies. But there are certain basic and fairly universal needs to guide us in setting priorities. These are the manufacturing and processing activities related to food, clothing, shelter, health, culture, about which detailed, practical information would go a long way towards filling the existing "information gap" and giving choices to people who at present now have none. The following list is not exhaustive, but serves to illustrate the range of new activities that could be developed in rural areas within the modest ambit of basic human needs (the procedure is broadly that adopted by ITDG's specialist panels).

---

1) See M.M. Hoda's article in this book, "India's Experience and the Gandhian Tradition".

RANGE OF BASIC TECHNOLOGIES REQUIRED FOR RURAL DEVELOPMENT

Agricultural production: tools and equipment for ground preparation, planting, weeding, harvesting, along with the basic tools and techniques required for their manufacture (metalworking and woodworking).

Water supply for agriculture: equipment for storing, lifting and moving water.

Crop processing: shellers, winnowers, mills, oil extractors, decorticators, fertiliser and feedstuff manufacture, and by-products (this would include processing of a wide range of products from biological resources).

Storage: storage equipment appropriate for different crops, using local materials.

Food preservation: metal and glass containers, cooking utensils, equipment for smoking and sun-drying; packaging for different foods.

Clothing: equipment for ginning, spinning and weaving cotton and wool; manufacture of dyes and finishing materials; tailoring equipment; leather tanning and manufacture of footwear and animal harnesses.

Shelter: brick and tile making, lime burning, cement substitutes, small-scale cement production; soil stabilisation; timber production and by-products; cast and forged metal fittings.

Other consumer goods: household utensils, equipment for pottery and ceramics, furniture, soap, sugar, domestic water supply (including water purification and sanitation), cooking stoves, fuel, toys, etc.

Community goods and services: school and medical clinic equipment, roadmaking, bridge-building, water supply, power sources and equipment, transport; data and equipment required to operate institutions such as health clinics and co-operatives, work-based education, and training-through-production programmes.

For each identifiable manufacturing activity - and there are obviously many more than this list suggests - we should aim at providing at least two or three levels or types of technology, to cater both for people who are wholly or partly outside the market economy, and for those who are already within a market system.

The first step is to mobilise the existing knowledge on intermediate technologies. The second, to indicate obvious gaps, or areas where new research and development work needs to be done; and the third to outline proposals for carrying out this work. The aim is to start with a 'state of the arts' survey, which should briefly describe the conventional high-cost 'modern' method, present existing alternatives, indicate their limitations or deficiencies and propose work programmes to improve them, or devise new technologies to fill the gaps.

This approach lends itself to dealing with the three types of situation which are familiar to the intermediate technology practitioner:

a) Those where there is a range of technologies which, on the face of it, are reasonably adequate to meet the needs of small, poor communities. There are for instance many types of water pumps and hand-looms already in existence. The task is to identify the pros and cons of each type of equipment, indicate how they are made or where they can be bought, and what they can and cannot do. Foundry work is a good example. The same applies to hand-made bricks; enough is known - on the basis of experience - to adapt this technology well down the size-scale;

b) Those where there are no obvious, reasonably efficient small-scale technologies, as cement manufacturing or cotton spinning. In these cases we have to 'point up' the gaps, and suggest ways of filling them. In cases like cement, the alternatives are to work on redesigning a plant to make it small as was done for paper-pulp machines, to provide a substitute (e.g. lime-brick mortars) or to combine the two approaches.

c) Those where the technology is relatively new and there is no body of current or recent experience to draw upon, as in the case of most forms of unconventional power sources and devices, or the application of modern scientific knowledge to old arts like chemicals from biological sources. In such cases, a state of the arts survey needs to be followed by specific programmes for design and field testing.

It is only through a sustained and systematic effort along these lines that an adequate flow of self-help technologies can begin to reach the potential users: people whose problem is not that of the rich - how to get best value for money - but quite a different problem - how to turn their labour into something useful.

SOURCES OF KNOWLEDGE AND METHODS OF MOBILISATION: ITDG's EXPERIENCE

Ways of acquiring the necessary knowledge of low-cost technologies will obviously depend, among other things, on the history of the technology and the pattern of development in the country where the work is started. At ITDG, we have developed a structure of Panels and Working Groups of advisers. Within each we try to get a mix of expert knowledge representing the professional and administrative, academic, industrial and commercial aspects of the particular subject under investigation. The aim is to appoint, as soon as funds permit, a full-time project officer - qualified and experienced in the subjec to carry out a work programme under the Panel's guidance and supervision. Currently, project officers are working under our Agriculture Construction, Building Materials, Water and Power Panels, and a small team of engineers operates the Group's Industrial Liaison Unit. The Panels on Chemistry and Chemical Engineering, Co-operatives, Forestry Rural Health and Transport are without full-time project staff, but are nevertheless pursuing work programmes which have resulted in useful publications and, in the case of chemistry, in consultancy work overseas.

Through the Panel membership, liaison with a number of universit research departments has been built up, as well as collaboration with British government research establishments, notably the Building Research Establishment, the Road Research Laboratory, the Tropical Products Institute and the Overseas Liaison Unit of the National Institute of Agricultural Engineering.

The systematic investigation and sifting of material already published is obviously our major source of practical data. Much of the Group's published information also derives from practical R and D work carried out by its own staff or associates: innovative R and D work, ranging from complete redesign and production to relatively minor modification, has been done on a small-scale paper-pulp manufacturing unit, brick-works, a wide range of agricultural and hospita equipment, ferro-cement boats, water catchment tanks and pumps. Original work concerned with organisation, training and business procedures has also been carried out - and the results published - on the construction industry, co-operatives and rural health. Most of the results of this work are already included in the Group's publications; a set of more detailed accounts of the work programmes completed, in progress and planned is currently being compiled and will be published as soon as possible.

THE SYSTEM OF COMMUNICATION

Low-cost technology lacks an effective communication and deliver system. Information on high-cost, energy-intensive technologies is promoted by government aid, by large companies, by the education system, and by media of all kinds. The task of communicating low-cost technology is much more difficult, not only because it is less famili to design-makers, but also because we are trying to reach people in the field where there are no clearly established channels of communication, and where more effective methods of communicating are unknown

Above all, and this requires particular emphasis, we are dealing with a situation where the demand for low-cost technologies cannot arise independently of the supply of information about them. All over the world, people simply do not know that there are low-cost alternatives. This applies almost universally to people in rural areas. As far as administrators and decision-makers are concerned, if they have been educated in the rich countries the chances are not only that they do not know about such alternatives, but they may well be prejudiced against them - the familiar 'second-best' argument. The primary task of organisations such as ITDG is to make it known that effective, low-cost alternatives do exist or can be developed. Until this has been done, it can hardly be expected that a demand for them will arise on a large scale.

While no one could claim to have discovered all the answers to the problem of communicating information and knowledge about low-cost technology, the methods developed by ITDG are probably of a fairly general significance. These methods fall roughly into four categories: the exploitation and diffusion of published information, field projects, consultancy assignments, and finally the development of closer links with intermediate technology groups operating in the developing countries.

a) Published information comes under three main categories:

- detailed specifications and drawings of equipment, e.g. agricultural machines and implements, or hospital equipment;
- guides to sources of equipment, 'step-by-step' manuals (e.g. for animal-drawn equipment, construction materials and water supply systems), annotated bibliographies and other source material, and finally industrial profiles (e.g. for foundry or leather works);
- detailed project reports of field operations, as in the case of ITDG's agricultural equipment projects in Zambia and Nigeria, water catchment and storage work in the Caribbean, Brazil and Ethiopia; reports of consulting assignments (for instance a small-scale chemical development project in Pakistan);
- reports on any of the lines above published by other organisations and distributed by ITDG's Publications Unit by arrangement.

The recently-launched Journal of Appropriate Technology is in a special category: it is intended to serve as an international forum for news, exchange of information, previews of major reports, and a means of linking practitioners in the field of low-cost technology.

b) Field projects. These have mostly taken the form of one, two or three-year programmes with clearly defined objectives, designed to identify needs, develop appropriate technologies and demonstrate their practical use. They are not intended to be, and are not, development programmes as conventionally understood. This must be emphasised since the Group is sometimes criticised for not undertaking extension work and for not launching major programmes to that end. ITDG is primarily a knowledge organisation with strictly limited resources. Even if we had the resources, we could not and should not attempt to do extension work or launch such programmes through the existing organisation. Nothing could be more fatal to the work of a knowledge centre than to be burdened by the pressures of implementing a programme on a 'development' scale (as distinct from a 'demonstration' scale). It hardly needs to be emphasised that extension work and the widespread diffusion of new methods and technologies is of crucial importance and that part of the communications task is to ensure that those aid and development agencies which are equipped to undertake such programmes get the fullest amount of information on relevant technologies. As far as ITDG is concerned, some involvement in implementation is in fact taking place but through separately-organised subsidiaries, 'Intermediate Technology Services' and 'Development Techniques', specifically concerned with consultancy and with technical development.

c) Consultancy. In addition to the routine answering of technical enquiries, this work includes the provision of technical inputs at the appropriate level to other agencies, governmental or non-governmental. During the 1973/74 period, ITDG has operated in this way in twelve different countries (Brazil, Canada, Ethiopia, Ghana, Iran, Jamaica, Kenya, Pakistan, Rwanda, Swaziland, Tanzania, and Zambia).

d) Overseas intermediate technology units. In the longer run, by far the most important way of communicating information is through the development of indigenous intermediate technology organisations in the developing countries themselves. Only through such centres can the knowledge and practice of low-cost technology become widespread and form an integral part of the development process. Focal points of this kind are now increasing in number(1) and ITDG has been associated with the growth of some of them, notably in Ghana, India and

---

1) Several of these centres were represented at the OECD meeting in September 1974.

Pakistan(1). The further development of such centres is high on our list of priorities. These may start as focal points for the assembly of information on low-cost technologies from outside sources but can rapidly develop indigenous resources for R & D, the dissemination of knowledge and the promotion of field applications. In other instances they can immediately begin to mobilise sources of information and innovation within their own countries and help to draw on outside sources. Ideally an appropriate technology unit of this kind should be brought in at the inception of every government plan for development.

THE POSSIBILITIES FOR CO-OPERATION

At first sight it is extraordinarily difficult to get effective collaboration between the various organisations now working on low-cost technology. They vary widely in size, degree of specialisation and form of organisation. Each has its own source of funds - always inadequate - its own lines of communication with other countries, and its own priority subjects.

The simplest form of communication is through the exchange of literature, and this is already proceeding. We hope that the Appropriate Technology Journal can increasingly help in this process, and contribute to the diffusion of knowledge as to who is doing what and where in the field of low-cost technology. But are there ways in which co-operation between the work programmes of each organisation could be made more fruitful? Are there ways of securing for each of them some of the advantages of a larger international effort without the constraints which attach to international organisations? Co-operation should not seek to merge the identities of the organisations concerned, but try to make them more effective.

A suggested list of topics on which more information and knowledge would be of particular use to practitioners in both rich and poor countries might include the following:

Methods of identification of needs and resources.
More comprehensive state-of-the-arts reviews.
Sources of relevant data.
Identification of gaps in knowledge.
New R and D work in hand and required.
Normalisation of ways of presenting data.

---

1) The Appropriate Technology Centre in Islamabad, Pakistan, has recently been started up with government support.

Locations for field testing and demonstration.
Joint publication of results of work done.
Joint approaches to funding agencies.
Joint approaches to 'user' agencies (governmental, non-governmental and international).

In operational terms, one could envisage for instance a meeting of practitioners concerned with small-scale water supplies, construction and building materials, or agricultural equipment. They could compare work programmes, identify gaps in knowledge, make proposals for filling these gaps, arrange for field trials, and aim at developing a comprehensive 'package' of new knowledge. Technologies appropriate to different climatic zones could probably lend themselves particularly well to this form of collaboration. Once such work programmes have been harmonised the proposals could be published in order to facilitate fund-raising and secure appropriate requests from developing countries to participate in the work.

CONCLUSION

Over the past twenty years, aid policies have treated the poor countries as if they were rich, and the introduction of modern technologies in the developing countries has contributed to widening the gap between the affluent urban elite and the great majority of the people living in the rural areas. The dominant view has been that the poor countries can only become rich by accepting the technologies, the institutions and the culture of the industrialised countries. Education has only reinforced this: most of the policy makers in the developed countries have been trained in Europe or North America. As a result, it is difficult to envisage that there may be different paths to development.

Over the last one hundred and fifty years, development has consisted mainly of sucking life from the rural areas. The time has now come to change this and to focus our intellectual and technological resources on the rural sector. This can be done by developing a wide range of new technologies which will have to be adapted and modified by the poor countries. The task of the rich countries is to help identify, in close contact with local innovators, the types of technology which are most urgently required. Most of these technologies will obviously have to be concerned with agriculture and rural development. No country after all has ever developed except on the basis of an agricultural surplus.

How this can best be done is not yet entirely clear. One of the big problems is to foster much closer contacts between the three different types of practitioners in intermediate technology: those who are concerned with the practical work of developing new pieces of equipment, those who focus on the field application of new technology, and finally those who, at the national or international level, are trying to mobilise all the available knowledge on more appropriate technologies.

In fact, the time is now ripe to start thinking of an overall strategy for intermediate technology. Five years ago, this would have been inconceivable: the number of people involved was much too small, and the whole idea of intermediate technology seemed absurd. This first stage then gave way to a much greater acceptance of the idea, but apart from a few rather isolated groups, there was and still is little action. We now seem to be entering a third stage, which should be marked by a large-scale and systematic attempt to mobilise existing resources, develop a wide range of new technologies, and establish an effective information network. The fourth stage, which has yet to come, should be that of widespread application and diffusion.

In terms of strategy, the crucial problem today is to make a successful transition from the second to the third stage. One tool would be to establish a really effective communication system in intermediate technology. What has been done until now has certainly been useful, but this is usually insufficient, largely for financial reasons. The communications system which must be built up should go beyond the existing community of practitioners and cover the people and institutions which in one way or another can influence the pace and direction of development.

II. BRACE RESEARCH INSTITUTE'S HANDBOOK OF APPROPRIATE TECHNOLOGY

by

T.A. Lawand, F. Hvelplund, R. Alward and J. Voss*

The Brace Research Institute is primarily an engineering organisation. However, from the experience gained in the field, we have evolved into an institute of practitioners of appropriate technology. This evolution led to our participation in the preparation of the first handbook on this subject. The departmental policy and history of the Institute are studied in order to identify the reasons why we were involved in the preparation of this handbook. This paper traces the transition of the Brace Institute from a technical operation to one concerned with the development and implementation of appropriate technology.

Why did this transition take place? What led us to adopt appropriate technologies? What lessons have we learned, and what new policies have our experiences created? And perhaps most important, what does an analysis of our evolution to appropriate technology suggest for future policy actions in developing areas? These questions are the real subject of this paper.

THE INSTITUTE'S DEPARTMENTAL POLICY

The Institute was set up on the bequest of Major James Brace, who was primarily interested in making arid or desert lands available and economically useful for agricultural purposes. It was his desire that the results of this research would be made freely available to all the peoples of the world. The Institute has fulfilled and is continuing to fulfil his wishes. The policy decision made in 1959 was to concentrate on the problems of water and power scarcity affecting the individual persons and small communities in arid, developing areas.

---

*T.A. Lawand is the Director of Field Operations of Brace Research Institute in Quebec, Canada; F. Hvelplund is a Visiting Economist at the Institute and a Research Fellow of the Handelshojskolen in Aarhus, Denmark; R. Alward is a Research Associate of Brace Research Institute; J. Voss is an Environmental Economist at the Institute for Man and His Environment of the State University of New York, Chazy, New York.

In the meantime, the Institute has built up a unique facility, the value of which is out of all proportion to its size. It maintains an active interest in the whole field of water desalination. It is now recognised as an international research centre for the application of solar and wind energy (especially with regard to solar distillation) and for the development of wind turbines. As an extension of its work on various methods of saline water conversion, the Institute has begun research into controlled-environment agriculture as a means of reducing the water requirements of plants in arid areas. It has undertaken studies on the use of greenhouses in colder climates.

The Institute has become a centre for information on appropriate technology, and one of its most valuable assets is the Brace Library, which contains collections of reference material on desalination, solar energy and wind power. These collections are considered to be among the most comprehensive and thoroughly indexed sources of information available in their respective disciplines. Extensive information has more recently been gathered on greenhouse agriculture and on appropriate technology in general.

Although the Institute concentrates primarily on the technological aspects of water and power problems facing small communities in arid areas, it is fully recognised that the 'tool' or 'system' is only one facet of the problem. A clear appreciation must be made of the cultural, social and political context in which the equipment is to function in order to establish its appropriateness to the community it will serve. It is essential to recognise that for a technology to be appropriate it must be directed towards the betterment of the community, in both its direct and indirect implications.

The basic philosophy has been to develop saline water conversion, pumping and other equipment which uses as much as possible local energy, material and human resources so that the technology can identify within the infrastructure of the local community. This policy was adopted in order to secure participation of the indigenous population in all phases of the construction and assembly of the equipment. This ensures continuity by developing their ability to handle its operation and maintenance. The equipment is characterised by its simplicity and ease of maintenance. Stocks of simple replacement components ensure continuity and dependability. The annual operating costs are comprised primarily of the amortisation of the capital investment, a fair proportion of which is made up of local labour and material charges.

This basic type of undertaking is generally referred to as appropriate technology. There are hundreds of millions of people whose everyday life is little affected by modern technological achievements and who lie outside the mainstream of development. The objective is to provide them with an option - or an alternative -

so that they may solve their own technological problems with systems, methods, energies and materials under their own control. In making use of what is by and large locally available, and in adapting the technology so that the individual villager himself feels part of the overall achievement, appropriate technology takes into account that facet of the human equation which is so often neglected - the dignity of man. For whatever we design, construct or discover, the final proof and justification of its merit is its acceptability by man.

THE EVOLUTION OF AN APPROPRIATE TECHNOLOGY APPROACH

Why did an institute working primarily in the field of engineering become involved in appropriate technology? There are several reasons. The first is financial: research and development and even field applications are relatively inexpensive. Hence, it is often possible, even with a modest outlay, to come up with simple technical solutions to the problems of the rural populations of the Third World. The second is scale: appropriate technology, by its very nature, deals with small villages and peasant farmers, and developments are themselves on a small scale. This is within the scope of a relatively small organisation such as Brace Research Institute. The scale of these activities also means that funding can often be more easily found so that in effect a wider approach can be realised with the direct involvement of the local inhabitants.

In 1960-61, a research test facility was built in the island of Barbados in the West Indies, where abundant quantities of sea water, sun and wind combine to provide an excellent proving ground for equipment development. This overseas mission which lasted until 1967, and which has expanded into other parts of the world, did more than just provide a convenient physical milieu for experimentation: it provided an insight into the real needs of the rural populations of the Third World.

In surveying the needs of these rural areas, it was evident that in maximising local resources, considerable attention had to be paid to the development of alternative indigenous energy sources, such as solar and wind energy. As a result, developments in the following fields have been studied:

- small scale desalination equipment - solar distillation units, vapour compression and reverse osmosis desalination, using windpower as a motive force;
- direct solar energy applications for heating water, heating air and drying crops;
- solar ponds for solar energy collection and storage;

- solar powered organic fluid rankine cycle engines;
- storage of thermal energy;
- the development of environmentally adapted greenhouses for arid areas to reduce water consumption, using saline water as a feed source;
- the development of environmentally adapted greenhouses for colder regions to reduce heating requirements through a more efficient use of solar energy;
- low-cost housing, and the integration of solar and wind energy sources directly into the structure for the provision of services;
- low-cost sanitary technology with a view to reducing water consumption;
- the development of a large windmill for water pumping, irrigation or electricity generation;
- the development of small-scale windmills using the Savonius rotor and sail mill principles.

After the initial R and D in the West Indies, and while working on some of the above problems, the Institute focused on a few specific applications such as solar dryers to process corn for a feed mill in Barbados, a solar distillation plant and some solar cookers in Haiti.

THE CULTURAL DIMENSIONS OF TECHNOLOGY

These applications in the real world pointed to the need for a more comprehensive approach: enthusiasm accompanied by good engineering design is not always sufficient.

We had reached a critical crossroads in moving from research and experimentation to the application of technology in developing areas. This required a move beyond the narrow confines of purely technical solutions to an approach based on a much broader range of inputs - cultural, social, political, and economic. Although we did not recognise it at the time, this led us to the adoption of what is now called the appropriate technology approach.

If the goals of the Institute were to be achieved, the following preconditions would have to be fulfilled:

- The technology must meet the fundamental needs of the community and be recognised as such.
- For the community to respond and accept the technology, a sufficient amount of 'animation sociale' of the local population has to be undertaken.
- The economics of technology has to be fully understood and appreciated.

- The cultural and social values of the local population have to be considered as an integral part of the innovation process.

In order to accomplish these tasks and achieve some measure of success, the Institute staff as practitioners have evolved, through trial and error, some basic principles of operation. Firstly, whereve possible, local technologists should become involved in all the phase of the development process, from research to application. The Institute has tried to help local technologists appreciate the importance of studying the fundamental problems facing their own rural populations. This is essential, since they can communicate in the same 'language' as these target communities and generally understand their culture and values.

Secondly, local social workers have a very important part to play in getting the indigenous population to appreciate and accept the technological innovation. The installation of a fresh water facility for example, decreases infant mortality rates, and creates new problems of birth control. The solution of the latter problem is often beyond the scope and capabilities of the well-meaning technologist. Thirdly, economists must be brought in to provide a more comprehensive evaluation of the costs and benefits of a given appropriate technology in a local context. In view of past development experience, it is obvious that both the short and long run consequences of a specific technology need to be considered. Economists can hopefully specify more comprehensive social welfare functions for each region of a developing area.

Our experiences as practitioners of appropriate technology has led us to appreciate that even the appropriateness of a given technology is not a sufficient condition for its widespread adoption. We have come to realise that no matter how simple, inexpensive or appropriate to the needs and resources of the local people, a new technology must be viewed within the cultural context in which it is introduced. Between the identification of the need for an appropriate technology and its successful application lies the critical problem of cultural adaptation.

Technology by its very nature is optimistic and presents what could be accomplished. From experience we discovered that the frustrating gap between what is technologically feasible, and what is adopted in practice, most often results from a basic scientific neglect of the critical role played by the other side of the cultural coin - society's attitudes and values. Values are shared beliefs about what is right and wrong. Technology determines what is possible, but values and attitudes determine what is socially acceptable. The criterion of social acceptability either limits or enhances the probability of adoption of a given technology. For example, even though population control is technologically possible, social attitudes toward family size or religious values with regard to birth control limit the range of technological solutions to the problem of over-population.

Our field operations throughout the world have led us to appreciate the critical relationship between technology and values in the development and application of appropriate technology. In our own fashion, we have come to realise that conceptually, appropriate technology is no more useful in the development process than an inappropriate technology if it cannot become acceptable to the individuals and the groups to whom it is proposed. Applied technological change implies social acceptability and the degree of acceptance and rate of adoption depend in turn upon a thorough knowledge of values and attitudes. 'Value competence' or 'cultural competence' must accompany the development of appropriate technology, and precede its successful applications.

THE TECHNICAL AND INSTITUTIONAL BACKGROUND OF THE HANDBOOK

The Institute has been collaborating for years with organisations such as the Intermediate Technology Development Group (ITDG) in the United Kingdom and the Volunteers in Technical Assistance (VITA) in the United States. This has given us access to great amounts of information and technology. Through our educational programmes and exchange visits, we have set up working relations with a number of groups and individual technologists working in similar fields in a number of countries.

It is not always possible to help the many technicians who write from Third World areas for information on the purely technical and scientific aspects of alternative energy sources. As a result we have embarked in the last few years on the preparation of a number of state-of-the-art surveys. These surveys deal with solar ponds, solar refrigeration and air conditioning, low-cost sanitary technology and commercially available and experimental windmills.

In 1974 we worked on the preparation of a handbook of solar agricultural dryers which includes an evaluation of the theory, design, construction and performance of such dryers, an assessment of the operating performance of the commercial and experimental dryers, and finally a bibliography and list of problems which remain to be solved. This handbook, which will be available in English, French and Spanish, has been an international effort: it has involved more than 30 technologists in all parts of the world who sent in their illustrated contributions. It will partially substitute for the lack of proper library facilities generally facing the technologist in developing areas.

The next step in the Institute's work has been the preparation of a more comprehensive handbook of appropriate technology. The institutional origins of this work can be traced back to the first Canadian seminar on intermediate technology organised in March 1972 by the Canadian Hunger Foundation. Interest in the subject had been growing in Canada for a number of years, particularly amongst the non-governmental organisations dealing with project-oriented development aid in the Third World. The guests of honour at this meeting were Dr.E.F. Schumacher and George McRobie of the Intermediate Technology Development Group with which the Brace Research Institute had been in contact since the mid-1960's. Following the meeting, a working group based primarily in Ottawa was set up to look into intermediate technology and recommend future action. One result was the submission in March 1973 of a proposal to the Canadian International Development Agency (CIDA) in Ottawa to provide counterpart funding for the publication of a loose-leaf handbook on appropriate technology

The non-government organisation division of CIDA accepted the proposal and gave financial support towards its execution. This joint undertaking shows that Canadian non-governmental organisations can work together and collaborate more closely in their development programmes. During the summer of 1973, the Canadian Hunger Foundation entered negotiations with the Brace Research Institute with a view to securing their collaboration in the preparation of the handbook. The main reason for approaching the Brace Research Institute was the fact that it has been an active practitioner in this field for the past 15 years. There were few other organisations in Canada, or indeed elsewhere, to which the Canadian Hunger Foundation could turn.

From its past experiences with interdisciplinary projects, the Institute recognised that it had to broaden its technological interests. Staff was expanded accordingly, and a number of contractual arrangements were made to enlist the services of outside specialists. Among these new collaborators, there were economists with varying backgrounds in development (socio-economic questions, values, etc.), sociologists specialised in social work and community development, technologists from Third World areas, and a political scientist dealing with the political implications of appropriate technology.

In addition, as part of the preparation of the handbook, questionnaires were sent to a number of organisations and individuals in different parts of the world. This has resulted in a considerable amount of information dealing with the experiences of other practitioners in appropriate technology.

OUTLINE OF THE HANDBOOK

The handbook has been designed to allow for a wide variety of uses. The first section presents the theory and philosophy of appropriate technology. It discusses the criteria by which a system can be judged appropriate or not, the various contexts within which appropriate technologies may or may not work and the mechanisms of implementation and diffusion. It also outlines the numerous factors which must be considered in order to evaluate the potential appropriateness of a technology in a given situation.

The second section contains a series of case studies in which an attempt is made to weave the theory and the technologies together into meaningful examples so that the reader can learn from the experience of others. Each of the case studies has been selected to illustrate important ideas from the preceding chapter. In addition, enough information has been provided to allow the reader to get started on a specific task or process, or to build certain tools.

Section three contains a catalogue of a number of tools that are available throughout the world to meet a wide variety of needs. This section is viewed mainly as an 'information exchange', and its aim is to make people aware of some of the devices others have used to improve their situation.

The final section in the book gives an indication of the new and experimental tools or systems that are currently under consideration in the field of appropriate technology. It also outlines possible new areas of research and future additions to the handbook. The appendices included in a reference section give the names and addresses of several groups and people involved in one way or another with appropriate technology so as to facilitate international exchanges particularly between the developing countries. There is also an extensive, though by no means exhaustive, bibliography, a suggested system of classification and a glossary of terms used throughout the text.

It must be stressed that this handbook should only be regarded as introductory. Appropriate technology is many things to many people and we have found ourselves in areas about which little is understood. Problems have been investigated that have not been thoroughly worked out. In order for this book to function as a dynamic and appropriate tool, it has to allow for the inclusion of new problems and new solutions. We are therefore asking for and relying on the comments, ideas and suggestions of the reader. The looseleaf format has been used to permit inclusion of any further information.

The amount of funding at the disposal of the Institute and its collaborators for the preparation of the handbook was extremely modest and the book is only a preliminary effort designed to familiarise the likely agents of change with the concepts and potentials of appropriate technology.

THE CRITERIA OF A TECHNOLOGY'S APPROPRIATENESS

The criteria of a technology's appropriateness presented in the first section of the handbook have evolved from an intensive state-of-the-art survey. Since the field is relatively new, viewpoints differ as to what constitutes an appropriate technology. Our selection has been eclectic and the following list is by no means exhaustive.

Appropriate technology should be compatible with local cultural and economic conditions, i.e. the human, material and cultural resources of the community.

The tools and processes should be under the maintenance and operational control of the population.

Appropriate technology, wherever possible, should use locally available resources.

If imported resources and technology are used, some control must be made available to the community.

Appropriate technology should wherever possible use local energy sources.

It should be ecologically and environmentally sound.

It should minimise cultural disruptions.

It should be flexible in order that a community should not lock itself into systems which later prove inefficient and unsuitable.

Research and policy action should be integrated and locally operated wherever possible in order to ensure the relevance of the research to the welfare of the local population, the maximisation of local creativity, the participation of local inhabitants in technological developments and the synchronisation of research with field activities

Obviously, in practice, it may not be possible to meet all these criteria. They do, however, provide general guidelines or goals to which appropriate technology practitioners should aspire.

CASE STUDIES INCLUDED IN THE HANDBOOK

The case studies included in the handbook deal with a variety of subjects and have been carefully selected to illustrate these criteria of appropriateness. The following examples will give an idea of the types of issues raised in the handbook.

a) Solar distillation

This case study traces the more recent historical development of a small fishing village on the island of La Gonave, Haiti. Defoliation of the land and subsequent climatic changes beginning in the early 1950's have brought unprecedented hardship. By the mid-1960's there were several major problems facing the community, one of which was the lack of potable water. Water for cooking and drinking purposes had to be obtained from wells, the closest of which was a day's walk there and back. In 1965, a community development programme was initiated in an attempt to raise the village economy to a level somewhat higher than its current subsistence level. As a part of the overall programme a solar desalination unit was constructed to supply the village with a continuous source of fresh water. The introduction, design, construction and utilisation of this solar still are discussed in the study. Also included are a basic description of the solar still and sufficiently detailed plans and specifications to permit the construction of a similar water desalination plant.

b) The integrated waste-fuel-food cycle

This paper is the result of many years of studies on integrated rural planning. Although it is not technical, it tries to clarify the problems faced by most developing countries in the world, where foreign investments have by-passed the large majority of the people in rural areas. The government of Papua-New Guinea has established an eight-point development programme, which is aimed at counter-balancing the growth of industry and cities through a healthier rural development. A very good set of recommendations are made to explain how small villages can enhance their economic activities through the use of local resources and better education. A fresh approach towards waste handling and processing is described. It shows that energy recovery goes hand in hand with better health and an improved quality of life.

c) Beekeeping development in Kenya

After a thorough study of traditional practices in beekeeping and a comparison with the methods and equipment used in the Western world, the authors have developed transitional types of bee-hives which lend themselves to the hardy type of local bees and to better bee-management practices. Widespread use of these hives has proven them to be very successful. Construction procedures and specifications are discussed and design advantages of this hive are analysed. A set of good photographs together with a brief bibliography of the subject complete the study.

d) Small bio-gas plants

This study is an introduction to the development of bio-gas plants as an alternative source of energy in rural communities in India. The anaerobic decomposition of animal waste is used to produce methane gas. The biological process, economics, installation and construction procedures together with clear and comprehensive calculations and drawings have been included. From these one can readily obtain an idea of the usefulness of this 'new' energy source in agriculture or for domestic purposes and how it can be used to raise the standard of living in some regions of the country.

e) Oil drum cupola foundry

This study describes a small-scale iron foundry built in Afghanistan to improve upon traditional techniques of steel production. A brief history of the project is given together with the reasons of introducing this improved technology. A description of the unit, costs and performance to date is discussed.

f) The Gujarat Industrial Investment Corporation

This study describes the activities of an investment corporation which has been set up in India to assist small-scale enterprises by providing loans for the acquisition or construction of buildings and machines, and for working capital. This government-sponsored institution has been remarkably successful in stimulating local small-scale industries. In spite of the fact that it gives loans to people without requiring collateral in the form of fixed assets, its success rate has been as good or better than that of the conventional commercial bank which insists on fixed assets as guarantees before granting loans.

g) Intermediate adaptation in Newfoundland

Appropriate technology can also be important for rich countries. This is particularly true for some of the less developed and mainly rural areas. Newfoundland, one of the provinces of Canada, is a case in point. It has one of the highest unemployment rates in the industrialised world in spite of significant industrial growth. The authors of this case study discuss the potential for an appropriate technology strategy in Newfoundland and analyse in detail two specific examples: 'longliner' boats for fishing, and controlled fish curing techniques. The final section of this study deals with the relevance to Newfoundland of appropriate technology and long-term planning.

h) Solar coffee dryers in Columbia

The solar dryers described in this case study are mainly used for drying coffee beans but they are also used to dry maize, beans and cocoa. The sun drying of coffee is widely practiced in Columbia. About 70 per cent of the national production of dry parchment coffee is dried this way. The reason for this is that the method is ideally suited to requirements of the thousands of small-scale producers. This study gives the operational characteristics and economic details on a variety of small-scale dryers as well as the necessary information on construction materials. Also included are a set of seven photographs of solar dryers currently in operation.

SUGGESTED POLICY ACTIONS

Our experiences show that practitioners and scholars of development have become disenchanted with the existing development policies. These policies are not only creating cultural and economic dislocations, but also by-passing the large masses of the population which stand little to gain economically from these programmes and which often suffer severe penalties in other aspects of their lives. The less fortunate groups of the population are left to fend for themselves, and foreign aid often does them little good. If economic development is to take place, and if cultural heritages are to be preserved, 'development from within' becomes a serious policy alternative. This requires working through local people, local resources, local structures, and using a technology that accommodates these low-cost indigenous inputs.

Practitioners who have gravitated towards this position are calling it the appropriate technology approach. Their problem has been one of conceptualisation and communication: how can successful but isolated experiences be shared and form an empirical basis for a viable alternative approach? This has led us to the development of an appropriate technology handbook, for there is today no authoritative book which adequately covers the totality of the subject. This handbook, which has resulted from the dedicated team work of a widely dispersed group of individuals and organisations, and in which Brace Research Institute has acted primarily as a secretariat, should be viewed as an interim step in this direction.

It is essential that this preliminary effort, so modestly funded, be continued on a more organised and rational basis. There exists in the world today a vast amount of knowledge on appropriate technology which needs to be collected and circulated. The task of exploiting this information is gigantic. With our handbook a modest effort has begun, and it would be unfortunate if the momentum developed in this initial phase were not carried forth. It is equally essential to recognise that the material collected should not be restricted solely to technical, economic and social data emanating from sources in developed countries. The subject is universal and it has universal implications.

The loose-leaf format of this handbook underlines the fact that we are dealing with a new and dynamic concept which cannot adequately be covered in a single book. It must be continually revised, upgraded and improved. It has also indicated the need for future activity on an expanded and more permanent basis. It is obvious that a journal, review or some other mechanism of communication, interdisciplinary in nature, and in a number of different languages is also required for the interchange of ideas. The subject matter must be neither purely academic nor essentially scientific. The language should be simple and not restrict itself to the technical terms of one particular branch of study.

If the problems of the underprivileged elements of the population in the developing areas are to be solved, the focus must decidedly be placed on the development of man. In order for him to progress, he must find out what processes, technologies, equipment or tools are used by other groups or individuals to improve the quality of their lives and raise their standards of living. The information must be made available in a form which is easy to understand and hence to apply.

Our handbook obviously has not been written for the ordinary citizen of the developing areas of the world: no first effort in dealing with such relatively new concepts could succeed in being so universal. It is hoped however that the message will reach those in an administrative, technical and catalytic capacity in both developed and developing regions who can act as the agents of change. The introduction of appropriate technology must necessarily be a slow but sure process. Only in this way can the overall awareness of the people be raised, and thereby help them to develop in an autonomous way.

III. THE IMPACT OF MICRO-DEVELOPMENT PROJECTS

by

Ross W. Hammond*

There are a great many organisations around the developing world working on micro-development projects. These activities are being carried out in agriculture, mining and extraction, and commerce, as well as in the manufacturing sector. However, this paper will selectively focus on micro-development projects related to small-scale industry development as opposed to these other economic activities.

The existing body of small industries in the developing countries exhibits great differences in organisation, capital structure, employment, mechanisation and product lines, in addition to operating in differing social, cultural, economic, and governmental milieux. Hence, it is difficult to generalise about small industry operations because of the diversity and complexity of the body of industries and the varying business climates.

Regardless of these problems, it is always desirable, and frequently required, that an evaluation be made of the impact of a micro-development project. Some of the most frequently used techniques to evaluate such projects include cost-benefit analysis, direct and indirect employment generation and income distribution, capital investment or value added.

The small industry sector in developing countries does however have a number of special characteristics which make accurate assessments of the impact of micro-development projects more difficult. Small industries are inherently high-risk activities and the proportion of business failures is considerable. By the time small industries seek assistance with problems, it is frequently too late for effective action. Their communications system is usually poor and ineffective. The small industry sector is geographically dispersed and lacks homogeneity; the small and weak units which compose it produce a great variety of goods, primarily for local markets and their only common denominator is small size.

---

* The author is chief of the Industrial Development Division of the Engineering Experiment Station, Georgia Institute of Technology, in Atlanta, Georgia, U.S.A.

The measurable results of micro-development efforts (e.g. managerial and technical assistance) with one company are miniscule when related to a whole country or even a region. Hence, an evaluatic of micro-development efforts must be an aggregate encompassing result with a great many enterprises which have been provided with assistanc generally on a short-term basis. This means that the results of a small industry development programme are frequently slow in appearing

These characteristics of the small industry sector raise a number of questions which will be discussed in this paper: Is micro-development prohibitively expensive? How can technology be effectivel disseminated? How can the visibility of micro-development projects be evaluated? What mechanisms are needed for more effective micro--development efforts? Can an original target group of small industries develop an innovative capability? Some discussion of these issues seems appropriate. In order to do so, I will draw from the 18-year experience of the Industrial Development Division of the Georgia Institute of Technology.

THE COST OF MICRO-DEVELOPMENT PROJECTS

Direct technical assistance to small-scale industries tends to be problem-oriented and short-term in nature. The problems may be important to the industry in question, but they are normally amenable to solution by a small expenditure of staff time on the part of the assistance organisation. The example of one Korean metalworking firm will illustrate our point. The technical problem facing the company was that two raw materials, small diameter steel rod and steel pipe, frequently were not exactly round. As a result, the quality of the products was poor, and the spoilage high. A simple hand-held shaving die was designed to permit the raw material to be made truly round before processing. The prototype die was designed and built in the mechanical engineering laboratories of Soong Jun University and it is now available to other metalworking companies in Korea. Problem identification, analysis, and the design and construction of the tool were done in few man-days.

Our own experience, both domestic and international, with about 4,000 companies is that small industry problem-solving may infrequently take considerable staff time, but that, on the average, an expenditure of two man-days is required per problem. This translates into a cost of approximately $200 in the US economy, but much less when the service is provided by a technical assistance organisation in a developing country where salaries are lower.

Other evaluation methods can be used to provide an approximate quantification of relative costs of micro-development projects. One analysis we did looked at 429 companies to which problem-solving assistance was provided in 1972. Direct jobs created or saved in these companies amounted to 6,415 at the end of the year. Total project costs were $251,580, which amounts to a cost of $39.21 per job created or saved. Full responsibility for creating or saving all these jobs could not be claimed by the programme since the amount of services varied with each company and other development organisations were sometimes involved. However, since neither indirect job creation nor multiplier factors were taken into account, it is believed that the $39.21 figure represents a fair evaluation of the cost per job.

A number of other evaluations of micro-development projects have been carried out and the results tend to vary. In the aggregate, however, the conclusion of the evaluators is that the return on the investment made in the project is many times the cost. In fact, small industry micro-development projects need not be prohibitively expensive if an experienced and well-motivated staff is available to implement the programme.

THE IMPORTANCE OF THE TECHNOLOGY DELIVERY SYSTEM

There appear to be three prerequisites for a successful programme of technology dissemination: reservoirs of information and expertise, a technology delivery system, and recipients of the technology.

Reservoirs of information and expertise exist in various public and private organisations in many developing countries. These include the information banks and professional staffs of public sector organisations such as ministries, state and local governments, libraries and other institutions. In the private sector, industries often have much information and expert staff personnel which may be tapped through arrangements of various sorts. The development banks can also often serve as sources of information and assistance for small industry. The extent and relative availability of such sources will vary from country to country, but in most cases access channels to technological information exist in some measure.

In all countries, there is also a small industry sector. These firms, with their variegated and fragmented problems, frequently require more information and there is no question that they form a body of recipients in need of technical information. Even when the technical resources exist and the small industry sector needs these resources, the delivery system which conveys the needed information to the recipients is usually the weak link. This delivery system may exist in full or in part, or it may even be totally absent.

In some developing countries, programmes of technical assistance to small industry have been set up by the government but the mechanism is often imperfectly developed. In one such country, technical assistance personnel were placed in a number of major cities to provide various problem-solving services to small industries. However, these people were not provided with automobiles or other means to visit the small firms in the relatively large areas assigned to them. Apparently the assumption was that the entrepreneur in need of assistance would find his way to the assistance office. In our experience this does not happen: entrepreneurs will not or cannot travel long distances to seek assistance. As a result, the industries in the immediate proximity of the urban offices receive advice and assistance, while those farther removed are as isolated from assistance as ever.

A technology dissemination system is only as good as its staff and in many places of the world good personnel are in short supply or even non-existent. The well-recognised need for training industrial extension personnel remains unfilled in many developing countries, thus perpetuating inadequate technology delivery systems.

While the situation varies considerably from country to country, the following conditions have to be met if a technology dissemination programme for small industry is to be effective:

- Technology needs and possible sources of information must be carefully identified;
- The authority and responsibility to act as a technology delivery system must be clearly assigned to one or more well-motivated organisations;
- The delivery system organisations need some separate and independent source of funding since the small-scale industry sector in most developing countries cannot support the costs of technology dissemination;
- The delivery system organisations must have developed or at least must be in a position to develop an experienced and trained staff which is results-oriented;
- A reliable evaluation procedure must be established in order to permit the assessment of results.

THE VISIBILITY OF MICRO-DEVELOPMENT EFFORTS

One of the inherent difficulties in increasing the visibility of micro-development efforts is that the results of any one technical assistance project for a small enterprise are rarely startling or momentous or even newsworthy. If, as a result of a micro-development effort, a company is able to hire 10 new employees, double its productive capacity, increase its profits or develop some new product, the impact is still small in the total picture of the developing country. It is only when the results of many such activities are considered that a substantial impact can be demonstrated.

By contrast, the funding and development of a large steel mill, textile plant, or shipyard is a more visible and glamorous project from which the government can obtain considerable press coverage and recognition. It is therefore not surprising if such large-scale projects, which usually involve foreign capital and technical assistance, receive preferential consideration by the governments of many developing countries.

Efforts to strengthen and build up small and medium firms, which are more likely to be locally-owned and managed by local people, tend to strengthen indigenous industrial and technological capabilities. Developing countries are beginning to realise the importance of a domestically-owned and operated industrial sector, and there is today a greater willingness on the part of governments to extend financial assistance and technical support to the small industry sector. However, the basic problem of visibility still remains, and it is only when many micro-development results are aggregated that impressive statistics related to the generation of employment, capital or income can be demonstrated.

THE CONDITIONS FOR A SUCCESSFUL MICRO-DEVELOPMENT EFFORT

The effectiveness of the mechanisms designed to stimulate the small- and medium-scale industry sector in the developing countries is very variable. Success or failure depend upon such factors as organisational structures, level of funding, staff capability and motivation, and ease of access to technological information. However, the truly effective micro-development programmes usually owe their success to the presence of four institutional elements.

The first is the existence of governmental programmes arising out of a recognition by governments of the importance of small industries, and aimed at providing incentives and support to small firms.

The second is the existence of institutional mechanisms, both public and private, to analyse and solve micro-development problems and to disseminate appropriate technology to recipients through industrial extension activities or other means. An essential element here is the presence of public or private lending institutions which have some responsibility for, and which are willing to work with, small industries.

The third is the existence of informational systems, which serve as data centres and provide access to R and D information from many sources, either as part of the institutional mechanism or available to it. An essential element of such informational systems is a 'current awareness capability' to ensure that information is available on recent developments in markets, products, processes, or equipment.

The fourth element is the existence of a personnel development system which provides continual personnel training (formal and informal) and upgrading, in addition to field experience. This is usually, but not necessarily, done through an institutional or technical assistance source.

THE DEVELOPMENT OF AN INNOVATIVE CAPABILITY

Innovative capabilities in the small industry sector do not appear as frequently as in larger industries. Occasionally, small entrepreneurs will demonstrate their innovativeness in the design of equipment, new products or new processes. This however is largely a reflection of individual ability and the nature of small industry activities is generally inhibitive to innovation. Most small firms are dominated by one entrepreneur who assumes all the managerial and technical responsibilities for running the company. The difficulties of day-to-day operations do not usually give the average entrepreneur enough time or opportunities to be greatly innovative. The larger firms by contrast employ staff personnel who have the responsibility for staying abreast of changing conditions and looking out for innovation.

This type of staff support is largely non-existent in the small industry sector. When an entrepreneur does develop an innovation, this innovation is generally not made available to other companies which might benefit from it. This may be due to the fear of competition, the absence of institutional mechanisms to disseminate information about innovation to other companies, or to any number of other reasons. However, when an innovation occurs in a wider institutional framework, the barriers to the diffusion of the innovation to small firms can be eliminated. For example, the low-cost rice machinery developed by the International Rice Research Institute has been made available to all interested manufacturers and is now in production in a number of rice-growing countries(1).

Another example is that of the 66 small metalworking industries in the Yong Dong Po area of Seoul, Korea. These firms are linked together in a tightly-knit association. In such an organisational framework, innovations developed by one company can be disseminated to others through the association. This, however, is an unusual arrangement which is not often encountered in developing countries.

Can a target group of small industries develop an innovative capability? In the absence of some special organisational arrangement such a question usually has to be answered negatively. An arrangement of this type would ideally reward innovation and provide a vehicle for dissemination of information about the innovation. It would also

---

1) See Amir U. Khan's paper "Mechanisation Technology for Tropical Agriculture" in this book.

assist in the adaptation of innovation to suit differing conditions.

In many developing countries the industry sector is an important contributor to employment and income distribution. It is also usually a training ground for indigenous entrepreneurs. For these reasons, it deserves more government support than it is presently given in some countries. Our experience suggests that properly motivated and staffed public and private assistance organisations can provide meaningful managerial and technical assistance to this sector at a relatively low cost, and using the existing infrastructure of technology and expertise.

IV. INDIA'S EXPERIENCE AND THE GANDHIAN TRADITION

by

M.M. Hoda*

INTRODUCTION

Whenever we think and talk of development in the poorer countries we must be clear in our minds. Are we concerned only with increases in per capita income, a more favourable balance of trade and higher production of steel, cement and electricity? Or do we want to develop the downtrodden masses living in the two million villages of the Third World? The real measure of development is the degree of well-being achieved by the 85 per cent of the people living in the villages and there should not be any controversy that the central aim of any development effort is the eradication of poverty and the provision of better living standards for the masses.

Most development plans however focus on the urban centres, reflect urban values and have largely by-passed the rural areas. There are very few effective development schemes for the villages and it is not surprising that the standard of living in the rural areas is declining, the village structures breaking-up, and migration to the cities increasing in a dramatic way. Unless this trend is reversed, real development in the poor countries cannot take place. The sophisticated and highly capital-intensive technology invented and used in the cities is unable to solve the problems of the poor people and we have to look for a technology which is less expensive, more labour-intensive and more appropriate to the situation of poverty

TECHNOLOGICAL CHOICES AND APPROPRIATE TECHNOLOGY

The technological choices facing the developing countries have generally been polarised between (a) modern technology imported from the industrialised nations and (b) traditional indigenous technology,

* The author is the head of the Appropriate Technology Development Unit of the Gandhian Institute of Studies in Varanasi, India.

which is primitive, inefficient and wasteful of skill and resources. There has been very little interaction between these two types of technology and they have evolved on two completely different planes.

Technology can be defined in terms of capital cost per work place. The indigenous technology may be called symbolically the '£1 technology', while the Western technology could be called the '£1,000 technology'. The gap between the two is so enormous that a natural, organic transition from one to the other is impossible. In fact, the introduction of £1,000 technologies has killed off traditional work places in villages at an alarming rate without providing any alternative employment for the millions who lose their livelihood. The village oilmen, potters, cobblers and weavers driven out of business by the competition from the oil mills, ceramic plants, shoe factories and textile mills, are joining the unemployed labour force in the big cities. Their £1 technologies are grossly inefficient and the Western £1,000 technology is too expensive for them. It is meant for those who are already rich and powerful. If effective help is to be given to those who need it most, it requires a technology which is more appropriate to the conditions of poverty, and which would range somewhere between the '£1 technology' and the '£1,000 technology'. This may again be called, symbolically, the '£100 technology'. It would be more productive than the primitive indigenous technology, but at the same time immensely cheaper and easier to manage than the sophisticated technology of modern large-scale industry. Intermediate technology can help to create much larger numbers of work places with existing capital resources and meet the needs of the poor in the villages. It should be small in scale, simple in use, rich in employment opportunities, sparing of natural resources and non-violent in spirit.

INDIA'S RURAL DEVELOPMENT EXPERIENCES IN PERSPECTIVE

The need to improve the technological level of traditional village industries, acute as it may be, is not an entirely new problem, and India has witnessed many attempts to meet it in the past hundred years. In the 1890's quite a few centres for rural development were started up in various parts of the country. They represented a systematic effort to promote the development of specific rural communities through the conscious application of technical knowledge. By the end of the 1940's, a number of such projects embodying important principles and approaches of community development were in existence all over India. Some of them had been initiated by provincial governments and princely states, others by private organisations or great individuals, including Christian missionaries. The missionaries who started their activities in the mid-19th century played a great role in bringing this development work (including education, hospitals and training in industry) to the remote and mostly tribal areas of the country. Later on, Gandhian workers and independent voluntary associations were to take the leading part in this effort.

Most of these experiences could have provided important guidelines for present-day planners, but no attempt has been made to know more about these plans and to learn the lessons from them. These projects were all systematically planned and scientifically conceived. They pioneered new methods of working with people in rural communities and made adequate use of industrial and agricultural technology. All the great Indian reformers, like Tagore and Gandhi, saw this very clearly. Before them (1885), Maharaja Sayajirao Gaekwad III of Baroda and his prime minister Raja Sir T. Madhav Rao had carried out their great experiment in Baroda for rural development. Tagore established a 'Shilpa Bhavan' (community) in Sriniketan and trained artisans in new technologies imported from far and wide. More than any one else, Mahatma Gandhi made it a movement, because he believed that "If villages perish, India perishes too". He organised the All-India Spinners' Association and the All-India Village Industries Association. He made the 'charkha' (the spinner's wheel) a symbol of new village technology and started a systematic study of all the village industries with a view to improving their technology and to giving them new dignity. Gandhi had appointed many experts to develop village technologies, which would help artisans and craftsmen to improve their productivity and efficiency. Men like Maganlal Gandhi, Satish Chandra Dasgupta and Jamnalal Bajaj carried out their research work and designed suitable machinery for spinning, weaving, oil extraction, leather work and the use of carcasses. 'Ashrams' (rural development units) were set up by Gandhi at Sabarmati, Ahmedabad, Wardha, Sodepur, Bardoli and other places.

After Independence (1947), Gandhi with the help of Dhirendra Ch. Mazumdar and others set up a chain of village industry complexes like Sewapuri, Benaras, Khadigram and Monghyr to train workers in improved technologies and make them qualified to set up small industries in their own villages. He also called on the State governments to try to make villages self-sufficient in their own needs for cloth, oil, shoes, etc. and thereby reduce their purchases of consumer goods manufactured by urban industries. As a result of Gandhi's effort, there are at least one thousand such units set up in various parts of the country. Men inspired by Gandhi carried his torch alight after his death. J.C. Kumarapa and Dr. Gadgil, known as Gandhian economists elaborated his ideas further and gave it a concrete shape. In 1956, a 'Sarvodaya' plan (community development plan) was prepared, which can be said to be a true village development plan. Acharya Vinoba Bhave and Jayaprakash Narayan provided the leadership for the work of village development through the Sarvodaya movement. In fact, the village development programme has become an article of faith for all Gandhian movements in the country.

In 1963, Dr. E.F. Schumacher, a British economist and former adviser to the British Coal Board, visited India at the invitation of the Planning Commission and Jayaprakash Narayan. He was influenced by the Gandhian ideas of industrialisation and technology, adapted them to modern needs and turned intermediate technology into a world-wide movement. In 1966 he set up with other like-minded people the Intermediate Technology Development Group in London to collect information on such technologies which would be really beneficial to the rural areas of the developing countries. This was the first organisation of its kind in a developed country which advocated cheap, inexpensive and labour-intensive machines and equipment for the developing countries, instead of sophisticated, modern and highly capital-intensive machinery.

Schumacher's movement of intermediate technology gave a new lease of life to the concept of village development and the Gandhian movement, reinforced as expected by Schumacher's ideas, took a lead in giving a new meaning and a scientific backing to the rural development programme. The Gandhian Institute of Studies, set up by Jayaprakash Narayan to conduct research on social science problems in a Gandhian perspective, advocated the view that technology could not become meaningful for the poor masses if it was not confronted with the general social and economic questions facing the country and made aware of its deep-seated rural problems.

The Gandhian Institute of Studies has taken an active interest in intermediate technology and organised many seminars on the subject. Later on, it decided to establish an Appropriate Technology Development Unit in the voluntary sector. This dream was eventually realised late in 1972 when such a unit was set up at Varanasi in co-operation with the Intermediate Technology Development Group of London.

THE GANDHIAN INSTITUTE'S APPROPRIATE TECHNOLOGY DEVELOPMENT UNIT

This Unit was set up with the aim to develop, crystallise and make visible the appropriate technologies that will really solve the problems of the poor in India. One has to identify what the problems of poverty are and find solutions which fit into the conditions of poverty. Technology must have a social outlook. In India modern technology has now reached the point where the establishment of an average work place in industry costs between 20,000 and 50,000 rupees (£1,000 - £2,500). We have to develop technologies which require somewhere between 2,000 and 5,000 rupees (£100 - £250) for setting up a work-place. This means ten times more jobs for the same capital expenditure.

The purpose of the Unit is to re-orient all such agencies, institutions and leaders which are engaged in the tasks of social and economic development of the rural areas to take up this new programme of work. It co-operates in this task with all those concerned, be they government departments, the Planning Commission, industrial corporations, big and small, or international agencies; but its priorities are clear: it is the voluntary sector and the voluntary movement in the rural sector which are of central concern. The Unit is interested in the improvement of indigenous industries, village technologies and small industries, and it is concerned with all the spheres of village activities (agricultural tools and implements, food processing, material handling and transport, water and irrigation, decentralised sources of power and energy, construction, animal husbandry, health and hygiene, education and training, community living and culture). Emphasis will also be put on the introduction of appropriate technology into the curricula of primary and secondary schools.

The Unit proposes to become a 'knowledge centre' where information on such technologies can be pooled and farmed out to those who require them to promote research, design and development in this field. One of its first priorities is to motivate the scientists and the technologists, the students and the teachers of the universities, engineering institutions, polytechnics, Institutes of Technology and other research and scientific institutions to carry out the work on appropriate technologies to help the poor rural communities in India.

The Unit also intends to set up a university liaison unit, which will include besides universities all technical institutions, polytechnics and research institutes. The teaching staff in these organisations will have to draw up original topics for student projects. With its contacts in the field, the Unit should be able to submit to them interesting and valuable challenges that are directly related to the real problems of the poor.

With the active help and co-operation of the Intermediate Technology Group in London, similar types of work are being done in many other developing countries. The Unit at Varanasi will be a chain in this international system and through communication with other centres will try to obtain the most recent knowledge and bring it to the notice of the Indian people. But to be useful as a link in the international chain, it will have to get the work done in India in order to influence research and development work in the country.

THE RELEVANCE AND IRRELEVANCE OF RESEARCH TO DEVELOPMENT

The record of foreign assistance to the less developed countries in the last two decades is so inadequate and lopsided that the developing world would have been better off without such assistance. The present level of aid is only of marginal significance and comes with so many project conditions, tying of aid, foreign consultants and sophisticated technology that it saps the initiative and freedom of action of the developing world. The developed countries are only interested in selling their turn-key projects, nuclear power stations, supersonic aircrafts and of course armaments and defence equipment.

Science and technology in the industrialised countries are now progressing at an ever faster rate due to the huge resources at their disposal. This activity is naturally directed towards the interests of the developed countries themselves. Only to a very minor, almost infinitesimal, extent does it have a bearing on problems of direct importance to the less developed countries. Scientific and technological advances in the West are having an impact on the Third World countries that is detrimental to their development prospects. Most of the technical innovations of the late 18th and 19th centuries were mechanical inventions, which were simply the result of harnessing the traditional skills of blacksmiths, clock-makers, millwrights, etc. Nowadays inventions grow increasingly out of basic discoveries concerning the structure of matter and energy, chemical processes, metallurgy and so on. The successful adoption and adaptation of present-day technology requires a much greater knowledge of general science, which in turn requires enormous expenditures in equipment and instruments.

Research in the developing countries is too much under the spell of Western science and technology and often its ambition is to produce results more in line with the Western tradition than with the needs of the developing countries.

The scientists and technologists of the poor countries are engaged in solving the problems of rich societies. This phenomenon affects the poor countries in two ways. In the first place, highly educated and trained persons migrate to the West and all the money spent on their education and training is lost to the nation. The responsibility for this situation lies with the kind of training and education they are given, for it makes them misfits in their home country. Secondly, even if they do not migrate they are mostly engaged in a highly sophisticated research work, which is relevant only to the industrialised countries and the modern Westernised sector of their home country. The poorer people, who in India constitute 90 per cent of the population, do not benefit from this research and this is one of the reasons why development of the broad masses is not taking place.

If the industrialised countries are seriously interested in helping the less developed countries with more appropriate technologies, they will have to increase their research activities and direct them towards the problems which seriously concern the less developed countries. This would imply aid of another type and on a larger scale than anything that has been done previously or contemplated until now. The real needs of the people will have to be studied in depth, then techniques and processes evolved to help them to work more efficiently and effectively. In agriculture, for example, studies are needed about the nature of tropical soils and their reactions to different rainfall patterns, about drainage, irrigation and fertilisers, about improved seeds, animal stocks, grazing pastures, cropping patterns, the use of agricultural implements, the prevention of plant and animal disease the storage and preservation of perishable products. Much more research should go into the planning of irrigation schemes, and greater efforts should be directed towards harnessing solar energy and wind power instead of concentrating on large thermal, nuclear or hydraulic power stations. Similarly there is a need for research into small village industries, low-cost housing, food preservation, grain storage, textiles, animal husbandry, etc.

FOUR STRATEGIES FOR INTERMEDIATE TECHNOLOGY

The technical approach to appropriate technology must be based on the fundamental questions: "What are we really trying to do?" "What are the real needs of the community?" "What are the obstacles that prevent solutions to these problems?" "What, then, is the most appropriate way of acting?" Appropriate technology should be neither a second best, nor an outmoded technology but a solution that fits best the local requirements. Four solutions can be envisaged, and a successful innovation policy would probably include some elements of all four. These solutions are:

- the reviving of an old technology,
- adapting a current one,
- inventing a new one,
- improving the traditional indigenous technology.

a) Reviving

Some people may think that the methods of the 19th century engineers, when labour was plentiful and the large-scale industrial development in the Western world just beginning, are suitable to the needs of less developed countries today. But, of course, the requirements are changing fast and the conditions in the developing countries are not exactly similar to those of Western countries at any past stage of their development. However, some of these are marvellous technologies, even from modern standards, but they were by-passed by other technologies. By-passed technologies can make a contribution and show unusual ways of solving a problem. It may be desirable to go for the simplicity of servicing and repair of 19th century engineers without blindly copying their technology. Water wheels and wind mills, for example, were abandoned in the 19th century as a result of the introduction of steam engines, electricity and internal combustion engines. But with today's energy crisis they can be made to serve a useful purpose again.

b) Adapting

A current technology can be made more appropriate simply by removing the labour saving elements. Substituting a hand lever for an electric motor represents not just a financial saving but also an employment opportunity. More important, it extends the application of the equipment to unelectrified areas and considerably simplifies the problem of maintenance. The appropriate technology answer is not to be found exclusively in small-scale industry, although this might appear at the upper end of our spectrum.

Scaling-down of a few important key industries such as cement, sugar or newsprint, can also be immensely beneficial to the less developed countries. This would help decentralise industry and reduce transport and distribution costs. The problem of depletion of resources and pollution could also be minimised.

c) Inventing

There is a considerable scope here for both new research and a blending of past and present technologies to evolve new designs. The statement of needs and definition of the problem will have to come from the developing countries themselves, but the solution can come from the industrialised countries and their highly sophisticated research institutions. This approach would direct the best of modern scientific knowledge for the benefit of the poor people of the less developed countries. This could include information on new materials, techniques for working them or the ways in which work is organised. New inventions can sometimes be directly adopted in developing countries on a small-scale, because large-scale introduction would need enormous amounts of money and the scrapping of costly machines. Biogas plants and utilisation of solar energy can be included in this category. Balasundram is working on open-end spinning which is the latest technique developed in the world.

d) Improving the indigenous technology

This approach will be most productive in the developing countries themselves. All the industries and crafts which exist in the villages and small towns could be studied systematically, and an organised effort be made to improve their efficiency and productivity. This should be done with the help and assistance of research institutes using modern equipment and the best of modern scientific knowledge. The Western research organisations can also help in this to some extent. This approach has been given great attention in India by the Khadi and Village Industries Commission. Individuals and some private organisations have also worked in this field, like Balasundram in Coimbatore, Mohan Parikh in Surat, Manibhai Desai in Poona and Anna Sahib Shastrabudhe in Wardha.

THE DESIGN CONSTRAINTS OF INTERMEDIATE TECHNOLOGY

Technology should be designed according to the needs and abilities of the poorest people living in the villages, the small towns and the slum areas of large cities. The people in the developing countries are often not used to the constraints of industrial discipline and regulated life. Other factors to be considered are the educational level, cultural norms, religious susceptibilities and habits of the population, the local climate and geography (humidity, temperature, rainfall), communication problems, energy and water resources and availability of raw material and spare parts. A lot can also be done by making use of the local traditional skills.

Bearing these constraints in mind, one can try to elaborate the concept of a '£100 technology' and envisage the development of industry on three levels:

a) home industries with an investment per work place of some £20, using local materials and operated essentially by individual entrepreneurs in the villages. Information about such technologies must be given in local languages and through pictures;
b) village industries with an investment per work place between £100 to £200. Such industries can be set up with the help of village co-operatives and government extension workers. Information can be supplied in local languages and Hindi;
c) small industries with an investment per work place between £1,000 and £2,000. This group probably represents the upper limit of the appropriate technology spectrum and includes industries working for large urban markets.

INFORMATION, COLLECTION AND DISSEMINATION METHODS

One of the very important problems today is to establish a communication channel at the level of appropriate technology. Any small improvement in modern technology, like steel manufacturing, air transportation or nuclear power is immediately known all over the world, but there is no channel through which improved technologies for villages developed in one part of the world is known to another. Such technologies exist in thousands of places, but there is no communication. It is useless for devoted village level workers to reinvent the wheel. If in a village someone is doing absolutely first class work and has invented some new tools or processes, the chances are that twenty miles away, no one knows about it. We therefore need a communication system with 'knowledge centres'. There must be an organisation acting on an international level as a knowledge centre to collect and redistribute information on a world-wide basis. But there should also be such knowledge centres in every developing country to obtain information from abroad and disseminate it to field operators. This must be a two-way traffic, and the problems encountered in the field must be brought to the knowledge of the national and international organisations. The field operators must be trained in the art of identifying 'knowledge gaps' and 'information gaps' and then the central organisations can be more specific in their search for new technologies.

The dissemination of information about what is available in machines, tools, plants and equipment, processes and new techniques could be done through government agencies. Information could be supplied in local languages, with properly illustrated charts, pictures, posters, diagrams, and pamphlets. Extensive use of mass media (television, radio and movies) must also be made. In addition to that, fairs and exhibitions can be organised to demonstrate the uses of new equipment, new techniques and processes.

LOW-COST TECHNOLOGY AS A POST-INDUSTRIAL TECHNOLOGY

The meaning of intermediate or appropriate technology has been aptly summarised by Gandhi. Soon after the independence of India, Prime Minister Nehru went to the small West Bengal village in the Noakhali District where Gandhi was trying to bring about communal harmony. Nehru went there to seek his advice and guidance on how to run the administration and the government. "Just keep one small thing in mind", said Gandhi, "when you are taking any action or making any decision, try to judge how this action or decision is going to affect the poorest of the people in the country".

Intermediate or appropriate technology should similarly be a technology to serve the poorest of the people and its impact on the masses should determine the appropriateness of the technology. For the developing countries, in many cases it provides work and also dignity to work. One of the lowest of untouchables cast in India are the 'Chamars' who deal with hides and skins. They have to do the dirty work of removing the dead bodies of the animals, disposing of the flesh, flaying and tanning their skins, and making various articles out of the skin, bones and horns. If some improved technology is provided to them, for instance pulley blocks for lifting and loading (instead of carrying the carcass on their shoulders), wheelbarrows for carrying, gloves and gumboots to deal with the dirt and filth, it would immensely raise their dignity in addition to making their difficult task easier. Similarly in every field, some simple technology should be provided to the poorest men of the developing countries.

The governments of the developing countries should therefore take a major policy decision to support the village industries and its technology by providing them with the following concessions:

a) All the village industries should be assured a regular supply of cheap raw materials and they should be protected from the large industries which are devouring raw materials from the village and forest areas.

b) Research institutes should be set up exclusively for developing appropriate technologies for the village crafts and industries.

c) A very intensive scheme of education and training for the technologies of village crafts, agriculture, irrigation, communication and culture, etc. should be started. The whole system of education should be oriented towards achieving this objective.

d) Protection should be given to the village industries: the large-scale industries operate with the help of cheap electricity, cheap transport, cheap foreign exchange, skilled personnel trained at government expense and high tariff protection against foreign competition, but none of these facilities are available to the village industries.

e) The village industries should be given support for the preparation of feasibility studies for market research, marketing and the development of a service infrastructure.

Due to the present energy and resource crisis, intermediate or appropriate technology has become a survival technology for both developed and underdeveloped nations. It is probably the only means through which we can raise the general standard of living of the poor communities and it will also help to restrain the industrialised countries from overspending the meagre resources of energy and raw materials we are left with. I think the bicycle exemplifies this very clearly. For poor countries where people have to walk miles on foot and carry heavy loads, it represents an enormous advance. For the rich countries, particularly in the overcrowded and over-polluted cities, it also becomes a major advance to save them from congestion and also pollution from fumes of the automobiles. Thus in a number of areas, intermediate technology represents a sort of converging point for rich and poor countries and also for a life of equality, peace and permanence in the whole world. In this sense we might view it as a post-industrial and post-modern technology.

V. THE SCALING-DOWN OF MODERN TECHNOLOGY: CRYSTAL SUGAR MANUFACTURING IN INDIA

by

M.K. Garg*

Prior to 1905, the most commonly used sweetening agents in India were two products manufactured by the age old indigenous technology: 'khandsari' - a sort of raw sugar which has to be further purified for direct consumption - and 'gur' or 'jaggery' - a concentration product of the whole cane juice without separation of molasses. In 1905, imports of crystal sugar started from Java in the then Dutch East Indies. Initially, there was some resistance to this new product but gradually imports began to grow, reaching a maximum figure of half a million tons annually valued at £10 million sterling.

The modern large-scale vacuum pan technology for manufacturing crystal sugar was introduced during the First World War. Up to 1932 its progress was rather slow due to the very keen competition with the imported crystal sugar from Java. In 1932, the Sugar Protection Act was passed and the excise duty on imported crystal sugar substantially increased. Thereafter the industry grew rapidly and in 1975 there were 222 large-scale vacuum pan factories in operation in India, producing close to five million tons of crystal sugar.

This modern sugar industry occupies an important place in the national economy. It employs some 200,000 persons and contributes about £50 million sterling to the national exchequer in taxes. To meet its requirements in machinery, six companies are manufacturing equipment in co-operation with foreign firms; their technical linkages however are limited to a few special fields.

THE IMBALANCES CREATED BY LARGE-SCALE TECHNOLOGY

In 1931-32 the large-scale sugar plants were consuming 6.6 per cent of the total cane grown in India. At present they consume on an

* The author has been working as a specialist on rural industries in the Planning Research and Action Division of the State Planning Institute in Lucknow, India. He is currently the head of Garg Consultants, a consulting firm specialising in appropriate technology

average only 28 to 30 per cent of the cane crop. The main reason for this low coverage is basically the agricultural pattern of India. About 85 per cent of the land holdings are below one hectare. Sugar cane is the most highly paying cash crop; not only does it bring in higher returns per hectare but it can withstand the vagaries of nature much better than any other crop. As a result, the small land holders strive to use part of their land for cane. In 1971-72 there were 20 million farmers producing 129 million tons of cane on 2.6 million hectares. The cane crop is widely scattered and the opportunities for establishing the intensive growth areas required by the large-scale mills are limited.

The modern mills get their supplies of cane at a price which is fixed by the government. This price normally gives a 25 to 40 per cent higher return to the farmer than the other means of disposal of cane crop, i.e conversion to gur by the farmer himself or the sale to the declining khandsari industry. The cane growers in the mill areas are much better off than those in the non-mill areas. Naturally, this has led the latter to ask for more remunerative arrangements for the disposal of their crop.

With the diffusion of intensive cropping cycles (two to three crops a year in place of the normal one or one and a half) the farmer no longer has the time to crush his cane for making gur by bullock power and wants to save his bullock power and his labour for intensive farming which is more remunerative than gur or sugar processing. This has created a strong movement in non-mill areas for the cash sale of cane.

In the mill areas the situation has become equally complex. In order to meet the fluctuations in cane yields, slightly bigger areas were allotted for the sugar mills to obtain their cane supplies. There have been increases both in the area under cultivation and in the yield per hectare: the land area under sugar cane has nearly quadrupled during the last 40 years (from 685,000 hectares in 1931-32 to 2.6 million hectares in 1971-72) and the availability of fertilisers and better irrigation have increased the average yield from about 30 tons per hectare to 45 tons.

The increase both in area and yields has been proportionately higher in sugar mill areas than elsewhere due to the intensive cane development work done by the sugar mills. As a result the cane available in these areas is at present about 175 to 180 per cent of the installed capacity of the mills The mills are forced to extend their working days from 120-140 to 200-220, well into the hot summer season when recovery rates go down. Even then, the complete crop often cannot be entirely crushed and in some years even has to be partly destroyed. This has resulted in a strong demand by the cane growers of mill areas for alternative means of disposal of their crop.

THE DEMAND FOR A LOW-COST MINI TECHNOLOGY

The imbalances brought about by the modern crystal sugar industry have created a demand for an alternative technology. Four groups are directly interested in such an innovation: the cane growers in areas where there are no mills, who want to get a better return from their cane crop and sell their cane instead of crushing it by bullock power; the cane growers in sugar mill areas who want a means of disposing of their surplus; the rural khandsari manufacturers who want to stay in business; and finally the government which is trying to find an outlet for the periodic surplus of cane.

The problem of developing an alternative technology which would eventually allow the cane growers to get the same price as that paid by the large-scale mills was referred to the Planning Research and Action Institute in 1955. The aim of this Institute, which was set up by the Government of Uttar Pradesh in 1954-55 with a grant from the Rockefeller Foundation, is to carry out action-research for improving and developing techniques for rural areas both in production activities and in home living.

A team was set up in 1955 under the leadership of the author to study the three main aspects of the problem, namely product selection, technology and organisational patterns.

The inherent strength of a large-scale technology is that its products meet the requirements and needs of present-day society. By contrast, the old indigenous technologies in developing countries tend by nature to produce only a particular type of product which is going out of use and which does not fully meet the new requirements of the market. This fact is not usually given enough weight in the various schemes for renovating and improving indigenous technology Most of the attempts in this direction on the part of research agencies seek to improve workability, introduce new tools and increase productivity. As for the government, it tries to improve marketing methods and provides financial measures for survival. Needless to say, such efforts have not been as successful as expected in putting traditional technology on its feet. To give an example, no improvement in lime technology can make lime take the place of portland cement. Unless the traditional low-cost lime technology can be geared up to produce portland cement, its future remains in doubt. Old indigenous technologies can, however, be one of the most promising bases for new low-cost technologies, provided proper attention is given to product selection.

Jaggery is a concentrated product of the whole cane juice; it contains about 80 per cent sucrose or cane sugar. Khandsari sugar is a powdery, yellowish product containing 94 to 98 per cent sucrose. The increasing preference of the Indian consumer for white crystal sugar (which contains 99.9 per cent of sucrose) led to the decline of the khandsari sugar industry, but the annual consumption of white crystal sugar in India is still very low: only 5.8 kg per head, as against some 50 kg in developed countries like the United Kingdom, the United States and Canada. The low consumption of crystal sugar is partly due to the fact that a sizeable part of the rural population uses jaggery, not only as a sweetening agent, but also as a food. However, even in remote rural areas, the use of white crystal sugar as a sweetening agent for drinks or for making sweets is increasing steadily.

The problem as viewed by the team of the Planning Research and Action Institute was not to improve the existing indigenous technology but to find or develop a technology for manufacturing crystal sugar on a small-scale, comparable in quality and competitive in production costs with large-scale sugar technology, if not on an absolute basis, at least in special circumstances or locations.

THE COMPARATIVE EFFICIENCY OF MODERN AND TRADITIONAL TECHNOLOGY

The manufacturing of sugar consists of four stages: the extraction of juice from sugar cane (crushing), clarification of the juice, its evaporation and concentration into masscuite and finally the formation of crystals and their separation from the masscuite to obtain the final product. The team made a detailed study of the various types of plants and equipment available for carrying out cane processing on a small-scale and compared their efficiency with the large-scale plant and machinery. It also studied the principles and practices which help to grow crystal.

This work showed that with the large-scale technology, only 10 per cent of the sugar is lost at the crushing stage. With the existing small-scale technologies (bullock crushers and 3- or 5-roller power crushers) used in the indigenous khandsari industry, the losses are around 30 per cent, i.e. three times higher.

Clarifications in the large-scale technology is carried out by chemical agents, principally lime which is used alone or in combination with sulphur dioxide or carbon dioxide. The three large-scale clarification methods (defecation, lime sulphitation and lime carbonation) remove between 35 and 60 per cent of the non-sugar component of the juice. By contrast, the traditional system of bark vegetable coagulant clarification in the gur and khandsari industry can remove only 10 to 15 per cent of these elements which retard the crystallisation of the sugar and give the final product a powdery consistency.

The evaporation and concentration is done in large-scale factories in vacuum pans while in the indigenous industry, open pan boiling is used. In the first case, sugar losses are around 2 per cent; in the second, they reach 15 per cent, and in addition, there is a significant drop in the purity of the product. In vacuum pan factories, crystallisation is carried out partly in the pan and partly in crystallisers where the masscuite is kept in motion. Approximately 1 per cent of the sugar is lost in the process. With the static crystallisation method used in the gur and khandsari industry, the losses are between 4 and 6 per cent and the sugar is powdery.

The separation of sugar crystals from the mother liquor called masscuite is carried out in the large-scale vacuum technology by mechanically-operated centrifuges while in the khandsari industry, it was done partly by static pressure and mostly by microbiological action. The efficiency of centrifugal separation is considerably higher: only 0.1 to 0.3 per cent of the molasses are left in the sugar, as against 3 to 5 per cent with the traditional technology.

THE DESIGN PRINCIPLES OF A PILOT PLANT

After carefully considering the above factors, the team decided to design a pilot plant based on the following technologies. For the extraction of juice, a 5-roller power driven crusher was used. For clarification, there was no choice except the chemical process, since the objective was to manufacture crystal sugar.

The scaling-down of the lime sulphitation process used in the large-scale technology was experimented upon by a research unit set up by the Government of India in 1936. The process did not prove successful when commercially demonstrated. These research results were carefully examined, changes in the design of the equipment (especially for the lime and sulphur-dioxide reaction) were made and the lime sulphitation process was incorporated in the proposed pilot plant design.

As far as evaporation is concerned, the large-scale vacuum pan system was found to be too complicated and capital-intensive. Scaling-down was a big research problem in itself and was judged uneconomic as well. The only choice left was open pan boiling. Large numbers of designs of open pan furnaces were available in various regions of India but an examination revealed that hardly any of them were suitable for the type of pilot plant envisaged. Two of these designs did, however, partially meet the requirements of the pilot plant and it was decided to try both of them. Scaling-down the technology of crystallisation in motion was simpler and a scaled-down model was incorporated in the design. A suitable and well tested small-scale centrifuge was available, and we accepted it as such for the pilot plant.

THE TECHNICAL PROBLEMS OF THE PILOT PLANT

The pilot plant was established at Ghosi in the Azamgarh district of Uttar Pradesh in 1956-57. Its capacity was kept at 30 tons of cane crushing per day. It operated for three months and produced white crystal sugar closely resembling the sugar manufactured by large-scale plants. The product fetched a price about 10 per cent lower than that of mill sugar, while khandsari quality sugar used to sell at a 25 per cent lower price. The sugar recovery rate on cane was 6.7 per cent as against 5.5 per cent in khandsari and an average 9.5 per cent in the large-scale sugar mills of Uttar Pradesh. The price paid to the farmer for his cane was 80 per cent of that offered by the mills and the working of the plant did not give any loss.

These results were discussed in a technical seminar organised in Lucknow in May 1957. The participants were technicians from the large mills, khandsari entrepreneurs and research agencies. An intensive programme of action research was then engaged to improve the technology. This programme focused on the design and development of a crushing unit comparable in efficiency with those used in the modern mills and on a better co-ordination between the various pieces of equipment used in the clarification process, while keeping instrumentation to a minimum. Development of better kinds of open pan boiling furnaces was undertaken to cut down the sugar losses and check the colour formation which is another handicap in open pan boiling.

In spite of the development of a crystalline structure, the crystal was not a true crystal but rather a conglomerate of tiny crystals. If a true crystal could be grown, losses would be reduced and the sale price of the sugar would be higher. The masscuite prepared on a small-scale had a higher viscosity than with the large-scale method. The separation of molasses was incomplete, and changes were made to slow down the acceleration time and increase the centrifugal force. If the travel time of the juice is too long, some sugar is lost and the recovery rate falls. The design of the pilot plant was accordingly modified to reduce juice travel to a minimum.

The process time required from the extraction of juice to its boiling into masscuite in the pilot plant was 12 hours. This was found to induce inversion of the sucrose. As a result it was decided to adjust the size and capacity of the equipment so as to remove the time lag between various processing stations and reduce the processing time to four hours only. Sun drying of sugar was adopted in the pilot plant. But it was found that about 0.25 per cent of the sugar was lost due to wind currents. For this reason, hot air drying was adopted.

FURTHER IMPROVEMENTS IN TECHNOLOGY

Further research and development work was carried out between 1957 and 1962. A new crushing unit with two mills of three rollers was developed in co-operation with the National Sugar Institute in Kanpur. The juice extraction rate increased from 62 to 68 per cent and the milling efficiency from 70 to 80 per cent. The frequent breakdowns of the crushing unit were completely overcome and a regular crushing rate of 4 tons per hour was achieved. The clarification equipment and practices were standardised so that the operations could be easily taught to semi-skilled workers from rural areas. Visual and simple chemical tests were evolved, thus minimising the risks of poor clarification.

A new type of furnace cut down the boiling losses from 15 to 10 per cent and increased the concentration rate. Not much headway could be made in developing a true crystal and the problem still remains. The changes made in the centrifugal design worked satisfactorily. Two new types of layout (semi-open and closed) were developed, and the juice travel time was cut down by 60 per cent. The readjustment of sizes and capacities, and the design changes in the equipment resulted in cutting down the whole process time from 12 to 4-5 hours. Progress in the development of artificial sugar driers could not be achieved due to the deterioration of the colour of the sugar. This problem should, however, have been solved by 1975.

To improve the economics of the plant, its capacity was raised from 30 to 60 tons per day. A comparison between the Institute's mini-technology and the other methods for producing sugar show that the recovery rate (i.e. the percentage of sugar recovered and bagged out from the total amount of sugar available in the cane at the time of the harvest) is between 57 and 64 per cent for the intermediate technology, as against 42 to 45 per cent for the traditional khandsar technology and 75 to 80 per cent for the modern technology of the large-scale mill. Further efforts are being made by the Planning Research and Action Institute to narrow this gap between the intermediate and the modern technology.

THE PATTERNS OF ENTREPRENEURSHIP

From an organisational point of view, the application of a technology could be divided into two categories: the entrepreneur-ownership type and the producer-ownership type. Large-scale technolog belongs almost entirely to the entrepreneur-ownership type. The responsibility for raising capital for management and for day-to-day operations rests on a strong central group which has the required expertise and incentives. Such an organisation hardly takes into consideration the interests of the supplier or even the consumer and at times an element of exploitation enters into it.

In the producer-ownership pattern, ownership is divided into a number of levels between the worker, the supplier and the entrepreneur. Capital investment is low and the management system is simple. The surplus arising from such an activity is shared at various levels in place of being collected in a centralised pool, and can be reinvested locally. This type of organisation can work at low efficiency and survive even when the odds are high. This type of organisation has allowed many of the inefficient indigenous technologies to withstand the heavy competition from the sophisticated large-scale modern sector.

In our small-scale crystal sugar plant, we considered the possibility of a producer-ownership type of organisation, but there were two handicaps: the process we selected was of a type which could best be carried out under one roof. Efficiency would be lost if the process were divided into a number of levels. Management also had to be centralised. It was therefore decided to try out both types of ownership rather than focus exclusively on the producer-ownership type (co-operatives).

THE FAILURE OF THE CO-OPERATIVE MOVEMENT

The cane growers were formed into a co-operative society whose membership was limited to cane growers of the locality. However, in order to foster leadership, a proviso was made to allow 5 per cent of the total shares to be sold to other interested persons of the region. 25 per cent of the capital was to be raised by the cane growers, 25 per cent contributed by the government as share capital, and the remaining half obtained as short and medium term loans from co-operative banks. Management of the society was vested in a board of directors partly elected by the shareholders and partly nominated by the government. Top management personnel were drawn from governmental and semi-governmental agencies, while the lower management personnel were recruited locally.

The members of the co-operative would supply their cane in proportion to the shares they held. No fixed price for the cane would be set at the time of supply; however, advances could be made. The price of the cane would be determined by the working results of the co-operative at the end of the season. The available surplus, if any, would be divided between the suppliers after deduction of a 6 per cent dividend on the shares.

The first co-operative unit was set up along these lines at Ghosi in 1957-58. Although the technical operation was only a marginal success, many demands to install such units came from cane growers in other areas. By 1962, eight units in all were built by P.R.A.I. Each of them was run initially under the administrative and technical control of P.R.A.I. for two years. Thereafter, the operational responsibility was transferred to the Co-operative Department. In 1962 the installation and construction part was also transferred to the Co-operative Department where a technical cell was formed. Whenever needed, guidance was given by P.R.A.I.

To date, four of these units have closed down and four are still working. The efficiency of three plants in this second group is quite high. Although the Co-operative Department made a year-to-year programme of expansion by providing capital loans, hardly half a dozen units in addition to the above eight were started in the last 12 years

This failure of the co-operative sector is due to a variety of reasons. The first is that the supply of cane received by the co-operative from its members was to be proportional to the share capital contributed by each member. In general, cane growers could not find sufficient capital to entitle them to sell their complete crop to the unit. The membership therefore had to be increased to collect the necessary capital. The increase in membership led to a situation where hardly 25 to 40 per cent of the members' crop could be processed. The farmers therefore had to look for other outlets for their crop or continue to make jaggery, and thus lost interest in the co-operatives.

The second reason is that the cane growers did not feel confident about getting a good price for their cane on the basis of the operational result of the unit. They insisted on selling the cane outright and to meet this requirement, the practice of direct purchase had to be introduced. In fact, the attitude of the cane growers vis-à-vis the co-operative was rather lukewarm.

A third reason for failure was the problem of motivation. When the programme was started, the administrative and executive responsibility for these units rested with the Planning Research and Action Institute which quite naturally had the motivation to implement the programme successfully. When the programme was transferred to the Co-operative Department, the Department passed on most of the responsibilities to the cane growers. The skill and expertise of the cane growers was not sufficient and the efficiency of the units began to fall. A fourth reason is that the management responsibilities remained in the hands of the Department's workers who were more or less forced upon local society. They did not feel responsible towards the cane growers and were more interested in working according to the departmental rules.

In two of the four successful units, the administrative and executive functions were carried out by the Board of Directors in a competent manner without any management personnel from the Co-operative Department. In the other two units, the Board of Directors was strong enough to control the Department personnel in the interests of the units and it built up a strong local cadre of workers.

PRIVATE ENTREPRENEURSHIP AND THE ROLE OF INDUSTRIAL EXTENSION

Efforts to attract rural investors to take up this new technology were made after the first plant operating on a co-operative basis showed success. As a matter of fact, when the working results of the pilot unit were published, some of the khandsari manufacturers expressed their keenness to take up the new technology. In order to promote the diffusion of this innovation a number of steps were taken by the P.R.A.I.:

- Financial and other working data on the various co-operative were made easily available by the P.R.A.I.;
- Facilities were established for training prospective entrepreneurs in the co-operative plants;
- Designs and drawings were supplied free of charge;
- Free technical guidance and supervision was given during the erection and installation of new plants;
- In the first six years of the programme, turn-key jobs were carried out for some selected plants;
- Operational advice was supplied during the working of the plant in the season;
- A training programme was organised to create a pool of skilled workers from which the new entrepreneurs could engage personnel;
- Technical literature in simple language was published to give operational help and insight into the working of the plant;
- The plant, machinery and equipment were standardised and the manufacturing of the equipment was kept under strict supervision of the Planning Research and Action Institute;
- Visits of entrepreneurs to each other's units and free discussions with regard to the handicaps and difficulties encountered were organised along with demonstrations of the solutions to technical problems. Technical experts from the National Sugar Institute in Kanpur and from other agencies were invited to these demonstrations;
- Technical seminars were organised to discuss the working conditions, handicaps and inefficiencies of plant design and processing techniques and to develop new solutions. Four seminars have been held so far and each of them has been a landmark in raising the efficiency of the plants;

- An association was formed to look after the interests of the entrepreneurs and to obtain facilities from the government and the development agencies dealing with the promotion of small-scale industries.

These extension methods have been highly successful in establishing the new technology. By 1973-74, 935 units had been set up by the small entrepreneurs in the State of Uttar Pradesh. The P.R.A.I. was also requested to help in establishing such units in other states. 280 plants so far have been set up in these other states where P.R.A.I. did a turn-key job for the first units. The programme is expanding every year, and on the average growth rate of these units has been around 40 per cent a year.

TECHNOLOGY EXPORTS

Six plants were imported by Pakistan between 1962 and 1964. Nepal and Sri Lanka have also imported a few plants. A team of experts from France came to study this technology in 1966-67. In 1974 teams from Tanzania, New Guinea and Papua came to explore the possibility of establishing this technology in their countries Two plants are under export to Ghana.

THE ECONOMICS OF INTERMEDIATE TECHNOLOGY

The first pilot plant was built in 1956-57 at a cost of 80,000 rupees (£4,000 sterling) and had a crushing capacity of 30 tons per day. In the next three years, due to some changes in the design and the introduction of some new machinery, the cost went up by 30 per cent. The second design of the plant, worked out after 1962, cost 250,000 rupees (£12,500 sterling) and its capacity was 60 tons per day. At present the plant cost has gone up to 1 million rupees (£50,000 sterling) for 80 tons per day crushing capacity. The major factor responsible for this has been the high rate of inflation in the last two years.

Initially, the capital for establishing such units was provided by rural and urban entrepreneurs. After Independence, changes in the Indian fiscal structure had made a substantial amount of capital come out of unproductive activities, and some of this idle rural capital was invested in the new technology. Government financing agencies also came forward and provided 75 per cent of the capital as loans. At present, the commercial banks have started providing loans on easy terms covering up to 80 per cent of the capital required. The technology has been accepted and is now included as one of the approved industries for the technically educated unemployed. If they invest

10 per cent of the capital required, the rest is provided by Government funds under this special scheme.

a) Investment and employment

In January 1973, C.G. Baron from the International Labour Office in Geneva, made a comparative study of the low-cost technology (the open pan sulphitation process) and the large-scale vacuum pan technology. His findings are summarised in Table 1.

Table 1

COMPARISON BETWEEN THE MODERN AND THE INTERMEDIATE TECHNOLOGY

	Large-scale modern mill	Small-scale low-cost mill
Capacity (maximum cane crushing (in tons per day)	1,250	80
Output of sugar in an average season (in tons)	12,150	640
Total investment required (in million rupees)	28	0.6
Total employment (seasonal and permanent)	900	171
Average investment per ton of sugar of output (in rupees)	2,305	940
Investment per worker (in rupees)	31,100	3,530

These figures can be normalised: for the same initial investment of 28 million rupees, one can build one modern plant or 47 small mills. In this case, the former will produce 12,150 tons of sugar for a total employment of 900, but the 47 small plants will produce two and a half times more sugar (30,280 tons) and employ 11 times more people (9,937 as against 900). These ratios obviously depend upon the assumptions about an average mill's performance in an average year. However rough they may be, they do show that the small-scale technology is more efficient in terms of capital output and employment generation.

b) Labour intensity

Another measure of efficiency is labour intensity, i.e. the ratio between the man-hours worked and the amount of sugar produced. Labour intensity varies considerably, even within the same region and with mills using the same technology. A 1962 study of the Indian Productivity Team for the Sugar Industry gave an average figure for the whole of India of 10.42 man-days of work per ton of sugar. The comparable figure for the Philippines was 2.10, and for Puerto Rico, it was only 0.64. Within India, there are wide disparities: in the State of Maharastra, it varied from 10.3 to a low 2.6, and in Western Uttar Pradesh from 20.7 to 9.3. For our small-scale intermediate technology the average labour-intensity is 31 man-days per ton of sugar, i.e. three times more than for the average modern mill.

c) Cost of processing and profitability

There are a number of variables which make it difficult to analyse the cost of production on a standard basis. Two major variables are common to both small-scale and the large-scale technology: the recovery rate, which varies from year to year and from mill to mill, and the number of working days, which extends from 100 to 120 in the case of small-scale technology and from 100 to 200 in the case of large-scale technology. Then there are other factors like the wage structure, labour benefits, management and other overhead expenses.

With the large-scale vacuum pan factories, depreciation is very variable for the reason that older mills were installed at a much lower cost. The capital cost in 1965 of a vacuum pan factory with a capacity of 1,250 tons per day was 18.9 million rupees but in 1973, it had risen to 28 million. For these reasons, it is easier to make a comparison on the basis of processing and converting a unit of 10 tons of cane into sugar. The data are summarised in Table 2.

For 10 tons of cane, the average recovery rate in the large-scale vacuum pan factories of North India is 9.5 per cent. With the small-scale technology it varies from 7.25 to 8 per cent. The cost of producing 100 kg of sugar thus works out at 235 rupees with the modern method, and between 223 and 236 rupees (depending on the recovery rate) with the small-scale technology.

The sugar made in the large plants has to be sold in near and distant markets which means additional costs for transport and local taxes. Further, there is always a delay in the disposal of the sugar since a much bigger region must be covered by the marketing system. These factors add 5 to 7 per cent to the cost and raise it to about 250 rupees per 100 kg. The mini sugar units which sell their product in the local market do not incur these expenses. The Government of India buys 60 per cent of the sugar produced by the 222 vacuum pan factories for sale through the state distribution system. The price of this levy is fixed on the basis of a production cost of 260 rupees per 100 kg. The remaining 40 per cent of the sugar is sold on the open market at a price which ranges between 320 and 350 rupees.

If the research currently carried out is successfully implemented the recovery rate in the small-scale technology will rise to about 8.5 per cent and the ensuing reduction in fuel expenditure will definitely tilt the economic balance in favour of the small-scale technology.

Table 2

COST OF PROCESSING 10 TONS OF CANE INTO SUGAR IN 1971-1972 (in Rupees)

	Large-Scale Modern Mill	Small-Scale Low-Cost Mill
Salaries and wages (1)	164.35	151.00
Fuel and power (2)	57.70	66.50
Stores and lubricant	103.23	62.80
Repairs and renewals	48.77	12 00
Depreciation	200.00	90.00
Overheads	39.23	10.00
Taxes: Excise duty	123.50	59.70 (3)
Purchase tax	50.00	50.00
Cost of cane	1,200.00	1,200.00
Transport charges on cane	47.50	- (4)
Capital cost	200.00	90.00
	2,234.28	1,792.00

Notes:

1) The labour in the large-scale mills comes from the organised sector and works on a year-round basis. After the season, a retention allowance is payable. In the mini sugar mills, labourers come from the agricultural sector and have some slack time during the winter months when the mills are in operation. These labourers used to go to the cities, but now prefer to seek employment in the small plants.

2) Fuel economy in the mini sugar mills is poor due to design factor as well as to the non-continuous nature of their operations.

3) The excise duty has now been doubled and the tax advantage is only marginal.

4) The mini sugar units are located near the farmers' fields and the cane is brought by bullock cart. For the large-scale mills, the cane is carted to the weighing station and from there it is carried by truck and railway to the factories; hence the high transport costs.

SOCIO-ECONOMIC FACTORS

The small-scale sugar plants currently crush 10 per cent of India's cane crop, and in 1974-75 produced 1.3 million tons of sugar, as against 4.8 million tons for the large-scale factories. This mini-technology based on local resources has thus increased India's sugar production capacity by a quarter. Some £24 million sterling have been invested in the rural areas. A seasonal employment potential has been created for 100,000 people who would otherwise migrate temporarily to the cities for 4-5 months or remain unemployed during the slack season. The additional tax revenues to the central and state governments amount to £1 million sterling.

A machine manufacturing industry has been set up. Its annual turnover is in the neighbourhood of £5 million sterling. The amount of iron and steel needed to manufacture these small sugar plants is only about 60 per cent of that required by the large-scale industry for producing the same quantity of crystal sugar. The machinery design furthermore does not require any imported components thus saving foreign exchange.

More than 60 per cent of the cane consumed by the large-scale units is transported by truck and railway. Practically no such transport is required by the small-scale units: they are within close distance of the cane growers, and transportation is done by bullock carts. These units also act as 'centres of technical radiation' by providing, among other things, repair services for the new types of agricultural implements which are now being introduced.

These units which employ about three times as many labourers for the same capacity as the large-scale mills, require only 40 per cent of the capital for the same output. The price paid to the cane grower is at par with that offered by the large-scale mills (or 25 per cent higher than the traditional means of disposal of cane). The total income thus added to the agricultural sector is around £10 million.

Some of the technological ideas developed for the small-scale plant have filtered down to the gur and khandsari industry. The gur industry has adopted the power crusher and crystallisation in motion. Both industries now use improved boiling furnaces and their efficiency has been substantially improved.

VI. THE UPGRADING OF TRADITIONAL TECHNOLOGIES IN INDIA: WHITEWARE MANUFACTURING AND THE DEVELOPMENT OF HOME LIVING TECHNOLOGIES

by

M.K. Garg*

In 1900, 89 per cent of the Indian population lived in rural communities. Village artisans, which accounted for some 18 per cent of this rural population, were employed (or more usually self-employed) in seven basic trades or community technologies: potters, carpenters, blacksmiths, weavers, tanners, leather workers and vegetable oil crushers. They provided basic household goods and agricultural tools to the community, served as repairmen and sold some of their goods to the urban areas. Farmers living near the cities and small towns often supplemented their income with small cottage activities like spinning, flour grinding, basket making and the preparation of milk products.

The villages were practically self-sufficient; when cash was required (e.g. for the payment of taxes), it could easily be obtained by selling a few surplus goods to the urban centres, and the capital base of the villages was generally strong. This position changed gradually with the growth of large-scale mechanised technologies which were to make a major impact in three directions: on product selection, on technology and on organisational patterns.

THE PROCESS OF DECLINE OF RURAL INDUSTRIES

The modern industrial firm, unlike the village artisan, had the capacity to change and expand its product range to meet the new requirements of the market. When the kerosene-fired hurricane lantern was introduced by large-scale industry, neither the village potter who made earthen lamps nor the village oilman who produced the oil to light it could meet the challenge and modify their product range.

* The author has been working as a specialist on rural industries in the Planning Research and Action Division of the State Planning Institute in Lucknow, India. He is currently the head of Garg Consultants, a consulting firm specialising in appropriate technology.

Industrial firms could also manufacture goods at a much cheaper price The reason for this is not so much the often repeated argument of scale of production, as the economies stemming from the use of modern sources of energy. Mechanised power in the form of steam, diesel or electricity is much cheaper than the muscle power, both human and animal, of village industry. A bullock can generate only 0.5 to 0.8 hp for 4 to 6 hours in a day. On the present basis the cost of electricity is only 0.55 rupees per hp as against 2.66 rupees for the equivalent amount of bullock power.

Mechanised power can be applied more intensively and in greater quantities in the manufacturing process and hence increase process efficiency. Muscle power has its own physical limit. In the village oil seed crushing technologies for instance, one bullock or at most two can be used, and only 30-32 per cent of the oil is extracted. With the same type equipment, 3 to 5 hp of mechanical power give an oil yield of 34 to 36 per cent, and the cost of energy is lower than with the bullocks. Mechanised power, unlike muscle power, has a great versatility and this greatly facilitates the development of new applications.

Aside from the inability to develop new products and the emergence of better technologies based on mechanical power, the third reason for the decline of village industries was the weakness of their organisational structures (ownership patterns, supply of raw materials, marketing, financing and availability of technical know-how including research and development facilities). Large-scale firms are generally owned by entrepreneurs who have substantial funds to hire any kind of technical skill, to innovate and to adopt new processes. The village industries were owned on a family basis by skilled workers or master craftsmen and employed the whole family. Labour costs as a result were small but their weak capital basis was a major obstacle to innovation.

The raw materials came mostly from local sources and were available partly on a barter basis in exchange for community services, and partly on a money basis. The large-scale firms, in a bid to get raw materials in sufficient quantity for their own use or for export, offered higher prices and thereby destroyed the supply organisation of village industry. They also disrupted the marketing system: the local traders, who used to sell the goods of the village artisan, went to the cities and started to sell the new industrial goods for which demand was growing rapidly. The artisans were left to carry their goods to the cities themselves, sell them in open markets, often at throw-away prices, and lose a lot of time which could otherwise be used for productive work.

This disruption of the links between the artisans, the traders and the suppliers of raw materials created a high degree of financial scarcity in the villages. The artisans had to turn to money lenders who charged a very high rate of interest. Large-scale industry, by contrast, could obtain easy loans and overdraft facilities at nominal rates of interest. They also had their stocks of raw materials and manufactured goods to offer as collateral.

The greatest advantage enjoyed by the large-scale firms came from the machinery manufacturing sector. The latter carried out much R and D work, and could offer new technology packages to the user. The government also supported R and D in the modern sector. By contrast, the technology of the village industries remained static. Village artisans could not finance any R and D, and hardly any government agency carried out any such work for them.

The process of decline of rural industry continued on a low key until the Second World War, but after Independence increased rapidly (the percentage of artisans in the village communities fell from 18 per cent in 1901 to 7 per cent in 1971). The main emphasis of India's Five Year Plans was to adopt as quickly as possible large-scale mechanised technology in all industrial activities. This required an infrastructure which was available only in the urban areas. Productive activities as a result concentrated in the cities and changes in fiscal policy contributed to driving the available capital to the urban areas.

The development of agricultural technology in recent years, which has brought about higher yields and allowed double or triple cropping has increased the income of the villages. However, the ultimate wealth left to the farmer after paying for fertiliser, diesel oil, implements (including tractors), irrigation water and domestic needs is proportionately much smaller. Though apparently there is a lot of money in circulation in rural India, the capital base of these areas is still very weak and production activities remain low. Hence the mass migration to the cities which has created very serious economic and social problems.

Another aspect often neglected by the planners is that large-scale technology has destroyed the pattern of self-employment and given rise to a class of organised professional workers who can hold society to ransom. Self-employment has an inherent flexibility to absorb fluctuations in the economy. A professional labour class does not have such a strength and this leads to instability and inefficiency, especially in a developing economy.

THE POTENTIAL FOR IMPROVEMENT

Despite the decline of rural industries in the last seventy years, there is still a large number of village artisans. The following table, based on the 1961 annual survey of industries of Uttar Pradesh, gives a rough idea of the economic and technological potential of these traditional industries in a typical Indian state (Uttar Pradesh then had a total population of some 74 million).

THE ARTISAN CLASS IN UTTAR PRADESH (1961)

	Total number of artisans	Percentage self-employed
Spinners, weavers, knitters and related crafts	444,700	86
Leather shoe makers	53,700	98
Leather tanners and related crafts	107,500	97
Potters	127,600	99
Blacksmiths	70,400	83
Oilmen	93,500	100
Carpenters	93,900	99

Any scaling-up of these technologies will directly benefit hundreds of thousands of people and will have a tremendous effect both on employment and the development of rural areas. However, if only big industrial units are established, they will provide employment at best to about 1,000-2,000 people per unit and for the most part in the cities. The number of plants required will be limited and the employment potential much smaller and considerably more costly.

There are two facets of the problem of upgrading traditional technologies: one may call them the 'current aspect' and the 'pilot aspect'. The main function of the first is to 'hold the line' and prevent a further deterioration of traditional cottage industries. Most of the organisations, both national and international, which deal with small industry are engaged primarily in this type of work through changes in machinery, financing, the building-up of marketing systems and efforts to increase productivity. This approach suffers from inherent limitations and its effectiveness is limited.

By 'pilot aspect' we mean a concentrated effort on all the problems of rural industry - product selection, manufacturing technology, organisational patterns (including the supply of raw material marketing and finance - with the objective of establishing sound manufacturing units requiring no subsidy. The products of cottage industries should be of the same quality and comparable in price to those manufactured by large-scale firms. This can only be done through a systematic action-research programme aimed at introducing mechanisation on a selective basis and developing processes which can be viable on a small-scale. Hardly any such processes are currently available. Some basic elements do exist but they can serve as a starting point only, and a complete system has to be built up through pilot projects The story of the whiteware manufacturing industry shows how this can be done.

THE EARLY ATTEMPTS TO UPGRADE THE POTTERS' TECHNOLOGY

In 1901 there were 3 million village potters in India but by 1971 only 1.2 million were left and more than half of these are only partly employed The traditional product manufactured by the village potter was a nonporous ware made from local clays which after firing at 800° to 900°C gave a red-coloured product. These were used for cooking, eating and storage as well as for structural purposes (roofing tiles, floor tiles and drainage pipes). Today, the whiteware articles made from China clay by the large-scale firms have replaced these traditional table-wares, while cooking and storing needs are now met by metal-wares. In the structural field, cement concrete has mostly replaced the red clay articles of the village potters.

Three alternatives for scaling-up the village potters' technology can be envisaged. The first is to develop a glazed type of red clay ware which would have nearly identical properties as the whiteware or China ware and which would partly replace the unglazed red clay wares. The second is to initiate the development of whiteware at the village level; and the third is to evolve on the basis of market needs new techniques and improved products, especially in the field of structural articles which can be manufactured by the village potters.

A number of attempts in these directions were made by several agencies and individuals, including the author. Glazed red clay wares require a technology almost similar to that of whiteware manufacturing. The glazed red clay wares developed as early as 1940 at Hala in the British Indian Province of Sind (now in Pakistan) did not find acceptance. The articles, though improved in looks, were poor in use. They developed defects and became dirty after 3-4 months. The cost of production was about 70 per cent of that of whiteware, but the price obtained was only 60 per cent.

Between 1937 and 1945 an effort was made by the Uttar Pradesh government to develop the manufacturing of household glazed red clay in Chunar, a region which had been producing glazed red clay toys since the 1900's (this activity still continues today). This effort did not succeed either. Another attempt was made by giving a white coat to the red clay ware and then glazing it. The product was very similar to the whiteware and had some of its important properties. This work was done at Khurja in Uttar Pradesh between 1940 and 1942 by a family of hereditary potters. A special kind of bowl was manufactured in large quantities for troops in the field but when the process was applied to other goods, it did not work. Productivity was low, consumer acceptance limited and the economics of the civilian market unattractive.

The only alternative which appeared to stand a good chance of success was to initiate the manufacturing of whiteware at the village potters' level. The work was carried out in four stages at four locations.

a) The Pottery Development Centre in Khurja

The first attempt was made in the township of Khurja in Uttar Pradesh in 1942. The local government decided to mobilise the village artisans to manufacture goods required by the defence forces. The small colony of potters at Khurja was organised to produce whiteware hospital goods, but the quality of the products did not meet the requirements of the defence forces. Later the decision was taken to organise a small factory. The potters were given piece wages for shaping the articles in their own cottage workshops and the raw materials were supplied to them directly. Subsequently, the materials were sold outright, and the goods purchased at a fixed price in the unfired state. Firing was carried out in the factory kilns. Later on, the potters were asked to fire their goods in the factory kiln at their own risk, and the finished goods were purchased at a fixed price. As time went by, the potters were encouraged to build kilns in their own house and become completely independent, except for the purchasing of processed raw materials from the factory. This method has paid dividends and now there are more than 250 cottage workshops at Khurja manufacturing goods worth about $120,000 annually.

b) Chinhat

The work at Khurja created a sizeable cluster of artisans in what was a semi-urban locality. The objective of the project initiated in 1957 in the village of Chinhat near Lucknow was to test whether a small cluster in a wholly rural area could also be successful. The same methodology was adopted, i.e. starting with a centralised work-shop leading to fully independent units. At Chinhat two kinds of technological innovations were also carried out: easy-to-handle decoration techniques and the development of a family kiln with the same fuel efficiency as that of the kiln used in the large-scale factory. Both innovations were very successful. Decorated Chinhat ware now has a very wide consumer market, and the economics of manu-facturing have been substantially improved thanks to the new kiln.

c) Gaura

The objective in Gaura, another village near Lucknow, was to find out whether the installation of a centralised small factory could altogether be avoided and whether manufacturing activities could start outright in the cottage workshops. For this purpose, a service centre was set up to supply the raw and semi-processed materials, provide centralised marketing facilities, lend money for investment and working capital and develop the necessary technical know-how through R and D and advisory services. The work at Gaura developed well but the lack of proper training to the village potters and weaknesses in the financial structure of the co-operative were a big handicap. The activities remained at a low level but seen from the viewpoint of a village's requirement, were probably sufficient.

d) Phoolpur

Work in the village of Phoolpur near Allahabad started in 1970. The objective of the Planning Research and Action Institute was not to develop big clusters like Khurja, but to bring the whiteware manufacturing technology down to the individual potters by organising them in as small a cluster as possible in a number of villages. A government-owned service centre with better technical and financial facilities was set up more or less on the same lines as at Gaura. Two types of workshops were established. The first were built around the service centre with government finance and leased on a hire-purchase basis to those artisans who had successfully completed the training programme. The second type of workshops were in the homes of the village potters whose facilities were remodelled to manufacture the whitewares. A well equipped training centre was also built up in the service centre, thus alleviating the two main problems experienced at Gaura, namely lack of training and weak financing.

The results of these various efforts to scale-up the potters' traditional technology have been very positive. In the State of Uttar Pradesh, some 15 million rupees worth of whiteware products ($1.9 million) are manufactured annually on a small-scale in cottage units in the semi-urban and rural areas. Employment for some 10,000 people has been created. The programme has reached a stage of self-generation. New centres in other parts of the State are being opened up at the request of local development agencies and two such centres - one in the eastern district and the other in Bundelkhand are currently being set up. R and D work is being carried out at Phoolpur and in the near future a porcelain type table ware and sanitary ware are expected to be introduced for manufacturing. Low tension electric goods are also being produced.

In any scaling-up programme of village technology, proper product-selection and the improvement or development of an adequate technology are very important. But the most critical factor is the type of organisational set-up required to initiate, expand and support such a programme for at least a dozen years or so. The artisans do not have the necessary financial strength. Nor do they have the capacity to locate or purchase the technical know-how. The sponsoring agency should therefore be in a position to provide adequate funds as well as facilities for the dissemination of technical know-how, together with marketing and purchasing facilities. This can be done best by starting a service centre on a regional basis. Such facilities can be provided on a 'no loss - no profit' basis or at a nominal profit rate. Unless such an organisation is built up, the results obtained have every chance of being retarded and lost.

THE DEVELOPMENT OF HOME LIVING TECHNOLOGIES

The potters' technology, whose upgrading we have just analysed here, is concerned primarily with production. Appropriate technology also has what one might call a 'consumption dimension': many technologies are aimed not at the manufacturing of new goods, but at providing basic services and amenities, mainly for the rural populations in developing countries. In this second group, we find a wide range of 'home living technologies' which deal among others with house building and construction materials, water supplies, drainage, the disposal of human faeces, kitchen fuel supplies and home lighting. The Planning Research and Action Institute has been involved in the development of many such technologies, and we will focus here in two specific cases: the disposal of human faeces and the supply of kitchen fuel through bio-gas plants.

Village people in India have for generations been in the habit of easing themselves in the open fields near the villages Even in urban communities one can find, in the early hours of the morning or the late hours of the day, men, women and children easing themselves in the nearby open space on the fringe of human habitation. Human excreta often contains harmful germs, which cause diseases like cholera, typhoid fever and diarrhoea, and intestinal parasites like roundworm or hookworm. The practice of defecating in the fields and in the open places allows the germs and parasites to be carried to other persons by flies, drinking water or food. There are other problems connected with this habit. The inconvenience caused to women, children and sick persons when they go to the fields during rains or bad weather is often very great, and the side lanes and back corners of the villages are turned into stinking, insanitary places. The health and nuisance problems resulting from these conditions need no further elaboration.

DEVELOPMENT WORK ON RURAL LATRINES PRIOR TO 1956

One of the first organised programmes for latrine construction in the Indian subcontinent began with the Health Units set up first in Ceylon and later in India by the Rockefeller Foundation in 1930. Bore-hole latrines with precast squatting slabs were tried out in these centres in a number of states, including Uttar Pradesh. These efforts met with some measure of success but there were practical difficulties in getting this type of latrine installed inside the houses, and in moving it to a new location once the bore-hole was filled up. There were also other disadvantages, such as foul odours and fly breeding, which were due to the absence of a water-seal. In Uttar Pradesh, the idea of replacing the bore-hole by a simple septic tank directly under the seal was experimented in the Pratapgarh Health Unit.

Later on, the India Village Service, a voluntary organisation in Uttar Pradesh, evolved a junior septic tank but due to certain defects, it could not be popularised. It was felt that more work was needed on this problem. A number of agencies, both official and non-official, were engaged in designing a suitable type of latrine for rural homes. Some work was initiated at Sevagram by Mahatma Gandhi where efforts were made to popularise trench-latrines and the use of the 'khurpi' (a small iron spade) for covering the nightsoil after easing in the open fields. These practices, however, remained confined to Gandhi's 'ashram' (community) and the programme did not radiate to the neighbouring areas. A few private firms also developed some latrines but on account of their high cost, they could not attract the attention of the village people.

THE P.R.A.I.'s DEVELOPMENT WORK

The problem of rural latrines was taken up first for study and later on for development and extension by the Planning Research and Action Institute in 1956. In the first stage, various types of latrines were assembled and their working and operational conditions were studied, primarily from the point of view of sanitary conditions. This work showed that no latrine could be called sanitary unless there was a water-seal between the place where the person sits and the place where the excreta is subsequently disposed. Without a water-seal flies and foul odours will develop. Rural latrines would therefore have to be of the same type as the modern water closet and the excreta removed by flushing or pouring water. Flushing a water-closet requires one to two gallons of water; villagers, however, do not have running water and at certain times of the year, water is in short supply.

The next phase was to determine to what extent the water quantity could be reduced. The development and testing of a number of models showed that with an increased inclination of the pan and a smaller siphon, the latrine could be flushed with a maximum of half a gallon of water. Another problem was the disposal of excreta after it has been flushed from the pan. In the villages, there is no drainage system and a septic tank would have to be kept out of contact from the villagers. Various experiments showed that the alluvial and sandy earths in Uttar Pradesh have an average water soaking capacity of one gallon per 40 sq.ft of surface area practically the year round.

Human excreta ferments and emits foul odour when the solid concentration is between 9 and 18 per cent. If the concentration is decreased by adding water, as in the central drainage system or a septic tank, fermentation and odour formation is nil. Similarly, if the concentration increases (e.g. when the water percentage is reduced by soaking, drying or otherwise) the same conditions will obtain: fermentation and odour formation will not occur. This knowledge about local soils and the importance of concentration rates were combined in the design of an earthen 'kachcha' pit (or crude' pit). After being flushed from the pan, the excreta is led to the pit, and the wall starts absorbing water, leaving excreta in a state of higher concentration. The top of this pit was covered by a cement and concrete cover. To reduce the cost and to make the local manufacture possible, the pan and other parts were manufactured from cement and concrete rather than from the porcelain used for city latrines.

The kachcha pit could be expected to fill up after a few years and would require cleaning. It may then not be completely dry, and removal can present difficulties. To provide for this emergency, it was decided to build two pits side by side. The leading pipe from the siphon was curved: once the first pit is full, the curved pipe is turned over so as to lead to the second pit, thus leaving the first one to dry. The excreta inside the pit dries up within six to eight months and is turned into valuable manure which can be easily removed. The cost of the equipment including the pan, trap, leading pipe and pit cover was 16 rupees (or about $2.00) in 1962 and the complete system could be built up for about 30 rupees. The cost at present has increased to 40 rupees for the parts and 80 rupees (about $10) for the complete latrine.

The 'P.R.A.I. latrine', as this innovation was called, is well suited for acceptance by the rural communities. Its low cost puts it within the financial means of the villager. The amount of water required is not excessive. Sanitary conditions are good and there is no excreta disposal problem. No costly provision like central drainage or safety tanks is required. The latrine can be made locally, and the skills required for manufacturing the parts, erecting the latrine and operating it can be easily acquired by the villagers through simple training.

ORGANISING THE EXTENSION SYSTEM

The design and development of the P.R.A.I. latrine took two years. In 1958 it was ready for extension. The pilot project for extension was taken up as a part of a wider environmental sanitation project in collaboration with the World Health Organisation and the Medical and Health Department of Uttar Pradesh.

A manufacturing centre was set up in Lucknow. Training of masons, sanitary inspectors, overseers and primary health inspectors was arranged. Subsequently, manufacturing workshops were started in nine divisions of the state. Individuals were also encouraged to take up the manufacturing and installation work. To make the design known to villagers, audiovisual aids and publicity in exhibitions and in villages were taken up. Literature in simple language was distributed. Training camps for the villagers were organised.

As a result of all these efforts, the design has found very wide acceptance both in rural areas and in small cities Wherever a central drainage system is not available, government agencies have been instructed to put up the P.R.A.I. type of latrine, even in hospitals and public buildings. By 1968, 200,000 latrines had been installed in Uttar Pradesh and the programme is continuing to grow. Other states have also taken up this technology and the Government of India introduced it very successfully in the Bangladesh refugee and prisoner-of-war camps.

THE PROBLEMS AND HANDICAPS OF THE P.R.A.I. LATRINE

Village houses are located close to one another. As a result, the space required for the latrine and the pit is often insufficient and such households, in spite of their strong motivation cannot avail themselves of this design. In some villages, the soil is very loose and some cases of pit collapse have occurred. To remedy this, changes in the pit design were worked out. If the soil is very clayey, the pit wall cannot absorb water at the rate required.

On the basis of experiments, it was concluded that the pits should be at least 15 ft distant from the water source both in a horizontal and a vertical direction. This distance was found to be sufficient to prevent the bacteria from travelling to the water source. In many locations, this rule could not be respected and the installation of the latrine had to be abandoned. In areas where the water level is 1½ meters from the surface, installation proved impossible, since a pit depth of up to 2-2½ meters has to be kept for getting the required soaking surface.

The first pit fills up in four to five years. Once full, it is left for drying, which takes about six months. In some cases where the first pit was opened for emptying the excreta, it was found that it had not dried up but was in a semi-liquid state. Removal of the contents was extremely difficult. A number of devices have been tried out but the problem appears to be related to the soil's absorption capacity. If the area of absorption is increased and the surroundings of the pit do not get any moisture from outside the excreta dries up completely but this requirement limits the scope of application of the latrine.

THE ECONOMICS OF COW DUNG: FERTILISER OR FUEL?

In India both in urban and rural areas, wood was the main kitchen fuel until around 1930. It was supplemented by dried cakes made from animal dung. The availability of wood had gradually decreased, especially after Independence: the village forests have been cut down and the land has been put to cultivation to grow more food. As a result, cow dung cakes began to be used in larger proportion in the rural areas. It is estimated that at the beginning of the century, cow dung represented hardly 5 to 10 per cent of home fuel consumption. This total increased to about 25 per cent by 1930 and 40 per cent by 1950. Today, it represents about 70 per cent.

Burning cattle dung as fuel is wasteful since it deprives the soil of valuable and inexpensive natural nutrients. According to the estimates of the Planning Commission, the 46 million tons of coal equivalent of dried dung burnt every year correspond to about 460 million tons of fresh dung. This quantity of dung is the equivalent of 1.38 million tons of nitrogen and 0.69 million tons of phosphorous pentoxide. In 1972 the total Indian production of nitrogen fertilisers in terms of nitrogen was 0.89 million tons and that of phosphorous pentoxide was 0.13 million tons. If agricultural wastes like leaves, grass, twigs, sterm and kernels are added, the amount of available nitrogen and phosphate could be increased by another 20 per cent.

The treatment of the soil with farmyard manure containing 0.3 per cent nitrogen increases the production of grain by 5 to 15 lbs per lb of nitrogen, depending upon the nature of the soil, irrigation and other factors. Assuming an average increase of 10 lbs of grain, the 1.38 million tons of nitrogen contained in 460 million tons of fresh dung could give an additional grain output of 14 million tons annually, apart from other benefits like the general improvement of soil structure and higher fertility. Farmyard manure has a residual effect: its beneficial effect on the crop is not confined to the season of application but persists over a number of years.

The habit of using dried cow dung as fuel is deeply ingrained in the villagers and it cannot be changed easily unless there is an attractive alternative. There are two ways of saving this dung for manurial purposes: one is to provide a cheap alternative fuel, and the other is to ferment the dung in a bio-gas plant. This yields a sufficient quantity of combustible gas which can be used for cooking and lighting and the residue can be used as manure without any loss of its fertilising value.

At this stage it is too early to think of replacing dung by any other primary source of energy (for example soft-coke) in the villages because of the economic difficulties involved. In order to release the whole of the cattle dung for manurial purposes an additional 46 million tons of coal will have to be raised. This would require a capital investment of about 3 billion rupees ($375 million). The carbonisation plants would probably need another 4 billion rupees ($500 million). The additional rail traffic thus generated would call for a 25 per cent increase in the current rail capacity. Distributing the fuel from the railway centres to the villages would probably call for a 50 per cent increase in the present transport capacity of the roads.

In this context the supply of rural domestic fuel is essentially a local problem which necessarily has to be tackled at that level. The development of gobar gas or bio-gas plants in rural areas is the only appropriate or low-cost way to provide the villages with a kitchen fuel which involves no health hazards, which is convenient to use and which also provides the much-needed manure. The problem of house lighting can also be solved: the gas from such a plant gives a brilliant light with the mantle type of lamp. The problem of electrifying the village could thus be given a low priority. For these reasons, the Planning Research and Action Institute decided to take up this problem and established a research station which operated first at Lucknow under the Rural Industries Section of the Institute and later on at Ajitmal.

THE GERMAN CONTRIBUTION TO BIO-GAS TECHNOLOGY

The application of anaerobic digestion to cow dung and farmyard manure was developed mostly in Germany. During the Second World War, three types of plants were used. The Schmidt-Eggergluse plant was developed for farms ranging in size from 570 to 1,300 acres with a cattle population between 100 and 300. The Weber plant was designed by a farmer for his own 50 acre-farm with 26 cattle. It consists of four cement tanks - two for fermentation, one for the storage of digested slurry and one for the gas. The Kronseder type of gas plant is very simple and cheap. It is meant for providing gas for cooking purposes in small farms. A tub for liquid manure, split in the longitudinal axis, floats in the cesspool once constructed on such farms for excreta. The liquid is covered by the tub and ferments. The gas produced is caught by the tub and a rubber tube conducts the gas to the kitchen. Nothing more is required.

An analysis of the above technologies shows that they are not really suitable to Indian conditions. These bio-gas plants require 25 to 150 cattle. They are fully mechanised and need an energy input in the form of electricity or diesel (this energy input comes to nearly 33 per cent of the total energy generated in the shape of bio-gas). The cost of the plant is substantially higher than what a poor village can afford. Such plants also require a high degree of organisation and suitable mechanical and managerial skills. In India their adoption would not only be difficult but practically impossible.

More than 80 per cent of the agricultural holdings in India are one hectare and below. A farming family of five people on the average has not more than three animals - one cow or buffalo and two bullocks. These families live in scattered villages with about 150 hectares of cultivable land divided into 50 to 100 families. Some farmers have up to 5 or 10 hectares but the bulk of the cow dung is produced in these scattered holdings.

DEVELOPMENT WORK IN INDIA

Social agencies have been very actively involved in the development and extension of gobar gas plants. The main ones are the Indian Institute of Agricultural Research in New Delhi, the All-India Khadi and Village Industries Commission in Bombay and the Planning Research and Action Institute in Lucknow. Useful work on cost reduction and the popularisation of these plants among the village masses has been done by organisations like Messrs. J.P. & Co., in Bombay (Gram Lakshmi Plant), Khadi Pratishthan Sodepur in Calcutta and the Rama Krishna Mission, Belur Math, in Calcutta.

The various agencies working on gobar gas have concentrated on a small and simple family-size plant which requires no energy inputs and no specialised skill to operate. The main emphasis has been put on cost reduction so as to put the plant within the financial reach of an individual family.

EXTENSION AND IMPROVEMENT OF THE FAMILY GAS PLANT

The P.R.A.I. took up the extension of the family gas plant. About 30 units were installed by giving part subsidy and part loan to the farmers. A survey made after a year indicated that hardly 20 per cent of the plants were working. A subsequent investigation showed that there were a number of reasons for this failure.

The volume of gas available was insufficient to meet the entire cooking needs of a family throughout the year, and after six to eight months of use the generation efficiency of the gas plant went down. The burning quality of the gas was not constant and the flame sometimes went out. More time was required to cook food with gobar gas than with other fuels. The gas contained moisture, thus clogging the pipe line which then required frequent clearing. Moisture spoiled and cracked the porcelain ring of the mantle in the gas lamps. Finally the cooking of 'chapatis' (pancake-shaped bread) proved to be difficult. The survey also showed that most of the farmers were not feeding the required quantity of cow dung because of its shortage.

It was decided to examine these problems in detail and a team was set up for this purpose in P.R.A.I. The group found out the reason for these defects. It was also asked to develop gas plants working on a feed of 50 pounds of cow dung - the quantity usually available to the farmer with three cattle. Over a period of two years, the team reported a number of findings.

a) Domestic gas requirements

For cooking purposes 12 to 15 cu.ft are required per person per day. A gas lamp (40 candle power) needs 2½ cu.ft per hour. On this basis, a family of five needs 60 to 75 cu.ft of gas for cooking and 20 cu.ft for lighting purposes (two lamps operating for four hours). The total daily requirement for a family is thus between 80 and 95 cu.ft. The 100 cu.ft gas plants which were installed could have met this but actually the generation of gas was found to be only 30 to 50 cu.ft from November to February and 50 to 80 cu.ft from March to October, with the peak in June.

b) Generation efficiency

The above gives a generation efficiency of 0.3 to 0.5 cu.ft of gas per pound of cow dung in winter, and 0.5 to 0.8 cu.ft per pound in summer. In a controlled experiment conducted by C.N. Acharya of the Indian Agriculture Research Institute, a generation efficiency of 0.5 to 0.8 cu.ft per pound in winter and 1.1 to 1.5 cu.ft in summer was reported. The various plant designs were based on that expectation. The variation of gas generation from month to month is directly related to the inside temperature of the digestor. According to C.N. Acharya, the inside temperature is a function of atmospheric temperature and it is 1 to 3°C higher than the atmospheric temperature. The best generation was reported when digestor temperature was about 34°C. This condition obtained for six months in the gas plant run by Mr. Acharya.

c) Digestor temperature

In the experiment conducted by P.R.A.I. the inside temperature of the digestor never went beyond 24°C irrespective of the atmospheric temperature. It was nearer to 22°C, hence the lower efficiency. In an effort to raise the temperature of the digestor, insulation on the outside surface was provided from local materials like straw and ash. No improvement in the inside temperature of the digestor was noticed. The reason appears to be that cow dung before feeding is made into slurry by mixing an equal quantity of water. The temperature of the water drawn from the well or by hand pump is always nearly 16°C, irrespective of the season, and the temperature of slurry was nearly the same when it was fed into the digestor. The heat generated during digestion raises the temperature of the slurry by 2-4°C. The effect of atmospheric temperature appears to be not more than two degrees in winter and about three degrees in summer. This means that the final temperature of the slurry will never be higher than 24 degrees in summer and 22 degrees in winter.

d) Ground temperature

The digestors are constructed inside the ground with cement and mortar bricks. In India, ground temperature below 1 foot is 16°C. Such digestors will naturally draw very little heat from the atmosphere since the top only is exposed and that too is covered by the gas collector. To benefit from the atmospheric heat, a digestor was made above ground. The cost was higher but it did show inside temperatures up to 28 degrees in summer. In June, it rose to 31°C for about two weeks and the gas output increased to 0.9 cu.ft. The cost of this digestor was 1½ times that of the underground one.

e) Preheating of slurry

An attempt to heat up the cow dung slurry was then made, but apart from adding another task to the farmer's family, it consumed 33 to 40 per cent of the gas produced. The idea of a solar boiler was then worked out, but such boilers, which are not available in India, had to be imported from Japan, and as a result were beyond the pocket of the average farmer.

f) Adjustment of the digestion cycle

Another attempt to raise the gas output was made by adjusting the digestion cycle. The plant originally designed had a 28 day cycle (the cow dung slurry stays inside the digestor for 28 days). During this period, 90 per cent of the gas was reported to be available. In fact, the cycle was found to be too short and no more than 68 per cent of the gas was available. The digestion cycle was raised to 56,

90 and 106 days. 92 per cent of the gas was available in 106 days as against 73 to 75 per cent obtained in the 56 day cycle. The increase in the digestion cycle required a larger size, thus raising the cost of the plant. The increase in the output of gas was nominal.

g) Development of mini-digestion

Development work was carried out on smaller digestors requiring a cow dung feed of 50 to 60 pounds which is available from 3 cattle. The smallest digestor had a diameter of 3 feet and a depth of 4 feet, against the 100 cu.ft standard plant with a diameter of 4½ feet and a depth of 12 feet. Its feed requirement was 25 pounds of cow dung. Two other digestors with a feed capacity of 50 to 80 pounds were also constructed The efficiency on all the smaller digestors was found to be at par with that of the 100 cu.ft unit. As a matter of fact, in winter the gas generation per pound in the small digestors was higher than in the standard digestor of 100 cu.ft. These digestors were placed on the ground, properly insulated and exposed to the sun throughout the day. Internal temperature was found to be 3°C higher than in the standard plant.

THE ALTERNATIVES : FAMILY PLANTS OR COMMUNITY PLANTS

If bio-gas plants are to be widely diffused, two alternatives can be envisaged: the installation of sophisticated community plants at the village level, or the development of more efficient small family plants. Community plants must be fully mechanised and require a heating mechanism. Two such plants have been built on a pilot basis in India with Hungarian collaboration - one at the Pusa Institute in Delhi, and the other at the National Sugar Institute in Kanpur. The investment costs were about 500,000 rupees ($65,000) in 1962 and it will now be about the double. Energy in the form of electricity or diesel is required to operate the plants. Installing and operating them calls for an efficient organisation. Such plants may not be workable in a subsistence economy based on the barter system, and the transition to a monetary economy will create further complications.

The other alternative would be to further improve the efficiency of the family gas plant, by increasing the gas generation efficiency to 1.2 to 1.5 cu.ft of gas per pound of dung and by getting 90 per cent of the total gas out within a week or two. Achievement of such high efficiency can probably only be done through biological means.

Anaerobic digestion is carried out by a colony of millions of bacteria, mostly of the 'verticalla' class of protozoa, which are naturally germinated in the cow dung solution in the absence of oxygen. The germ pushes away all the undesirable materials in the cow dung by its cillai (hair-like projections), selects and consumes the suitable material through an aperture and suddenly whirls round giving a corkscrew motion of its stem, during which process the gas is liberated. The verticalla resembles a lotus stalk with a flower on top. There are also other bacteria like 'cynocripta', 'englema', 'epistylis', etc.

Maybe a new species of bacteria could be developed They would have an active stage of working between 16° to 22°C in place of the present 30° to 38°C, and a multiplication speed 5 to 10 times higher than the present species. If such a species could be developed, the gas plant based on three animals or 50 pounds of cow dung will provide 75-80 cu.ft of gobar per day, sufficient to meet the needs of the typical farmer family. The increase in multiplication rate will lower the digestion cycle; this would require a much smaller gas plant whose cost will then be within the means of the farmer. Such development will constitute an appropriate or low-cost technology. Discussion with a number of bacteriologists indicate that such a possibility exists.

VII. THE ROLE OF TECHNO-ENTREPRENEURS IN THE ADOPTION OF NEW TECHNOLOGY

by

P. D. Malgavkar*

THE STORY OF PRAKASH: ENTREPRENEURSHIP INSTEAD OF BRAIN DRAIN

"Please send it by special truck positively today. Take comprehensive insurance for 100,000 rupees on our account". Prakash Jagtap, sitting in his glass enclosed office in his two-storey factory in Poona, was giving instructions on long distance telephone to the principal of an engineering college situated some 160 km away. The much fondled and valued cargo was actually a twenty-five year old 800 hp dynamo meter which had not been used for years and needed thorough overhauling and servicing before it could be commissioned.

Prakash graduated in June 1969 with distinction in mechanical engineering. Amongst five brothers and three sisters, he is the youngest. His father is in the printing and publishing business. Twenty-five per cent of his classmates after graduation went to the United States for further studies and/or employment. Many advised him to go abroad. His father left it to Prakash whether to go abroad or start a business. If he decided against leaving, the money earmarked for sending him abroad would be placed at his disposal for starting a business. His brother-in-law Salunke, who was also a graduate in engineering and had about ten years of experience in running an industry, advised Prakash to decide once and for all

---

* The author was formerly the principal director of the Small Industry Extension Training Institute (S.I.E.T.) in Hyderabad, India. He is currently an industrial consultant in Poona and a Visiting Professor at the National Institute of Bank Management (Bombay). The present paper is the outcome of the active co-operation given by Mr. V. Salunke of Accurate Engineering Co., Mr. T.P. Vartak of Mark Elektriks and Mr. Prakash Jagtap of SAJ Steels Pvt. Ltd. from Poona and Bhadu, a village lad from Assam. These entrepreneurs willingly submitted to hours of questioning at short notice, patiently explained their points of view and went through the draft. The sole criterion in reducing the voluminous material presented by them was whether it was pertinent to the adoption of technology. The author would like to thank all of them for sharing their experiences in the cause of technology transfer to less developed countries.

whether he would go abroad or start a business of his own. After a month's deliberation, Prakash decided to start a business. Having burnt his boat, he was advised to select an item for production which could substitute for current imports, which was not consumer-oriented and which had a high technology content. He noted that in Poona and its surroundings a number of prime mover manufacturers were located. Prakash therefore selected water brake dynamo meters which are used for general testing, inspection and research by these manufacturers. In September 1969, he established the Dynamic Engineering Co. His assets: a letterhead and a 'factory' - a room in his father's house.

The criteria for product selection was based on the reasoning that import statistics would easily indicate the demand within the country and that consumer products required sophisticated marketing skills. Capital goods, by contrast, meant contacting a selected market where the buyer was more convinced by performance than by brands. The high technology content would ensure product specialisation, thus reducing the risk of mushrooming competition. Another reason for selecting a product which required a high technology was to get away from mass production techniques where high material costs always weigh in favour of large companies.

A search then began to find out more about the dynamo meter. He first consulted the textbooks which helped him graduate in mechanical engineering. The books gave a general description of about one and a half pages and nothing about design, parts and component drawings. Prakash approached his engineering college professor who authorised him to open the dynamo meter used in the college laboratory. He found that it consisted of a small number of castings and that there was nothing very complicated about its construction and assembly. Thereafter, he called on foreign consulates and in a period of one week in Bombay collected addresses of about 250 manufacturers from all over the world. Back in Poona, the letterhead of Dynamic Engineering was shot out requesting information on products, types of machinery and designs from all these manufacturers. Literature started pouring in. He noticed that there was not much of a difference between the college laboratory dynamo meter he had stripped earlier and these new designs.

Simultaneously, he visited the diesel engine manufacturers association, got the addresses of manufacturers in India, asked what type of dynamo meter they were using, what improvements they would like to have and the possibilities of their buying his dynamo meter with improvements incorporated in its design. Some eighty manufacturers replied to his letter: the dynamo meter had become a necessity for testing and certifying the performance of diesel engines, especially if these were to be exported. He further contacted personally some 40 local diesel engine manufacturers.

In December 1969, in the process of contacting manufacturers, Prakash came across one of the leading producers of diesel engines who had a dynamo meter of a well known foreign make relegated almost to the scrapyard. His offer of servicing it was willingly accepted by the factory manager who promised to buy eight dynamo meters provided Prakash could guarantee their performance. Proud of the possibility of an order and requiring space to strip the dynamo meter, he shifted his factory to his father's garage. He had seen a Japanese catalogue referring to this particular type made in the West and he argued that if the Japanese found it good enough to copy, it must be really good.

During this preparatory time, he used to spend every day two to three hours in his brother-in-law's factory to acquaint himself with the machine shop practices and organisation. Simultaneously he wrote to a number of machine tool manufacturers in India for their catalogues and prices which gave him an idea of the relative performance of different types of machine tools.

He was now in a position to submit a proposal for getting financial assistance from a bank. He took his proposal to a number of banks (at that time 14 banks in India had just been nationalised). He wanted financial assistance to cover land and building, machinery, working capital and bills facility to the tune of 180,000 rupees, out of which he would contribute 30,000 rupees which is father had set aside for sending him abroad(1). After discussions with a number of banks, one of them showed interest in his proposal but at the last minute, the manager asked him about his experience in this particular line. By then he had spent about 7,000 rupees in developing the prototype dynamo meter which was ready in the garage and he could convincingly reply to the bank. The proposal which was submitted in November 1969 was accepted by the bank eight months later.

Of course, manufacturing the prototype was far from smooth. The first machine was way off the mark as it indicated 5 hp whilst absorbing only 2 hp. Prakash checked all the castings, the patterns, assembling, etc., till at last he succeeded by June 1970 in turning out a dynamo meter performing to international standards.

The first year was a difficult one. In 1970, with two engineers helping him, his turnover was only 5,000 rupees. In 1971, it went up to 86,000 rupees and in 1972 to 140,000. By 1973, it had reached 400,000 rupees, and he was employing 40 people, of whom 12 were engineers. He started with dynamo meters of 5 hp capacity. In 1973 his range of dynamo meters varied from 5 hp to 400 hp and he supplied 60 factories. The old 800 hp dynamo meter about which he was phoning the engineering college at the beginning of our story, was loaned to

---

1) Approximate rates of exchange: £1 = 20 rupees, $1 = 7.5 rupees.

help him ensure that the drawings and design he had developed for a similar dynamo meter (to be supplied to the Defence Department) were of the right kind. This dynamo meter is to be sold for 100,000 rupees His target for 1974 is a turnover of one million rupees and he is now recognised as an authority on dynamo meters.

Prakash went abroad in 1973 when he visited over 40 factories making dynamo meters. This gave him confidence that his production facility and product were as good as any in the world.

He owes a great deal to Salunke, his brother-in-law, for introducing him to financial discipline. Even though he started as a small proprietory unit, his accounts were audited by a qualified chartered accountant from the beginning. Further, he was advised never to sell on credit but to send the documents through the bank. He was afraid the big customers and factories would not accept documents through banks, but having committed himself to such a policy he went on to explain to the customers that being small and new, his bankers would not accommodate credit sales. As the equipment was badly needed by the big factories (including some public sector projects) they accepted his conditions.

The future for him is all exciting. Having established his mastery in dynamo meters, he has gone into making special alloy steel The stainless steel strip required for the vernier callipers made in Salunke's factory has to be imported in India, which poses problems and delays. This was a challenge to Prakash. During his 1973 visit abroad, he had seen in Germany small manually-operated furnaces turning out special-purpose alloy steels. He has assembled a 200 hp rolling mill and a direct arc furnace of 100 kg capacity capable of heating up to 1700°C. He has improved upon the furnaces he saw abroad (his furnace has an electrical control panel) and employs two graduates in metallurgy for this project. He looks forward to supplying special purpose stainless steel to Salunke who will have a ready supply of stainless steel strips to make the vernier callipers.

He is now helping five other engineers to set up their own manufacturing units, and he is restructuring his firm into a private limited company, since the new scheme of alloy steel which he has undertaken requires long range institutional finance.

THE STORY OF VARTAK: FINDING THE MARKET NICHE

Vartak graduated in 1965, specialising in electrical engineering With the massive programme of electrification in the country, he thought of manufacturing electric motors since these are the common motive power for any other equipment. He therefore established in 1967 a partnership firm called Mark Elektriks with another graduate in electrical engineering. Between 1967 and 1969 he manufactured standard types of motors but he could not compete on production costs: the bigger manufacturing units bought their materials in bulk which gave them a considerable advantage in price since materials are the major portion of the cost of standard motors. Nor did he have the advantage of a well known brand to penetrate the market.

At that time a U.S. Peace Corps volunteer specialised in industrial management was attached to his firm. He suggested to Vartak that the partners should take advantage of their knowledge by going in for special purpose motors which are generally not manufactured by bigger companies, either because their collaboration agreements with foreign firms do not cover special purpose motors, or because development costs for special purpose motors are too high relative to the size of the market. After 1969, Vartak started to manufacture custom-built special purpose induction motors required by the original equipment and machinery manufacturers making cranes, hoists, washing machines, etc. He also developed torque motors. The yarn winding machines which use such motors require a reduction in winding speed as the yarn is wound. One way of achieving this is with eddy current couplings. A client lent them one such coupling which could no longer be obtained in India because of the ban on imports. Vartak is now successfully manufacturing the eddy current couplings.

To know more about special purpose motors, Vartak collected addresses of foreign manufacturers from the foreign consulates in Bombay and asked for specifications and quotations from the manufacturer. In one catalogue, he noted that 42-volt motors were manufactured by a foreign firm for driving mercury pumps in the alkali industry. He got in touch with alkali manufacturers in India to find out whether they would like to have 42-volt motors. The alkali manufacturers were at their wits end since they could neither find these motors in the country nor import them from abroad. They readily responded by placing orders for such motors.

Later on Vartak went abroad to visit manufacturers of special purpose motors. These trips to America, England and other countries made him confident that there was nothing extra special or formidable with the foreign manufacturers which he could not provide in his factory. In fact he bagged a good order for torque motors from a foreign manufacturer as he could offer them much below the price of the foreign firm. The torque motor requires considerable skill and attention, where the Indian workers score, and the wages in India are very much lower than in England.

As a service to industries in the Poona region, he took up the rewinding and repair of electric motors, an activity in which the big manufacturers are not interested, since rewinding services have to be as close as possible to the users of electric motors. He trained eleven people from small towns where intensive electrification had taken place. Most of them have prospered and two of them now have their own jeeps.

Vartak is now full of confidence with his good turnover and high profits. After running into losses during 1967-68, he turned the corner in 1969 and his net profits in 1971 amounted to 71,000 rupees.

Having achieved success in special purpose electric motors, he began to look for new avenues. In 1969, he learnt that the principal of an agricultural college was working for quite some time on develop ing a culture for nitrogen-fixing bacteria. The university was approached to let the principal act as adviser for the commercial exploitation of the cultures he had developed. The nitrogen-fixing bacteria called 'rhizobium' and 'azotobacter' were developed and tested in a number of government agricultural centres. With the use of this culture, the crop production in the test farms went up by 10 to 15 per cent. The cost of this culture works out to less than a dollar per acre. Vartak's target for 1974 is to cover half a million acres.

His contribution to this new venture was to get the principal of the college to act as the adviser and later as the research director for the scheme. He had the bacteria scientifically tested on more than a hundred farms, obtained direct reactions from farmers to this innovation, turned the laboratory experiment into a commercial scheme, and got a local firm to manufacture the shakers which are required to keep the culture continually shaking. Vartak admits that the shakers made in the United States are better, but if the American manufacturer were to supply the technology for making the shakers to the Indian firm, the latter would be able to manufacture them at much below the U.S. cost and supply them to the American firm for marketing either in the United States or in foreign countries.

Vartak strongly advocates labour-intensive technology if, in the process, the overall cost does not go up. This is particularly important since demand in the developing countries is limited, at least in the initial stages. He suggests that a fetish of quality need not be made. Low-cost equipment which may not give the same quality and output as abroad may well serve the purpose of a developing nation, and is at any rate better than no equipment at all.

THE STORY OF SALUNKE: STIMULATING NEW ENTREPRENEURS

Salunke, a graduate in electrical engineering, is the managing director of M/S Accurate Engineering Company which manufactures gauges, instruments and tools. His slip gauges are made to a precision of 1/1000 of an inch. During his visit to an American firm manufacturing similar gauges, he noted how they were finishing them and offered to supply gauges with the same specification at much below the U.S.cost. Today he is supplying that firm with one million rupees worth of gauges a year. Salunke now has 60 qualified engineers in the age group of 25 to 30 on his total payroll of about 200.

A dedicated person with a desire to adapt modern technology, he established in February 1970 a 'Forum of Industrial Technologists' in Poona with the support of the local Chamber of Commerce and Industry and a State Government Corporation. The objective of the Forum is to exchange views and ideas regarding industrial technology, to discuss the latest technological developments in the advanced countries and the ways and means of bridging the technological gap between the developed and developing countries, to assist and guide members in economising scarce resources and materials and to encourage young entrepreneurs to break new technological ground. The Forum has six panels: mechanical engineering industry, chemical industry, electronics industry, metallurgical industry, electrical industry and management problems.

The average age of the 137 technologists in the Forum is 35, and 34 of the companies on its roll have developed new products. The Forum encourages and supports innovation through advice and facilities. Its main focus is on the adoption and adaptation of technology rather than on research. The diffusion of technology takes place through a direct feedback from the Forum members, who share their problems and difficulties in the use of existing equipment, study imported machinery and equipment with a view to developing an indigenous manufacturing capacity and encourage fresh engineers to take up developmental challenges.

Salunke has refined a method for technology development and transfer: he entrusts a technical entrepreneur with the full responsibility for developing the specific machines or components which are required for his own manufacturing facilities. Once this equipment is successfully developed, he allows the entrepreneurs within his firm to produce it on their own on a commercial scale. One example of this is the development of a precision watchmaker's lathe similar to a Swiss model by three technical entrepreneurs in his factory. Once the lathe met the accuracy and performance tests, the engineers who developed it were given the opportunity to start their own concern and manufacture it on a commercial basis. The first precision lathe shown in a recent exhibition brought forth over 1,700 enquiries and 50 firm orders. These engineers have been given a shed on a temporary basis in an industrial area and within a couple of years they will move out to a growth centre in a backward area which Salunke is developing as chairman of a public sector project. Similar units have been set up for making dial gauges and the wooden boxes which contain the slip gauges.

The process of development of the watchmaker's lathe will lead to the establishment of small firms making the chucks, motors and attachments required for the lathe. Efforts are being made to select ten engineering students who will be encouraged to start up these new industries.

IMPROVING THE TRANSFER PROCESS: THE SUGGESTIONS OF THE ENTREPRENEURS

These three techno-entrepreneurs made the following suggestions for accelerating the process of transfer and adoption of technology.

a) The problems of information

"There is absence of information within the country about new products, machinery, equipment and plants being manufactured in other parts of the world. If such information were readily available, individual entrepreneurs would not have to search for addresses of outside manufacturers and collect information in a haphazard way before going into manufacturing". (Vartak)

"Local information centres whose main task would be to give help in locating sources of specialised information should be set up throughout the country". (Vartak)

"Cheap editions of technical books and journals will encourage entrepreneurs to have a reference library of their own". (Vartak)

b) Learning from foreign designs

"The equipment, machine tools and components which are now being imported should be analysed with a view to finding out the possibilities for manufacturing them locally and meeting the country's requirements in the coming years". (Prakash)

"An agency recognised for its technical competence should have a certain allowance of foreign currency to import a few samples of foreign machinery and equipment which is likely to be in demand in the country". (Salunke)

"An institution should be created to import foreign machines with a high innovation content. This equipment should be placed on display for about a year, after which it should be sold to interested manufacturers at full cost for stripping it apart and developing new equipment". (Vartak)

"A more liberal trade policy would allow technologically innovative industrialists to strip foreign equipment and components with a view to manufacturing them in India. Vibrator motors with an amplitude from 10 mils to 60 mils and frequency from 10 to 80 kcs is a good example: knowledgeable manufacturers in India claim that they can produce such motors at half the imported price provided they are given opportunities of stripping one or two such motors". (Prakash)

"The production of certain types of equipment often requires specialised components. These are generally made abroad on a large scale and it would be uneconomic to produce them on a small scale in India. The permanent infinitely variable drive, to take one example, requires a special chain which would be easier to import than to manufacture since local demand is too small to justify a full scale plant. The same is true of the special purpose bearings and seals needed for certain types of machinery and equipment. Import of components which account for less than 5 per cent of the cost of the final equipment should be allowed liberally". (Prakash)

c) Picking up ideas from abroad

"Visits to foreign factories and exhibitions by an experienced manufacturer gives him ideas about the possibilities of improving his products and developing new types of equipment for producing high quality goods". (Salunke)

"Concessional foreign tours on the lines of Group Travel Scheme should be devised for techno-entrepreneurs to visit technologically developed countries". (Vartak)

"Qualified engineers from developing countries should be given opportunities to visit plants in developed countries and, if required, they should be given some in-plant training in their own firms beforehand". (Prakash)

d) Support for risk-capital

"It takes two to three years to develop a machine after stripping an imported one. During this period, techno-managerial counselling through voluntary agencies is more meaningful than through governmental agencies. However, the support for developmental finance will have to come from the financial institutions supported by the government". (Salunke)

"The development of new technology is 'a shot in the dark' and a risk capital agency should be established to support it". (Vartak)

d) Co-operation with foreign firms

"A survey should be made of the material-saving, labour-intensive and low-volume manufacturing industries in Poona and later in India with a view to finding out their potential for manufacturing machinery and equipment for export. Manufacturers from industrialised countries could pass on their specialised technology to these industries with an understanding that marketing outside India will remain with the foreign companies. The latter will then get the products at cheaper prices and make more profits whilst the Indian manufacturers will be able to use their production capacity, skills and knowledge to the full". (Vartak)

THE INFLUENCE OF THE ENVIRONMENT: POONA AS AN INDUSTRIAL CENTRE

The techno-entrepreneurs whose story has been told here come from the city of Poona, about 160 km from Bombay. It was earlier known as an educational and administrative centre. In 1941, it had a population of 237,000 and there were some 135 factories. Thirty years later, the population was over 1.1 million and there were close to 1,500 factories. Industrial development started quite early: the British Government established an ammunition factory in 1869 and in 1885 the first private industry - a paper mill - was set up. Later came a copper and brassware factory (1888), a textile mill (1893) and a glass blowing factory (1898). By the early 1950's Poona had several large industries along with conventional small industries making consumer articles such as 'beedies' (local cigarettes), bricks, coarse cloth, brass and copper utensils, stationery, perfume and soap.

The development of an engineering industry gave a fillip to small-scale firms supplying components and semi-finished products as in the case of the two engine factories set up between 1946 and 1954. This growth was further accelerated by the proximity of Poona to the main Indian market in Bombay and by the State Government's restrictions on industrial expansion in Greater Bombay. The growth of industry in the Poona region from 1940 onwards can be seen from Table 1.

Table 1

THE GROWTH OF INDUSTRY IN THE POONA REGION - 1940-1971

Year	Number of New Industries established in the Poona Region	Cumulative Total
Before 1940	135	-
1940-1945	60	195
1946-1950	128	323
1951-1955	64	387
1956-1960	188	575
1961-1965	391	966
1966-1970	465	1,431
1971-1972	42	1,473
Total	1,473	1,473

Source: Data collected by Mahratta Chamber of Commerce and Industries and published in B.R. Sabade, "Poona: The new industrial city", *Commerce* (Bombay), 23 June 1973.

These data show that about 60 per cent of the existing firms were established in the 1960's. A further analysis indicates that about 41 per cent of them are engaged in engineering. Metal processing, chemicals, textiles, printing, electrical and electronics instrumentation each account from between 9 to 10 per cent of the total number of firms(1).

Besides the 76,000 workers in these industries, there are another 30,000 people working in the defence factories around Poona. Together with their dependents, about 425,000 people (38 per cent of the population in Poona) are dependent on industry. If one takes into account transportation, banking and other industrial services, about half of Poona's population can be said to depend upon industry and ancillary activities.

Table 2 brings out the changes in Poona's industrial structure between 1961 and 1970. Food and beverages, tobacco and textiles declined to 13 per cent of factory employment in 1970, as against more than 27 per cent in 1961. The machinery, chemicals and allied industries gained ground and accounted for nearly 50 per cent of factory employment in 1970 compared to 28.5 in 1961.

Table 2

THE CHANGING INDUSTRIAL STRUCTURE OF POONA DISTRICT

	Percentage share in total factory employment	
	1961	1970
Food and beverages	13.6%	5.8%
Tobacco	3.7	1.6
Textiles	9.9	5.6
Chemicals & chemical products	5.3	6.2
Basic metal industries	1.7	6.2
Machinery (except electrical)	16.7	30.0
Electrical machinery	4.8	8.5
Other industries	44.3	36.1
	100 %	100 %

Source: Commerce Research Bureau data, cited in K.V. Desai,"MIDC's Activities in Primpri Industrial Area", Commerce (Bombay), 23 June 1973, pp. 9-11.

In 1960 the State Government developed a large industrial area near Poona. Many new areas have been added and the present total industrial area in this region is about 3,500 hectares. To quote from a recent study: "Behind this industrial effort and supporting it in terms of research and training, stands the city's complex of

1) For further details, see the source of Table 1.

institutions - its social overhead capital. Thus there are the College of Engineering, Government Polytechnic, the Sir Cusrow Wadia Institute of Electrical Technology, the Industrial Training Institute and an Institute of Paints Technology. The National Chemical Laborator one of the 35 national laboratories - is also located in Poona. The State Government has put up its Industrial Research Laboratory and Small Industries Research Institute at Poona"(1). For small-scale industries there is an extension centre of the Small Industries Service Institute. The Central Mechanical Engineering Research Institute has started functioning a few years ago in Poona.

A number of professional organisations like the Poona Management Association, the Institution of Engineers, the National Institute of Labour Management and the Institution of Production Engineers have opened up chapters in Poona and they extend special assistance to industries in this region.

Young graduates in physics, engineering, chemicals and electronics are starting up new industries in increasing numbers. Formerly these graduates would have sought employment in existing firms, but with the industrialisation in Poona, industry has acquired respectability and an ambitious young man today aspires to have an independent enterprise of his own. This upsurge of ambition in men wanting to be self-employed is the motive force which has brought Poona to prominence as an industrial city.

THE STORY OF BHADU: TECHNO-ENTREPRENEURSHIP AT THE VILLAGE LEVEL

The scene: a village of about fifty huts near Mangaldai - a small town with a population of about 12,150 some 60 km. away from Gauhati, the capital of Assam, in the North-Eastern region of India. We landed in this village with the Entrepreneurial Development Officers on the morning of May 26th 1974 at 10 a.m. The village is on a cart track which could be approached with difficulty by a jeep. There is silence all around whilst a few farmers watch us from a distance. Das, the Entrepreneurial Development Officer, leads us to a farm the size of a hectare, where fresh transplantation had taken place only the previous evening. The paddy seedlings were transplanted in regular rows and are standing in about 5 cm of water. An old lady comes hurriedly and talks with Das in the local Assamese dialect. Das explains to us that the lady and the villagers are frightened because of our presence, especially as we had come in a government jeep and are dressed in shirts and trousers. She was wondering if anything was wrong. The soft and intimate manner of Das convinced the lady that we meant no harm and that we had come to see how her son Badreshwar Barua ("Bhadu") was progressing in his experiment with IR8, a new variety of paddy.

1) Source: special issue of Commerce (Bombay), 23 June 1973.

Bhadu, a young boy aged about 22, was sleeping like a log in his hut. Was he drunk? No, he had been working the whole night in the light of a hurricane lantern lifting water from the pond with the aid of a bamboo basket to water his freshly transplanted crop. The previous evening he had finished the transplantation, given the necessary dose of fertiliser and then saw to it that the crop would be adequately watered by working the whole night. Tired and exhausted he was now asleep.

The neighbouring farmers by now had gathered around us. They were closely watching this experimental introduction of a new hybrid variety of paddy in their village. They were men of experience who had worked all their lives with traditional seeds and had waited for monsoons for sowing their crops. They cultivated paddy by the broadcasting method, never went near chemical fertilisers, and scoffed at the idea of double cropping. Bhadu learnt from the Entrepreneurial Development Officers of the new hybrid varieties, the fertiliser dosages and the possibility of double cropping. He had therefore sown his seeds on raised seedbed in April and had just finished transplanting the seedlings. The elders were sure that this whole business was a mad venture and that Bhadu would come to his senses after meeting with losses and failure.

The Assam Government launched the Entrepreneurial Development Programme under the guidance of S.I.E.T. (Small Industries Training Institute) in November 1973. It placed inter-disciplinary teams in different parts of the State, inviting people for self-employment schemes, screening the respondents on the basis of their risk-taking abilities, need for achievement, and sense of efficacy, together with their family background, past experience, knowledge, skill and aptitude to start up a new business. Each team consists of four to five people from different disciplines (agriculture, animal husbandry, plant protection, horticulture, industrial extension, civil construction, road building, etc.). After screening the respondents, those showing possibility of self-employment are given a three-week course on the social and psychological aspects of entrepreneurship, the economics of self-employment and the management of new enterprises. After the course, the respondents are given practical field experience for the appropriate length of time in the lines they have chosen. Bhadu was one of the respondents who showed entrepreneurial characteristics. He went through the three-week training programme and was given field experience in one of the demonstration farms in Assam.

With all the agitation going around, Bhadu woke up. He changed over from loin cloth to trousers and bush shirt, and rubbing his eyes open, came to meet us. He told us that he had been working for forty-eight hours to ensure the success of his transplantation as timely water was considered to be one of the important elements of success for the crop. We asked him why he did not hire a pump. Small portable diesel pumps are hired out by the Agro-Service Centre for about one dollar an hour, but the farmer has to bring it from the city in his bullock cart and deliver it back to the Service Centre. About 2 hour's run of the pump would give enough water to the field. But that was 2 dollars, and Ghadu did not have any money left. He would therefore have to depend on his brawn and muscles till the monsoon would start in a few days.

Bhadu matriculated in 1972, with great difficulty. He then spent two years looking for a job ranging from "peon in the office to primary teacher in the village school", but with no success. Early in 1974, he read an announcement about training for self-employment in the local daily newspaper. The advertisement published by the Senior Special Officer for Employment of the local Entrepreneurial Motivation Training Centre aroused his interest and he went to visit their office which was only six miles from his home village. After completing the motivation training in February 1974, he selected agriculture as his main field. His father however tried to dissuade him from going into what Bhadu calls 'business-like agriculture' on the ground that these practices were very costly, too theoretical, and unsuitable for the region and that the government would harass him for the repayment of loans. The village elders were equally sceptical, but Bhadu was sufficiently strongly motivated to withstand these pressures against innovation and induced his father to give him 3 'bighas' of land (approximately one acre, or 0.4 hectares) for this experiment.

Das, whose background was in agriculture and plant protection, gave him an idea about the new varieties of paddy and about chemical fertilisers and convinced the local bank to give him a loan of one hundred dollars. Ten per cent of this amount was given by the State Government as margin money to the bank, and the remaining 90 per cent came from the bank itself. Against part of this money, the bankers and Das provided Bhadu with hybrid seeds and fertilisers. The balance of the money was spent in getting the land ploughed, since Bhadu had neither bullocks nor plough of his own.

Bhadu started cultivating summer paddy, which was unknown in the village. This innovation became a major topic for discussion in the community but Bhadu kept out of the discussions and transplanted his seedlings. The first challenge came when the rains failed to materialise. The villagers laughed when they saw the paddy going pale, but Bhadu solved the problem by spending two days and two nights watering his paddy with a bamboo bucket. When we came along by jeep from Mangaldai on the morning of May 26, the villagers anticipated some trouble. They thought that Bhadu had committed a big mistake and that his property would be auctioned to pay back the loan he had taken. But when they realised that we had come to see his progress, their attitude changed: "If government officials are taking such an interest in Bhadu's experiment, there must be something unique about it; let us observe it closely".

In the following weeks, the rice continued to grow, and the villagers were happy when Bhadu counted twenty-five tillers per plant on the average. Everything was going well until a new problem appeared. One day, Bhadu observed some insects in his field. The neighbouring farmers explained that these were 'gundhi bugs'(1) and that they would suck the milk from the grain. He contacted the agricultural extension officer who suggested dusting the crop with an insecticide. Bhadu bought the powder but could not get a duster. He treated the crop manually with a home-made muslin sieve, but was somewhat sceptical about the outcome. He therefore followed his father's suggestion and used the traditional practice: he caught a number of fish, chopped off their heads and fixed the heads on sticks planted in the field. Villagers believe that the insects prefer fish-heads to sucking the milk from the grain.

A few days later, Bhadu observed that several 'gundhi bugs' were indeed sitting on the fish heads. He also set up a light trap for the insects on two consecutive nights. By that time, the grains had become hard enough not to be damaged by the insects, but the agricultural extension officer who visited the field in August could not stop Bhadu from going ahead with his light trapping programme. The crop was now out of danger, but Bhadu had developed the habit of constant caution.

Bhadu expects that the rice crop will bring him 200 dollars. With that money he will repay the loan and use the rest for further improvements. He has never been late on paying the interest on his loan and wants to go on with his work. His father has given him more land for cultivation, and he is now growing the local Tingri variety of paddy with fertilisers. He has also left part of his field without fertiliser for the purpose of comparison. He has raised seedlings of 'Manohar chatti', another improved variety of paddy and hopes to grow wheat at some later stage. One of his next plans is to raise a loan to buy a lift pump.

The villagers are no longer as sceptical. The educated young are arguing with the elders that Bhadu's experiment is worth following by others and there are strong indications that if the yields are sufficiently high, the whole village will adopt the new technology. But if he fails, the effects will be disastrous. It is therefore all the more necessary to ensure that this pioneering effort will succeed at all costs. Failing this, the village will continue with its old

---

1) Or paddy ear-head bug, known in India as 'gundhi bug' on account of its offensive odour. The scientific name is 'leptocorisa'.

methods for the years to come. If a number of farmers take up a new technology, it is not so very important if a few of them fail. But if the first person to innovate does not succeed, the consequences can be disastrous.

CONCLUSION

The three young techno-entrepreneurs from Poona were well qualified as engineering graduates and deliberately chose to start a new company. They came to the conclusion that they should select a product with a high technology content which was badly needed by the industrial sector, and for which the demand could be assessed from the import statistics and the import substitution possibilities. Prakash deliberately selected the dynamo meter on these considerations, and Vartak shifted from standard motors to special purpose motors produced in small quantities and which require a lot of labour and limited amounts of raw materials.

Having identified the product, the entrepreneurs found out more about manufacturing through books, literature, foreign manufacturers and a careful study of existing machines. The search for foreign manufacturers is a time consuming process which could be made more efficient if there was an Information Centre in India where all this information is readily available.

The college professors played a leading part in giving advice, offering opportunities to work in their laboratories and allowing the opening-up of the machines in their workshops. This however is at the discretion of the individual, and policy decisions will have to be taken to ensure that techno-entrepreneurs are offered such facilities by colleges and laboratories on a regular basis.

The techno-entrepreneurs then located prospective users within the country, asked for suggestions about possible improvements, and found out whether they would be interested in local machines which met their specific requirements. This market research could clearly be handled more systematically by a specialised agency.

Not content with the theoretical training in the colleges, they got down to work in the workshops to familiarise themselves with workshop practices, cost estimations and production planning. They also saw opportunities in odd specifications. Vartak for instance, found that 42-volt motors were manufactured by a foreign firm for a special purpose and used this knowledge to identify buyers within the country for this odd voltage. They overcame the prototype difficulties with perseverance, kept in close touch with their clients and were on the lookout to widen their range of production.

They went abroad with the specific purpose of seeing factories making similar products, learning how to bring down costs and improving the quality of their products. They found out for themselves that they were not in any way inferior to the well-known foreign manufacturers and used the opportunity to sell their products to their counterparts abroad. Vartak is now supplying torque motors to a well-known foreign maker and Salunke is shipping slip gauges to an American firm. Equally important, they realised the need for financial discipline and tight budgetary control.

The ban on imports stimulated the adoption of technology by forcing industrial users to support indigenous development. However, the import of certain machines (for stripping them) and of some mass-produced accessories and components for final assembly would facilitate the adoption of new technology. Visits to exhibitions and factories abroad clearly stimulate innovative ideas. In-plant training in high technology factories in industrialised countries could certainly accelerate the transfer of technology.

Having succeeded in their original line, they did not sit on their laurels but became real entrepreneurs. They jumped their specialisation and selected a line which they felt was prestigious or profitable. Prakash went in for special purpose steel because it was a challenge which, if he succeeds, will allow him to supply stainless steel strips to his 'guru'. Vartak went for nitrogen-fixing bacteria. Salunke dedicated himself to developing new techno-entrepreneurs, and has an active interest in the development of backward regions. He made the Forum of Industrial Technologists more meaning ful and acted as chairman of a public sector development project. Their suggestions for accelerating the process of adoption of technology deserve careful consideration, for they come from their hard earned experience.

Bhadu's case on the other hand is from the rural environment, where the level of technology is low, the environment unfavourable and the extension agents far away. The introduction of a rudimentary new technology and its success poses a host of problems which are somewhat different from those of techno-entrepreneurs in an industrial region, but just as complex and challenging.

VIII. THE REORGANISATION OF THE LIGHT ENGINEERING INDUSTRY IN SRI LANKA

by

D.L.O. Mendis*

Sri Lanka's 1972-1976 Five Year Plan has presented the general framework for the country's economic development. This framework has been further elaborated in sector plans for agriculture and for industry. The Industrial Sector Plan stresses the need for a self-reliant approach to development. It recognises the existence both of a traditional sector and a modern sector. In both these sectors, the development of an indigenous machine-building industry has for a long time been retarded by the unrestricted import of even the simplest types of equipment. Very high priority has now been given to the development of a dispersed small-scale light-engineering industry to act as an agent for spreading technological skills in the rural areas. Implicit in this policy statement is the recognition of an existing traditional technology which is basically an indigenou low-cost technology. Although this recognition is itself without precedent, it will not serve the objectives of rapid economic development unless the progressive improvement of this indigenous technology is also planned and vigorously pursued.

In both the public and the private sector in Sri Lanka, there already is a considerable investment in modern technology which is comparatively highly productive and capital-intensive. The progressive development of indigenous technology in the traditional sector has been planned so as to gain from linkages with the high technology in the modern sector. This contrasts with earlier policies when new industrial technologies were introduced with little regard to existing low-cost technologies whose development as a result was retarded. Examples include the introduction of modern power-looms which competed with existing handlooms, the development of mass-produced aluminium household goods which almost destroyed the traditional pottery industry, the setting-up of a State Hardware Corporation to

---

* The author is Adviser (Techniques) in the Ministry of Planning and Economic Affairs in Colombo, Sri Lanka, and Visiting Lecturer at the Faculty of Engineering, University of Sri Lanka.

manufacture various agricultural tools and implements and building materials which were already being made by indigenous industries or the production of soap by the local subsidiary of a multinational corporation which almost eliminated the small-scale soap makers.

THE LOCAL LEVEL: DISPERSED PRODUCTION CO-OPERATIVES

The chief vehicle for implementing the new development policy based on low-cost technology and the improvement of indigenous industrial capabilities was the Divisional Development Council (DDC). The Council's projects are all in the form of co-operatives. One particular type is the Dispersed Production Co-operative. This is an organisation of individual entrepreneurs and workers in a given industry in a particular region or administrative division. This type of organisation offers a number of advantages:

a) The individual worker is not confined to fixed working hours, but may engage in productive work at any time of the day or night that suits him;
b) In some cases, he may be assisted by members of his family;
c) The training of new workers (not only family members but outsiders as well) takes place on a person-to-person basis in a comparatively small work-place;
d) The co-operative which is owned by its individual members takes the place of the middlemen, so that there is no possibility of exploitation by such people;
e) The individual worker, if he is the owner, has every incentive to develop his work-place and improve his technology and his productivity; and if he is an employee, he has the prospect of setting up one day his own work-place with the technical and financial assistance of the co-operative;
f) Financial assistance for an owner to improve his work-place or for an employee to start his own business, is more readily available from the People's Bank through the co-operative organisation than it would be for small independent individuals;
g) Because of his membership in the co-operative, the individual producer can tide over market fluctuations more easily than if he were alone;
h) Since the co-operative arranges for the supply of inputs and takes care of the marketing problems, its members can devote most of their time to actual production. This means a maximum utilisation of the available labour and technology.

THE REGIONAL LEVEL: MEDIUM-SCALE INDUSTRIAL CO-OPERATIVES

Dispersed production co-operatives of village blacksmiths were first set up on the basis of one for each Revenue Officer's Division. By August 1974 more than 100 of these Light Engineering Industrial Co-operatives, as they are called, had been established in a total of 158 Revenue Officers Divisions all over the island. At the next higher level of technology, co-operatives were set up for the owners and skilled workers of the small workshops which are very common in provincial towns. These are called Medium-Scale Industrial Co-operatives, and it was proposed to have one such organisation for each of the 22 administrative districts in the island. Later it was decide to also have Medium-Scale Industrial Co-operatives in each of the Revenue Officers Divisions in the Colombo district. By August 1974, some seven of these Medium-Scale Industrial Co-operatives had been established. A Union of Light Engineering Industrial Co-operatives was also formed as the apex organisation for the light engineering industry.

The Union, as the representative body of the organised light engineering industry in Sri Lanka, has broken into some new areas of activity, of which ship-breaking is the most noteworthy. The Union has also purchased a modern production factory which manufactures motor and tractor spare parts, weighing machines and other products.

Through the organisation of Light Engineering Industrial Co-operatives in the rural areas, the traditional low-cost technology of the village blacksmith has been made more productive, and the quality of production was improved. The co-operatives organised the supply of raw materials including scrap iron and steel, charcoal and small hand tools such as files. They also undertook the sale of finished products through a commercial centre. Orders for various tools and implements were obtained in advance from government organisations as well as from the plantation sector, an important consumer of such hardware in Sri Lanka.

QUALITY CONTROL

With the organisation of supplies and markets for the village smithy, the highest quality was demanded. Quality control is ensured by a combination of traditional and modern methods. Traditionally, the village blacksmith made and sold implements to the farmer on the basis of a person-to-person guarantee. This has been formalised in the co-operatives by giving each member an identification number which is stamped on every product, which is then guaranteed for one month from the date of purchase. If a defective item is brought back by the purchaser, the co-operative replaces it and returns the defective item to the manufacturer. If the defect appears to merit special attention, the problem is referred to an outside consultant. This consultancy service is provided by metallurgists in the University's Faculty of Engineering and the Sri Lanka Institute of Scientific and Industrial Research.

THE PRICING MECHANISM

Since the blacksmith obtains his raw materials from the co-operative, he is liberated from the clutches of middlemen who had previously supplied the inputs and purchased his products, and generally retained an unconscionable profit. The producer now enjoys a maximum profit. In order to give a fair deal to the consumer, a Price-Determining Committee has been set up in each co-operative. Usually this committee consists of three blacksmiths representing the producers and three local farmers representing the consumers, with an official of the Planning Ministry as the chairman. The committee examines the amount of raw materials and labour necessary for manufacturing the products on the basis of the blacksmith's low-cost technology and arrives at the prime cost of production. A reasonable profit margin is allotted to the producer and a further mark-up of between 10 and 15 per cent is allotted to the co-operative to determine the selling price to the consumer. Profits accumulated in the co-operative will in future be used for the progressive development of technology.

INVESTMENT AND EMPLOYMENT

With this type of organisation, productivity has been considerably improved. Today a master blacksmith and two assistants can produce more than 2,000 rupees (about $250) worth of digging forks per month using the simple technology of the village smithy which represents a total capital investment of some 5,000 rupees (about $630) at today's prices. New employment has been generated in two different ways. Firstly the master blacksmith will take on more helpers to increase his production. Secondly, skilled blacksmiths who are working as helpers will be able to set up their own workplaces. The co-operative encourages this type of development with individual loans from the People's Bank. Some co-operatives have more than doubled their membership through this process in the last three years.

THE ROLE OF THE TECHNICAL CENTRES

Once a co-operative has been established and has started to function, steps have to be taken to set up a technical centre. This centre will be equipped with tools which are not generally available in the small work-places of the co-operative's members such as a power hammer and power saw for reducing large scrap into the smaller billets which are issued to members. It will also include a lathe, a drilling machine and both electric arc and gas-welding equipment. This equipment will be used for certain specialised operations like the production of some component parts for small agricultural implements. An assembly unit will also be set up in the technical centre to make small agricultural machines.

THE PROGRESSIVE DEVELOPMENT OF TECHNOLOGY

The progressive development of technology in individual work-places has been planned through the technical centres. At the village level a blacksmith is usually equipped with a hearth and a pair of bellows, an anvil, a number of sledge hammers, a vice, several pairs of tongs which he usually makes himself, a few files and sometimes a grinding wheel, and finally a manually-operated drill. With this equipment he can make a wide variety of tools and implements. In general he was dependent on traders for the supply of raw materials such as scrap iron, steel and fuel (usually charcoal). The trader also sold him the tools of his trade that needed periodic replacement like the files and the grinding wheels. Likewise the blacksmith was dependent upon middlemen for the sale of his products. As a result, he invariably functioned at a low economic level and did not have the capacity to generate sufficient savings to develop his technology

With the organisation of industrial co-operatives, the village blacksmith is now able to put aside a certain percentage of his earnings and invest it in the development of his technology. The progressive development of technology can be described as step-by-step improvement of the tools of his trade. This improvement will in the first stage allow him to use the physical energy available to him in a more productive way. Thereafter, it will enable him to add mechanical energy to his production process. He will be able to pay for mechanical power with his own savings and with credit facilities made available through the co-operative. Next, using mechanical power he will be able to manufacture products on a more systematic basis and in larger quantities, and go into a production-line type of manufacturing if regular orders for his new products are forthcoming.

In practice, when a blacksmith produces more and increases his savings, he will probably have priorities other than the improvement of his technology. Settling old debts, improving the house, and purchasing movable property like furniture and jewellery generally take precedence over improvements in the work-place. The first blacksmith who takes steps to improve his technology is a pioneer whose example is watched closely by the others before they decide to follow suit. The interest of the co-operative is therefore to ensure the success of this and the consultancy service available through the technical centre must be used to maximum effect at this stage.

The technology of the village smithy can be systematically improved in a number of ways:

a) By introducing a more efficient hand-operated blower to replace the bellows;

b) By the better conservation of heat in the hearth and the use of waste heat for tempering and perhaps for annealing, and by the use of a specially-designed oven placed above the hearth;

c) By introducing a mechanical finishing hammer (manually operated in the initial stage) to achieve a better finish in a more economical way;

d) By using grinding wheels to save on filing;

e) By introducing mechanical power which may be used initially for the mechanical finishing hammer, the blower, the grinding and finishing wheels and a drill; power may be used later to operate a simple lathe and a power hammer.

It is preferable to introduce mechanical power when the blacksmith already has a certain amount of equipment such as a mechanical finishing hammer, a blower, and a grinding and finishing wheel: at this stage, mechanisation is an economic proposition. Mechanical power can also be used for other types of equipment, for instance, for the simple home-made hand-operated lathe which is introduced sooner or later in almost every smithy and which is used to turn out wooden handles for knives and other implements.

It has been found that such a planned step-by-step development of technology is well within the bounds of reality, when the productive capacity of the village craftsman has been liberated by the organisation of co-operatives, when the middleman has been eliminated and when regular markets have been made available. At a later stage it is expected that the village blacksmith will be able to purchase and use more sophisticated equipment, including welding plant and machine shop plant. He will then have qualified for membership in the Medium-Scale Industrial Co-operative at the district level.

INNOVATION IN THE MEDIUM-SCALE INDUSTRIAL CO-OPERATIVES

Medium-Scale Industrial Co-operatives were set up to harness the un-utilised and under-utilised production capacities in workshops dispersed all over the island. These include small machine shops, repair garages and foundries. It has been possible to produce small machines such as two-wheeled tractors by farming out the manufacture of component parts to members of these organisations. This particular project has been tried out in no less than seven co-operatives which have all successfully manufactured and tested prototype two-wheeled tractors (the engine, the chain and the main bearings of these machines are imported). The project has been launched in the teeth of opposition from those who believe that the only way to make a two-wheeled tractor in Sri Lanka is to set up a joint-venture with a foreign firm. The success of the project is therefore to be measured not merely in terms of the number of tractors manufactured in this way, but in terms of the number of individual entrepreneurs and skilled workers who have been given an opportunity to use their talents to create a commercially viable product. This is a very important element in achieving a self-reliant attitude to development

The first priority has been given to the development of the two-wheeled tractor, but the need for local manufacture of water pumps has been repeatedly emphasised by the members of several co-operative One such pump has been developed locally; it is mounted on the frame of the two-wheeled tractor and driven by the same power unit.

At the beginning of this project the Ministry of Planning set up ad hoc committees of individuals who were known to be interested in it in order to launch the manufacture of this machine through the industrial co-operatives. This group of individuals brought in their varied expertise and gave the necessary initiative for developing the prototypes. This method has been quite successful. Prototypes have also been manufactured in public sector organisations under the direction of these ad hoc committees. It is expected that some of these public sector institutions will also undertake batch production of these machines in the future.

Other machines which have been developed in small workshops as a result of the organisation of the industrial co-operatives include a bicycle and a water pump. Both are now in batch production. Several prototypes of single cylinder air-cooled internal combustion engines have also been made but none of them have yet been taken up for batch production.

The reorganisation of Sri Lanka's light engineering industry by the Ministry of Planning and Economic Affairs under the Divisional Development Council's programme has met with such success that the same approach is being used for other industrial sectors. Steps are being taken to organise village potters who have been unfortunately discriminated against: bureaucratic ignorance has promoted other industries that tend to suppress the traditional pottery industry. Similar work in organising the traditional textile industry can be expected in the future.

IX. MECHANISATION TECHNOLOGY FOR TROPICAL AGRICULTURE

by

Amir U. Khan*

INTRODUCTION

In tropical regions where farm income is low, farm holdings are small, and labour is cheap, many attempts have been made to mechanise agriculture with equipment from the industrialised countries. However, farm equipment from the advanced countries, having been developed for conditions of either large farm holdings and high labour costs, such as in the United States, or for small farm holdings and high farm incomes as in Japan, is not well suited to the agro-economic conditions that prevail in the developing countries. Consequently, mechanisation of tropical agriculture has been slow and limited mostly to the larger farm holdings which constitute a small part of the total arable land.

It has been argued that a wide range of farm equipment is readily available from the industrialised countries and that the problems of mechanisation in tropical agriculture lie primarily in the proper selection of farm equipment. A closer analysis reveals that this assumption is based only on the operational considerations and does not take into account the other aspects that have an important bearing on agricultural mechanisation. The economic, social, and cultural needs of the farmer, lack of foreign exchange to import equipment, local manufacturing capabilities, and the compatibility of the mechanisation technology with the resource endowments of the country are important factors and must be considered accordingly.

Because not enough appropriate farm equipment is available for individual ownership by small farmers, efforts have been made to introduce larger equipment through various forms of joint use. Such efforts have met with marginal success in only a few countries, mostly for equipment for the difficult farming operations of land preparation and harvesting. Chancellor studied the contract hire services in Malaysia and Thailand and found the sales of small two-

---

* Dr. Khan is head of the Agricultural Engineering Department of the International Rice Research Institute in Los Baños, Philippines.

wheel power tillers to individual farmers rapidly increasing in Malaysia, even when large tractor hire services were economically available(1). It seems that the desire to retain control of the production operations and the prestige of ownership are strong factors that favour individual equipment ownership by farmers. The power tillers provide the farmers with greater flexibility, for these machines can also transport persons and haul goods.

MECHANISATION CONSTRAINTS

Two distinct agricultural mechanisation technologies have evolved to suit different sets of agricultural and socio-economic conditions. The Western approach emphasises dryland farming with large, high-powered equipment. This capital-intensive technology is primarily aimed at replacing agricultural labour with machines. Its introduction is not desirable in the populated tropical Asian region and has been seriously questioned by many social scientists. Numerous attempts during the last 30 years to introduce this technology in many developing countries have had rather dismal results. In India, where large tractors of over 35 hp have been continuously marketed since the end of World War II, only one per cent of the total arable land is worked with such tractors today.

Mechanisation in Japan has not followed the Western approach. Rice, a major crop in that country, is grown under wetland conditions on small farm holdings. The high support price for rice and the rising standard of living, coupled with rapid industrial growth and widespread opportunities of part-time industrial employment in the rural areas, have resulted in the mechanisation of Japanese agriculture with small but quite sophisticated and costly farm machines. The equipment, developed to meet the requirements of the relatively rich Japanese farmer, is far too complex and uneconomical for tropical Asia. Recently introduced Japanese combine harvesters and paddy transplanters are excellent examples of functionally suited but economically unacceptable machines for the tropical regions.

Besides being complex, farm equipment in the industrialised countries is developed for manufacture with capital-intensive mass production methods. Production technology to manufacture such designs is not readily available in the developing countries. Since the design of a product is closely related to the production process, the scope for the production of imported farm machinery designs in the developing countries is severely limited. Unless appropriate farm machines

---

1) W. Chancellor, <u>A Survey of Tractor Contractor Operations in Thailand and Malaysia,</u> The Agricultural Development Council, New York, 1970.

are made available to suit the low-volume production technology in the less developed countries, low-cost machines would be difficult to produce and mechanisation will continue to be a luxury only a few rich farmers could afford.

With the introduction of new varieties, problems of drying and processing have assumed new proportions. Technology for drying and processing agricultural crops in large central plants is available from the industrialised countries. The establishment of such plants, however, requires a well-developed infrastructure that often is lacking in the tropics. The shortage of economically viable drying and processing systems for farm- or village-level operations is a serious problem in the developing countries.

Rapid technological developments in the industrialised countries are making it increasingly difficult to transfer mechanisation technologies to meet the needs of the farming communities in the developing countries. In the Western countries, the trend is toward higher powered machines with more sophisticated and automated control systems. To some extent, mechanisation developments in Japan are following a similar trend.

Generally most imported farm equipment in the developing countries sell at two to four times its price in the country of origin. Furthermore, since this expensive equipment must compete with low-cost local labour, the economic yardsticks on which the farm machine is originally based are no longer present.

The high cost of imported farm equipment limits its market to a small number of relatively rich farmers and hinders the development of an effective sales and service network in the country. For example, before power tillers were locally produced in the Philippines in 1970, less than 1,000 units of power tillers were annually imported and these were distributed among six different makes and over twelve different models. With such a limited market, manufacturers could not establish effective sales and service networks in remote parts of the country. With the recent production of locally designed simple power tillers, the market has multiplied several-fold and service and parts for such machines are becoming readily available all over the country.

Most developing countries are undergoing varying degrees of balance-of-payment problems. Large-scale importation of equipment to mechanise tropical agriculture is not possible because there is a lack of foreign exchange in most developing countries.

MECHANISATION STRATEGY

The author believes that the slow pace of agricultural mechanisation in the tropical regions is largely caused by the non-availability of appropriate mechanisation technologies to meet the overall requirements of the small farmers in the tropics. If mechanisation is to succeed, suitable technologies will have to be developed to suit the agricultural, industrial, and socio-economic conditions of the developing countries.

Small- and medium-sized farm holdings of up to 10 hectares constitute a large segment of the arable land in the tropics. Farmers with such holdings find it difficult to operate economically with animal-drawn tools, but the larger tractors of over 30 hp are too large for their needs. Ironically, this large group of developing country farmers have hardly any access to appropriate farm equipment. To be successful, agricultural mechanisation must be introduced on the smaller farms in the tropical regions and must be considered in this particular frame of reference.

Almost all past national and international efforts to mechanise tropical agriculture have been concerned primarily with the utilisation of imported machines. Agricultural mechanisation and indigenous production of farm equipment are two facets of the same problem and must be looked into simultaneously. The establishment of a viable farm equipment manufacturing industry is a prerequisite to widespread agricultural mechanisation in a country. The availability of farm machinery designs that could be produced locally will favourably affect the development of indigenous farm equipment manufacturing industries in the developing countries.

There is little doubt that the desire of the developing countries to industrialise and the socio-economic implications of agricultural mechanisation will require the local production of farm equipment. To maximise employment in the manufacturing sector, it would be desirable to produce farm machines in the small-scale industrial sectors through low-volume, labour-intensive production methods.

Social scientists have understandably pointed out the dangers of displacing farm labour with agricultural machines. Unfortunately their analysis is generally based on the field labour displaced directly through the introduction of large, high-powered agricultural machines. It is unrealistic to base such conclusions on a mechanisation technology that is completely inappropriate for most developing countries. Certainly the introduction of mammoth equipment, such as that used in the Western countries, will displace labour and would be highly undesirable in most developing countries. On the other hand, agricultural mechanisation through the introduction of small, individually owned farm machines, such as is done in Japan, does not necessarily displace labour but rather permits more intensive and timely operations, thereby improving land and labour productivity. It is interesting to note that among 11 Asian countries, Japan has the most mechanised agriculture (3 hp/ha) and yet is among the countries with the highest labour input (1,400 man-hr/ha) in the production of rice in Asia (see Table 1). An agricultural mechanisation strategy based on the indigenous production of small individually-owned farm equipment will provide additional employment in the production, marketing, and servicing functions.

LOW-VOLUME MANUFACTURING

One often hears the popular argument that farm machines cannot be economically manufactured in the developing countries unless a substantial local demand develops. This erroneous conclusion is drawn because the product designs and the production processes available from the developed countries are often not able to tap the full potential of low-cost labour that is available in the developing countries. Undoubtedly the economies of scale are important in the industrialised countries where labour costs are high. Economical production, however, can be organised in the developing countries with labour-intensive methods if the machines are appropriately designed for low-volume production.

The many interesting examples of low-volume production that are beginning to appear in some developing countries strongly substantiate this thesis. The low-volume production of the jeepney, a locally adapted version of the Jeep, by small metalworking firms is a good example in the Philippines. Most of the jeepney components are manufactured all over the country with manual methods using simple tools and jigs. The complete body grill and other sheetmetal parts are hand made; in quality they are comparable to, and at times better than, machine-stamped parts. While the engines and the transmissions are imported, numerous other components and the chassis are produced by large metalworking shops in the Greater Manila area and supplied to small jeepney shops all over the country. The production of the jeepney has resulted in the development of a sizeable automotive component manufacturing industry in the Philippines which has substantially helped in the local manufacture of other makes of automobiles. The jeepney is among the lowest-priced automobiles and is the most popular mode of transport in the country.

Another interesting example of appropriate product development is the motorised lowlift pump. This pump was developed by a village farmer-mechanic in Vietnam in 1963. In less than four years, nearly 43 per cent of the farmers in the village had purchased such a pump(1). Different versions of this lowlift pump are now manufactured in low volumes in Vietnam, Thailand and the Philippines.

---

1) R.L. Sansom,"The Motor Pump: A Case Study of Innovation and Development", Oxford Economic Papers (New Series), Vol.21, No.1, 1969.

Table 1

SOME AGRICULTURAL MECHANISATION INDICATORS FOR 11 RICE-PRODUCING COUNTRIES IN ASIA

Country	Arable land holding (ha)	Agricultural working population/ha	Horsepower per hectare			Total	Hp per agricultural worker	Labour hours for rice cultivation/ha	Net domestic agricultural production US$	
			Human	Animal	Mechanical				Per person	Per hectare
Sri Lanka	1.59	1.20	0.120	0.148	0.110	0.378	0.009	N.A.	293	352
Taiwan	1.11	1.95	0.195	0.164	0.164	0.505	0.074	1300	349	696
India	2.62	0.90	0.037	0.204	0.008	0.249	0.009	1000	148	133
Iran	6.17	0.37	0.090	0.048	0.154	0.292	0.418	N.A.	417	154
Japan	1.06	2.16	0.216	0.120	2.664	3.00	1.231	1400	626	1350
Korea	0.90	1.96	0.196	0.236	0.003	0.435	0.0013	830	244	477
Nepal	1.22	2.49	0.249	0.480	0.004	0.733	0.0016	N.A.	99	236
Pakistan	2.37	1.09	0.109	0.288	0.013	0.410	0.012	N.A.	154	169
Philippines	3.66	0.71	0.071	0.104	0.023	0.198	0.030	800	242	186
Thailand	3.64	1.10	0.110	0.184	0.054	0.348	0.050	N.A.	102	112
Vietnam	1.57	2.10	0.210	0.244	0.023	0.477	0.004	N.A.	203	421

Source: APO Expert Group Meeting on Agricultural Mechanisation, APO Project SYP/III/67, Tokyo, October 1968, Vol.II.

Another interesting example is the high-speed air-cooled engines in a rural area of the Chachiengsao Province in Thailand. The owner of the company that manufactures the engine is an innovative individual who incorporated into the engine design many ideas from popular makes of imported engines. He also has developed simple production equipment to produce these engines with labour-intensive methods. In 1974 this company was producing 1,500 engines a month in the 10-. 15-, and 20-hp sizes and was marketing them for use in motorboats, irrigation pumps, and power tillers. In addition to these examples, there are many interesting cases of low-volume production of diesel engines, water pumps, machine tools, motor rickshaws and other machines in India, Pakistan, Taiwan, Thailand, and Sri Lanka. These examples tend to indicate that fairly complex machines can be economically produced in the developing countries if the product and the production process are appropriately adapted to low-volume production.

DEVELOPING NEW EQUIPMENT

Product development is essential in the transformation of engineering research into a commercially useable form. This field is almost exclusively catered to by industry in the advanced countries. The academic institutions usually conduct basic research and industry often uses the results in developing a wide variety of products that eventually generate industrial activity. Product development requires risk capital which the struggling small-scale industry in the developing countries cannot afford.

Manufacturers in the industrialised countries have the resources to develop new farm equipment, but for many reasons, they do not find the development of farm equipment for the tropical countries very attractive. Understandably their strategy has been to develop new equipment for their domestic markets and to expand its sales later in the tropical regions. Funds for research are primarily available to the academic and research institutions in the developing countries, which are mostly in the public sector. Such institutions in the less developed countries have therefore the responsibility to cater to the product development needs of small-scale industry.

IRRI'S MACHINERY DEVELOPMENT PROGRAMME

In 1967 the Agricultural Engineering Department of the International Rice Research Institute (IRRI) initiated a programme on the development of low-cost farm machines for use and manufacture in the tropical rice-producing countries. Under this programme, engineering drawings and other technical assistance are provided without cost to manufacturers who are willing to produce such machines.

Six IRRI machines - the 5 to 7 hp tiller, axial flow thresher, rotary power weeder, batch dryer with rice hull furnace, seeder for pregerminated paddy, and extendible blade-lug wheels - have generated much interest and are now regularly being produced commercially.

a) 5 to 7 hp tiller

This tiller was developed to compete with the simplest of the imported power tillers. It was designed to make maximum use of locally popular standard machine components, such as engines, roller chains, sprockets, bearings and seals. It simplified the manufacturer procurement problems and minimised parts problems for the end-users. The other components of the tiller were fairly simple and could be produced by small metalworking firms in most developing countries. Care was taken to limit the production operations to simple cutting, bending, welding, and machining operations. A range of attachments has been developed for transporting, irrigating, and cultivating upland and lowland crops. The tiller, now produced in six Asian countries, sells for approximately half the price of comparable imported power tillers. As of December 31, 1974, over 7,000 IRRI power tillers have been commercially produced in the Philippines

The success of the 5 to 7 hp tiller and the continuing need for a larger power tiller has led to the development of another IRRI power tiller with a 10 to 12 hp diesel engine, multispeed transmission, and steering clutches. This larger tiller was recently released to six companies in the Philippines and these manufacturers are now fabricating their units.

b) Axial flow thresher

Experience with two earlier hold-on type threshers - the drum-type and the table-type threshers - indicated that farmers were interested in a high-capacity throw-in machine for threshing wet and dry paddy and other crops. This led to the development of the axial flow thresher which has rapidly gained popularity and is now being produced in eight countries. The threshing material moves in an axial direction in the threshing drum and this prolongs the threshing action Preliminary separation of the grain and straw occurs in a full-circle concave that extends to the full length of the threshing drum. This thresher has an output of about one ton per hour with a three-man threshing crew. It can thresh rice, sorghum and soybeans. With some modifications it can thresh wheat.

c) Power weeder

Since paddy fields in the developing countries are small and have no headlands at the end of the rows, turning of a ground-supporte machine is difficult. The IRRI power weeder is a portable lightweight (22 kg), three-row machine that can be lifted off the ground at the end of the paddy rows. One Japanese company that started producing this machine in 1971 has found an excellent market for it in areas where the soil is so soft that the fields cannot normally be worked with power tillers. The company has developed 14 attachments for this machine for a variety of uses.

d) Tractor lug wheel

Cage wheels with fixed lugs have been popularly used with large tractors in wetland conditions to improve traction. Since the lugs are welded at a fixed radius, the wheels cannot be adjusted to suit the varying soil and field conditions that are generally encountered in the tropics. The IRRI extendible lug wheel has 14-inch wide metal blades mounted on a wheel frame that can be radially extended to obtain better traction and improved mobility in difficult field conditions. When a tractor bogs down in soft fields, one or two lugs can be fully extended to extricate the tractor from the bogged condition under its own power. The bogging down of a tractor is a serious problem for tractor contract operators since it requires winches or other tractors that are difficult to obtain in the countryside. The IRRI extendible lug wheel has become so popular in the Philippines that almost all makes of large farm tractors sold in the Philippines for paddy cultivation are equipped with such wheels.

e) Batch dryer

Drying of paddy during the wet season is a serious problem in developing countries. Simple batch-type dryers are popular in Japan for farm-level drying; however, such dryers are expensive to import and difficult to fabricate in the less developed countries. The IRRI batch dryer was designed with features similar to those of the Japanese dryer but it can be easily fabricated by most small metal-working shops in the developing countries. A simple pot-type oil burner with an automatic fuel cut-off valve permits safe operation. The machine can dry one ton of paddy in 5 to 6 hours. The standard drying bin is made of sheetmetal; however, plans are available for wooden bins to reduce costs. Three companies are now manufacturing the IRRI batch dryer in the Philippines; it is also produced in Taiwan.

Because of increasing costs of burner fuel, an alternate rice hull furnace was developed for use with this dryer. The furnace consists of an inclined grate burner, a hopper for rice hull, and an ash trap. Hulls can be fed either manually or automatically by connecting the feed mechanism to an eccentric drive pulley driven from the blower fan. This furnace is produced by the three dryer manufacturers and is finding an increasing market in the Philippines.

The IRRI machinery development programme has demonstrated that public research institutions can play a significant part in developing local industry by channelling research efforts to solve specific research and development bottlenecks that small-scale industries are facing in the developing countries.

INDUSTRIAL EXTENSION

During its early years, the IRRI machinery development programme encountered considerable difficulty in encouraging manufacturers to produce the IRRI agricultural machines. Advertisements offering free agricultural machinery designs were placed in the leading Philippine papers with little success. Subsequently orders were placed with some selected manufacturers for the supply of a few IRRI-designed machines. This prodded some companies to produce these machines and helped in developing their interest. As the manufacturers gained some confidence in the IRRI machinery programme, commercialisation of the machines became relatively easier. After one of the IRRI machines achieved some commercial success, manufacturer interest increased rapidly and many companies now actively seek machinery designs from IRRI.

The commercialisation of the designs that originate from most industrial research institutions has been a chronic problem in the developing countries. Experience indicates that distribution of engineering drawings and technical information to manufacturers is in itself not sufficient for successful commercialisation. Most small- and medium-sized metalworking firms in the developing countries are not used to working with engineering drawings and prefer duplicating actual sample machines. These firms have very limited capital to undertake any risks on new unproven machines and they prefer to produce machines for which there is an assured order.

To encourage the introduction of IRRI machines among small manufacturers, a design-release procedure was adopted under which engineering drawings and prototype machines were loaned to the interested manufacturers. These manufacturers were requested to submit their quotations for supplying a few machines to IRRI along with their long-range manufacturing and marketing plans. Orders were then placed with some selected firms for the supply of a few, usually one to six, machines to the Institute. This approach provided the manufacturers with some additional business and permitted them to evaluate their production capabilities for producing IRRI machines without taking any risks.

Such a strategy was possible since most IRRI machines were so designed that no special tools, materials and processes were required in their manufacture. An added advantage of this strategy was that the design weaknesses and production problems were pinpointed and rectified early during the limited production phase. The manufacturers were encouraged to modify the IRRI designs as they saw fit either to suit their production facilities or to improve the machine's performance. Almost every IRRI machine underwent some degree of modification at this stage.

Each manufacturer submitted its first machine to IRRI to be evaluated and checked against drawings. The manufacturers were informed of the test results and given the appropriate recommendations for improvements wherever necessary. If the first machine performed satisfactorily, the manufacturer received approval to proceed with the production of the other machines ordered by IRRI. The manufacturers were encouraged to test-market some of the machines being fabricated for IRRI and any inquiries received by IRRI were channelled to them. Even before they could complete the IRRI order, the manufacturers were often able to sell a few of the more promising IRRI machines. Such sales provided the manufacturer with a strong incentive to enter regular production.

To facilitate marketing in the early production stage, IRRI provided leaflets, instruction manuals, sales literature and test reports to the manufacturers. Attractive decal labels, indicating that the machine was based on an original design from IRRI, were provided for attachment to the commercially produced machines, and these helped in the introductory sales. The machines received from the manufacturers were extensively field-tested in various part of the Philippines and in other countries through co-operating subcontracting organisations. Manufacturers were informed of any design or production defects during this stage.

Once satisfactory field performance was obtained, the manufacturers were given the go-signal to produce the machines regularly. This design-release procedure was quite successful and is now the standard method for releasing all IRRI designs.

Because of the limited resources available to the programme, it was difficult to organise a major industrial extension effort for commercialising the IRRI machines in countries other than the Philippines. Co-operative arrangements were made, through subcontracts, in nine Asian countries, with organisations that indicated an interest in extending the IRRI machines. These organisations were provided with those machines which had gained some commercial success in the Philippines and a modest amount of funding for hiring some technical staff. Engineers working with the IRRI co-operator were brought to the Philippines for short-term training in the use and production of IRRI machines.

Some useful experience has been gained in extending the IRRI machines to manufacturers in other countries through the co-operator approach. It shows that the agricultural and industrial research organisations in the public sector are interested in testing and evaluating machines but are not as effective in working with local manufacturers. These organisations seem faced with problems somewhat similar to those of the academic institutions. Lack of interaction with local industry and insufficient motivation to commercialise machines are common difficulties. On the other hand, manufacturing firms are primarily interested in the specific machines which suit their own production plans and are reluctant to devote efforts to encourage local manufacture of all the IRRI machines. A good example is that of a manufacturer in India who spent all his efforts unsuccessfully in installing his own make of diesel engine on the IRRI power tiller rather than in extending the complete range of IRRI machines to other manufacturers. In all cases the subcontractors' engineers, who obtained training at IRRI, were the key persons in the programme and success or failure was highly related to their motivations.

MANUFACTURING DEVELOPMENTS

Progress in the Philippines has been quite encouraging. A sizeable industry for the manufacture of small farm equipment has been established to produce IRRI machines. Eighteen companies are now manufacturing IRRI machines in the Philippines. The production of some IRRI machines has also started in 10 other countries. In the relatively short period of 4 years since the release of the first design, over 20,000 IRRI-developed agricultural machines have been commercially produced in Asia. Table 2 indicates the number of IRRI machines that were produced in Asia as of June 30, 1974.

Nearly half of the manufacturers of IRRI machines had not previously produced or marketed any agricultural machines. The rest had been either importing farm machines or manufacturing rice-processing or construction equipment.

In the Philippines, a sizeable ancillary industry has developed to supply components to the larger firms. The power tiller cage wheels, plows, harrows and trailers are produced by small metalworking shops and more production specialisation is beginning to occur. IRRI is also experimenting with a decentralised production approach by licensing one company to produce the power tiller transmission only. The strategy is to provide a low-cost power transmission assembly which is difficult for small metalworking shops. The small firms would produce the remaining components and market the tiller

Table 2

IRRI AGRICULTURAL MACHINERY DEVELOPMENT PROJECT, 30 JUNE, 1974

	Asian manufacturers reporting										
	1	2	3	4	5	6	7	8	9	10	Total
1. No.of machines manufactured till June 30, 1974:											
Power tiller	3,900	11	1,234	7	457	-	-	-	-	-	5,609
Batch dryer	-	-	-	35	-	-	4	-	74	-	113
Axial flow thresher	20	-	2	8	20	60	9	9	-	-	108
Table thresher	-	-	-	-	84	-	-	-	-	-	84
Grain cleaner	-	-	-	20	-	-	-	-	-	-	20
Bellows pump	-	-	-	100	-	-	-	-	-	-	100
Multihopper seeder	-	-	-	232	-	-	-	-	-	-	232
Single-hopper seeder	-	-	-	338	-	-	-	-	-	-	338
Power weeder	-	-	-	-	-	-	-	-	-	7,500	7,500
2. Production capacity utilisation:											
Current	85	NR	50	95	75	70	100	NR	95	NR	
Before start of IRRI machines	70	NR	20	90	45	50	75	NR	90	NR	
Percent change	15	NR	30	5	30	20	25	NR	5	NR	
3. No. of new workers employed	339	4	90	65	42	18	12	5	45	NR	680
4. Additional capital investment (US$)	76,000	NR	120,000	10,000	8,300	1,400	12,000	1,500	3,800	NR	233,000
5. Additional capital investment per worker (US$)	199	NR	1,330	154	197	78	1,000	300	84	NR	330

NR = Not reported by manufacturer.
US$1 = 6.60 pesos.

These manufacturers have received engineering drawings and technical assistance from the IRRI project. However, additional manufacturers are now producing IRRI or similar machines by indirectly acquiring the designs for which no data have been collected.

Prototype units of some IRRI machines have been fabricated by manufacturers in Indonesia, Korea, Pakistan, Sri Lanka, Thailand, and Vietnam. Commercial production has started in some of these countries. A total of 11 Asian engineers have received short-term training under this programme.

in nearby markets. It is felt that a decentralised production can considerably reduce the price of the power tiller.

In the Philippines, three larger companies, the Marsteel Corporation, the IGRI Industrial Sales Corporation and the Durasteel Industries Corporation, have developed tools, dies, jigs and fixtures to produce the power tiller and the axial flow thresher in larger quantities. These companies have modified the machinery designs to suit large-volume production. The original fabricated power tiller transmission casing, which was formed manually, has been replaced by a stamped casing to permit mass production. The first two companies have replaced the primary chain transmission with spur gears and are developing steering clutches for their machines. These change will no doubt result in improved machines although it is not clear how these developments will affect the smaller firms which may not be able to incorporate such changes.

Three companies, Oberly & Company, Kaunlaran Industrial Shop and C&B Crafts, have made many improvements on the axial flow threshe These improvements include the installation of oscillating screens for improved grain cleaning and the recycling arrangement for semi-threshed material.

Manufacture of IRRI machines outside the Philippines has started with the production of either the 5 to 7 hp tiller or the axial flow thresher. These machines are now being manufactured in 10 countries in Asia. A company in Ghana and another in Ecuador are also manufacturing the axial flow thresher and the grain cleaner.

In India one firm is manufacturing the axial flow thresher and the machine is being marketed for 5,500 rupees (US$ 733) without the engine. This company has recently modified the axial flow thresher for wheat threshing. In many countries tenderised wheat straw is used as cattle feed and the modified axial flow thresher does a good job of tenderising the straw. The use of this machine for wheat opens a large market for it in almost all developing countries.

In Indonesia, a company has fabricated the axial flow thresher and is currently modifying it to handle the panicle-harvested rice varieties which are quite difficult to thresh manually. Another company is now fabricating a prototype power tiller. A large fertiliser manufacturer in Indonesia is now setting up the production of IRRI machines in that country. Five engineers from this company who recently completed a short-term training in the production of our machines at IRRI have returned to start the production programme.

In Japan the power weeder has found a ready market. The Ohtake Company has been successfully producing this machine since 1970 and as of December 31, 1974, has produced 14,550 minicultivators. It developed 14 attachments for weeding and cultivation in soft paddy fields. Recently two other Japanese companies started producing similar weeders in Japan.

In Pakistan the axial flow thresher is being produced by Messrs. Habib Industries, Karachi, and is sold in the paddy areas in the state of Sind. The company is now adapting the machine for wheat threshing with the assistance of IRRI.

In Vietnam three manufacturers, the Vietnam Agricultural Machinery Company, the Than Nong Cong Ty, and the Binh-Duc Cong Ty, are producing the power tiller and the axial flow thresher. The manufacturers' response on the IRRI machines has been very encouraging in Vietnam and the information received just before the end of the war indicates that many new companies had started the production of IRRI machines.

In Sri Lanka a machine shop co-operative, the Nugegoda Medium-Scale Engineering Co-operative Society, which has 22 members, recently started producing the power tiller. Each machine shop produces some parts of the tiller and the machine is assembled and marketed by the co-operative. The Sri Lanka Government recently organised over 1,000 light industrial co-operatives and if this initial experiment succeeds, more co-operatives will soon be producing IRRI machines in the country.

In Taiwan, one firm is producing the batch-type dryer and is marketing it as the 'IRRI/TARI Dryer'. The Taiwan Agricultural Research Institute (TARI) is our subcontractor and has been responsible for introducing this machine in Taiwan.

In Thailand, two manufacturers, J-Chaorenchai in Ayudhaya, and the Anusarn Company in Chiengmai, are producing the IRRI tiller. Three manufacturers have started to produce the axial flow thresher and other manufacturers in Thailand are highly interested in it.

PROGRAMME IMPLICATIONS

The increased production of the IRRI machines has resulted in considerable new industrial employment in the countries where such machines are being produced. As of December 31, 1974, seven manufacturers in the Philippines had directly hired more than 700 additional workers in the manufacturing operation at an additional investment of US$ 200 per job. Many additional jobs in material procurement, management, marketing, and servicing operations have been created. Many of the IRRI machines are being used in contract operations and their overall impact on employment is difficult to estimate.

In the early stages, most IRRI machinery manufacturers were located in the Greater Manila area, but as the IRRI machines became popular, many firms in the smaller towns started production, thereby creating employment in the provincial areas. IRRI is now encouraging production in smaller towns by limiting the release of the machinery design to only a few manufacturers in the Metropolitan Manila area.

Some of the public policies in the Philippines were changed as a result of the local production of IRRI machines. The tariff on small air-cooled engines was reduced and a higher tariff was levied on imported power tillers to encourage local manufacturers. Many IRRI machines were approved for International Bank for Reconstruction and Development loan financing and local banks now provide loans to farmers for the purchase of such machines. Interestingly IRRI did not make any representation to the Philippine Government for these changes in policies. These changes occurred after some IRRI machines gained popularity in the local market and the government realised the need for encouraging local farm machinery industry.

While the machinery development programme was initially directed toward meeting the need of the intermediate-size farm holdings of approximately 2 to 10 hectares, it is becoming clear that the IRRI machines are being used for contract operations and are benefiting even the farmers whose holdings are less than 2 ha. The power tiller and the axial flow thresher are being widely used by farmer-owners for contract operations to provide services to neighbouring farms or by tillage or threshing service contractors. The contract operators are providing a much-needed service to small farmers who normally cannot afford to purchase the machines. This makes modern mechanisation technology available to farmers who would normally have continued to use traditional methods.

The farm machinery marketing structure in the Philippines is undergoing some rapid changes. Most large companies have their own sales organisations with dealership networks in the major rice-producing areas of the country. In the case of the small manufacturers, often the proprietor initially handles the sales function and markets the machines directly to farmers in selected areas. As sales increase, these small manufacturers start to sell through independent farm equipment dealers or develop their own sales organisations. Because of increased farm machinery sales, many independent farm equipment dealerships are now being set up in most small towns and a national network of independent machinery dealers is slowly evolving.

The project has been instrumental in stabilising the prices of some of the imported power tillers in the Philippines. One well-known 4.5-hp imported power tiller, now sold at a rather attractive price in the Philippines, competes very well with the 8-hp IRRI power tiller. Such competition with imported machines is healthy because it provides local manufacturers an incentive for improving the machines they produce.

The local production of agricultural machines has resulted in considerable savings in foreign exchange. In most IRRI machines, the engine, bearings, and chains are imported components. These components are often less than 20 per cent of the total cost of the machine.

Some Philippine manufacturers of IRRI machines have started to export their products to Malaysia and Indonesia and are looking forward to increased export markets.

CONCLUSION

The experience gained from the IRRI machinery development programme has many important implications for the technology transfer and industrial development process in the developing countries. The programme has demonstrated that it is possible to develop an indigenous industry in the less developed countries through carefully tailored research and development programmes in public institutions. Low-cost, demand-oriented products that could be economically manufactured with existing simple production processes should be developed. These are critically needed by the small-scale industry sector. The programme was sharply focused on the specific research and development needs of the small metalworking firms and it assisted such manufacturers directly without going through the normal industrial assistance institutions in the public sector.

The programme's strategy was to focus attention on the establishment of an indigenous farm equipment industry rather than on conducting a wide variety of research on agricultural mechanisation. The programme was highly market-oriented and in executing it, IRRI gave the needs of the farmers and those of the manufacturers appropriate consideration. The programme's leadership guarded against the tendency to conduct research for knowledge and continually maintained the focus on hardware development.

The programme emphasises the importance of pre-project evaluation and market studies. In the product development process, factors that bear heavily on the acceptance of new technology, such as the end-users' needs, production capabilities, and the economic and industrial structure of the society, were fully kept in mind.

In the selection of technology, many social scientists emphasise the socio-economic objectives but quite often disregard the most important factor of market acceptability. It would be unrealistic to introduce a technology, no matter how effectively it may meet the social objective, if it cannot become commercially viable in a society.

The popular socio-economic criteria of employment generation, resource utilisation, and income and wealth distribution are important, but secondary to the market factors. The significant role that the market signals play in the assessment, development and transfer of a commercially viable technology has been well demonstrated by this programme. Whatever the social and economic benefits were, the natural outcome was the introduction of a commercially viable technology.

The programme has demonstrated that agricultural mechanisation based on the Japanese pattern in which small farm machines are individually owned can be successful in the developing countries if appropriate machines can be made available to the farmers at reasonable prices. The indigenous production of farm machines can be instrumental in lowering the cost of mechanisation in the developing countries if existing low-volume production technology is used in the manufacture of such machines. Apparently there is considerable entrepreneurial potential in the developing countries but critical technical assistance is needed to develop it.

While metalworking shops did not develop new machines, when faced with competition they were able to substantially improve the designs originating from IRRI. As soon as a machine demonstrated some commercial potential, some manufacturers demonstrated the capability of developing fairly sophisticated machines comparable to imported ones. Providing local firms with simple machinery designs seemed to overcome a psychological barrier that had previously kept them from manufacturing some of the imported designs. The selective approach to industrial assistance offers considerable potential when applied to many other industrial sectors in the developing countries.

X. INTERMEDIATE TECHNOLOGY AND REGIONAL DEVELOPMENT IN THE PHILIPPINES

by

Rufino S. Ignacio*

THE UNIVERSITY AS A FOCUS FOR REGIONAL DEVELOPMENT

Mindanao State University (M.S.U.) is located about 500 miles from the capital city of Manila, in the midst of a culturally distinct and socially deprived area of the Southern Philippines. It is committed to the integration through education of the cultural communities of Mindanao, particularly the Muslims, into the national body politic. It is also committed to providing the necessary professional manpower for the development of the island.

The growth of M.S.U. has been phenomenal. It started in 1962 with 828 students and a budget of 700,000 pesos ($1.3 million), in 1975 it had some 7,000 collegiate students and 8,000 high school enrollees and its budget totalled 60 million pesos ($10 million). The main campus is located in scenic Marawi City, in the province of Lanao del Sur; external units were gradually established in Lanao del Norte, Cotabato, Sulu, Tawi-tawi, Misamis Oriental, and Davao. These external units have programmes geared specifically to the educational and cultural needs of the area they serve: the Iligan Institute of Technology for instance offers various 3-year technical programmes for the burgeoning industrial complex of Northern Mindanao, and the Sulu College of Technology and Oceanography in the province of Tawi-tawi gives courses in deep-sea fisheries and oceanography. A string of community high schools exist at strategic locations to better equip young Muslims for entry into college. The standard of scholarship is generally high since the students, Christians and Muslims alike, are well selected from the different high schools in the region. Various remedial and special programmes have been devised to assist the less gifted students.

Young as it is, Mindanao State University has proven its worth as a school of good standing and in some programmes like engineering

---

* The author is Vice-President for Academic Affairs of Mindanao State University in Marawi City, Philippines.

and education, could compare well with the best universities in the country, on the basis of criteria such as the quality of faculty and students, faculty pay and incentives, research facilities and scholarship, performance of alumni, and the flexibility and relevance of its programmes.

The university is developing very well as an instrument, or venue, for the direct development of the region. Its informal programmes, alongside and in co-ordination with the core academic programmes, are designed to meet the immediate needs and demands of the region. M.S.U. has informal programmes on tourism, adult education, executive development, public leadership, science training for elementary and high school teachers, manpower training in engineering, agriculture, forestry and fisheries, and others. An integrated programme for the development of small and medium industries was established in 1974 under the name of Regional Adaptive Technology Centre (RATC).

THE ORGANISATION AND PROJECTS OF THE RATC

The RATC is a university-based organisation which is directly concerned with the enhancement and growth of small and medium-scale industries in the region. The Centre's programmes focus on appropriate technology, entrepreneurship, and the formulation of public policies for industrialisation. These thematic thrusts have resulted from a long association and close interaction with the Technology and Development Institute (TDI) of the East-West Centre in Hawaii. The activities of the Centre include training, consultation, research and development, surveys, and the establishment of pilot and training plants to demonstrate the viability of a new industry. Organisationally, the RATC is divided into a software and a hardware division. The software division works on the literature used in the various training programmes, conducts surveys and helps formulate the necessary delivery and functional linkages with the community and the local agencies. The hardware division undertakes research and development on technology and conducts the various training programmes for specific industries.

One of the RATC's projects on the University campus is a ceramics centre which serves both as a training centre and as a demonstration plant. A brassware centre is being set up in Tugaya, 50 miles from the campus, in co-operation with government agencies like the National Science and Development Board, the Mindanao Development Authority and the town council of Tugaya. A low-cost housing project, initiated by the M.S.U. College of Engineering, is being promoted by the RATC through training and technical assistance for the establishment of low-cost housing villages on a self-help basis. An integrated coconut charcoaling and copra drying plant, invented by a local artisan, is being improved by the University through the RATC; in co-operation with private entrepreneurs, the Development Bank of the Philippines and the Philippine Coconut Authority, the RATC is currently working out the installation of similar plants in the most suitable areas of the region. The Naawan Fisheries Station, a marine fishery research unit of the University, which has made a breakthrough in the spawning and culture of prawns (jumbo-shrimps) is training the local fishermen along the lines of the RATC concept of technology dissemination and entrepreneurship development.

TRAINING PROGRAMMES AND THE PROMOTION OF LOCAL TECHNOLOGICAL TRADITIONS

As the RATC is a new organisation, different techniques and methods of training are being experimented on; only at a later stage will it be possible to evaluate and compare their effectiveness.

a) The Ceramics Centre

The province of Lanao del Sur, as revealed by a recent survey, abounds in family-owned and family-operated cottage industries like weaving and brassware. The villagers, both young and old are craft-oriented. The ceramics industry is being introduced by the RATC because of these basic artistic skills of the villagers, the presence of abundant sources of fine clay and the projected demand for ceramics products. A family-based skills training programme was thought to be reasonable. In the Ceramics Centre, groups are divided on a family basis: children and parents work together, with the latter playing the role of work leaders, and duties are distributed among members of the family. Training is conducted in the local dialect and when participants do not know how to read or write, illustrations are used.

The first batch of trainees have yet to 'graduate'. The RATC plans to loan out the potters' wheels which the trainees are now using in the Centre, and is envisaging the possibility of lending them some glazing materials to produce ceramics wares from their own local clay deposits. The artisans could then bring the un-kilned wares for treatment to the large kiln of the RATC, and the fees for this service will be paid after the sale of products.

The longer-range plan of the RATC is to get the villagers to organise production and marketing co-operatives and to build community kilns of their own. The interest shown by the families in their training and the quality of their products and designs give the RATC staff much hope for introducing the industry in the province.

b) The Brassware Centre in Tugaya

Brassware manufacturing is an age old industry among the Mananao Muslims but has never really developed on a large-scale because of the low quality of products (the artwork however is excellent), the lack of capital, faulty marketing and other problems. Using its own resources and a grant from the National Science and Development Board (NSDB), the RATC is conducting a research project to improve the manufacturing process and find solutions to the problems of the industry.

A Brassware Development Centre, aimed like the Ceramics Centre at training and production, is planned to be built as soon as possible in Tugaya, the town of the brassware artisans. The Centre will serve to demonstrate the improved yet simple new technology of brassware manufacturing. Similarly, marketing co-operatives are in the process of formation. Financial inputs into the Centre are coming from the NSDB and the Metal Industries Development and Research Centre. As in the case of the Ceramics Centre, training will be conducted in family groups.

THE DEVELOPMENT OF INDIGENOUS INNOVATIONS

a) The coconut charcoaling and drying plant in Iligan

Coconut charcoal and dried coconut meat (copra) are two major products of the Philippines and there is a tremendous domestic and foreign demand. In concert with the local inventor of an integrated coconut charcoaling and copra plant, the RATC has been working on proposals to improve the invention and diffuse it within the country. The RATC discovered that this plant is the only one of its kind in the Philippines and possibly in the world. It had the technology patented in favour of the inventor, an unschooled but very experienced man in his fifties. The invention, according to experts of the Philippine Coconut Authority (PCA) and the United Nations Development Programme (UNDP), could revolutionise the coconut industry in the country if used widely on the farms.

Manufacturing plants of this type could be installed in strategic areas and operated on commercial basis. Proposals to this effect are presently being prepared by the RATC in co-operation with the Department of Industries. The staff has prepared the financial projections and plans, and has even completed the incorporation papers for two groups of farm-owners who wish to have the plant on their own holdings. The PCA also plans to install a similar plant in its demonstration site in Leyte Island, 200 miles north of Mindanao.

Simple training packages on the maintenance, repair and upkeep of the plant, and lessons on the formation of co-operatives are programmed by the RATC in co-operation with appropriate government agencies. Local entrepreneurs are also given guidelines on the plant's economic and financial aspects.

b) Low-cost housing technology

The M.S.U. College of Engineering has done research on low-cost construction materials for a number of usages, including housing, feeder roads and irrigation. Building blocks made from soil mixed with cement and bamboo roofing have been explored quite successfully in the laboratory. The RATC, in co-operation with the College actually built a 10-house village on the campus using soil-cement and bamboo roofing. The idea was to encourage local people to use such materials. Students and trainees from various localities did the construction; both groups got trained in the technology, and at the same time the project saved a lot of money for the university.

A much larger village of 50 cottages is planned to be constructed on a self-help basis in the remote town of Lumba-Bayabao, 50 miles from the campus. Low-cost feeder roads using these new materials developed in the laboratory will service the area. The RATC staff and the researchers of the College are in charge of the project. Housing-related industries are made aware of the project.

c) The jumbo-shrimp industry

In 1973 the M.S.U. Fisheries Station in Naawan (60 miles from the main campus) made a major breakthrough in the spawning and rearing of prawns under controlled conditions. The experiment has tremendous economic implications to the country since the jumbo-shrimp is a major commodity food item with a demand that far exceeds supply. The Fisheries Station has graduated to commercial-sized hatcheries that are capable of supplying the fry requirements of the region, but the new technology must be taught to the people. What the RATC is doing, concurrently with the Fisheries Station personnel, is to spread the good news, convince pond-owners to convert their present fishponds to prawns or open new ones, and invest some of their money in the new technology. Eventually, marketing co-operatives will have to be formed, and storage facilities and other services will have to be developed.

One offshoot of this project is a much larger hatchery-research complex built by the Philippine government with financial inputs coming from six other countries in Asia. The complex is located on the island of Iliolo, 200 miles north of Mindanao.

d) Other projects

The RATC is currently looking into the manufacture of simple machines using junk vehicle parts, food processing equipment and low-cost scientific instruments.

THE GROWTH PAINS OF AN INTERMEDIATE TECHNOLOGY CENTRE

The above summary of the RATC's activities after one year of operation suggests that it is indeed an active organisation. But as with any new institution there are some growth pains, and a number of problems remain to be solved.

a) The place of RATC in the university

Some hard-core academics in the university tend to consider the RATC as a low-priority project and several deans of colleges think that the RATC's activities could very well be done by other government agencies and not principally by the M.S.U. It is true that there are government agencies responsible for the development of small and medium-industries in the country, but their resources are too centralised in Manila and there are not many meaningful projects in Mindanao. Added to this predicament is the precarious peace and order situation in the island. With this in mind, it can be argued that the university is duty-bound to do something for rural industrialisation and that the RATC is the most appropriate instrument. Linkages are being established with Manila-based agencies like the Department of Industry, the National Cottage Industry Development Authority and the UP-ISSI (Institute of Small-Scale Industries of the University of the Philippines). At the international level, the RATC has tie-ups with the East-West Centre in Hawaii and the International Development Research Centre in Canada.

b) Curriculum development

The integration or introduction of courses on small-scale industrialisation, entrepreneurship and similar problems into the curricula of the colleges is promising but not easy to achieve. Some headway has been accomplished in this regard. The research thrust of the College of Engineering for instance is on the development of appropriate technologies, and the newly-created Institute of Development gives courses on small industry development. What makes it difficult to really re-orient the educational system is the fact that the existing curricula follow the standard university patterns which for most part are western-oriented. Engineering, for example, presupposes studies on big industrial plants, their design and efficiency, and does not focus on the processes and technologies obtaining in the region; as for the curriculum on business administration, it deals with big companies and large-scale investments. The practice so far has been to plug in two or three courses on small-scale industry into the already heavy curricula of the colleges. In other words, small and medium-scale industry development, as a target discipline in itself, is not yet sufficiently wide to form the basis of a complete curriculum.

c) <u>The availability of qualified personnel</u>

Qualified staff members to develop, run and maintain the RATC's projects are not readily available mainly because of the newness of the programme. A dynamic programme of staff development has therefore been established, and ten people have so far undergone crash training in Manila and Hawaii. This has got to be supported on a continuing basis. Another problem is the rapid turnover of staff in RATC: one key staff member has become vice president of a bank, and another went to UP-ISSI in Manila.

d) <u>Financing</u>

One of the crucial problems is the availability of money to support the RATC's programmes. From the way it looks, the RATC is an expensive proposition indeed: publications, travels, seminars and conferences cost a lot. It is premature at this stage to expect the beneficiaries of the RATC to plough back some of their resources into the programme, but efforts are currently being made along these lines.

At this early stage, three preliminary conclusions could be drawn. The first is that there are positive signs that the RATC is making a direct contribution to development. The second is that the RATC is a viable and relevant programme based on the university. The third conclusion is that this experiment could prove to be a useful instrument for formulating a science and technology policy on a regional level.

XI. THE DESIGN AND OPERATION OF A WATER FILTER USING LOCAL MATERIALS IN SOUTHEAST ASIA

by

Richard J. Frankel*

An appropriate technology for Southeast Asia was developed to provide potable water for rural communities using local filter materials. A series filtration system using these materials produced a sparkling clear water from highly turbid surface waters without the aid of coagulants. Effluent quality is comparable to that obtained from the best slow sand filters.

THE SITUATION IN SOUTHEAST ASIA

In Southeast Asia, economic development is closely linked with improved social welfare of the massive rural population. The technological priorities appear to be transportation, community water supply, and rural electrification, in that order. These nations have imported conventional water treatment methods (coagulation, sedimentation, rapid sand filtration, and chlorination) as their approach for providing potable water to rural communities.

For several reasons this technology has proved to be a disillusioning experience. Capital costs are high, and each plant must generally be tailored to a local set of conditions. Thus design and construction are time-consuming and require well-trained personnel. Operating costs are likewise high and operational difficulties are numerous. Based on a recent evaluation of rural water supply projects, the following problems were found typical: laboratory equipment was not available for daily or weekly jar tests to determine the proper chemical doses; operators were not sufficiently trained to perform or understand the coagulation jar test results; chemical costs were high in rural areas and operators often tried to cut back on the use of chemicals to reduce water-treatment costs; chemicals ran short, and ordering in advance or obtaining additional chemical deliveries

* Dr. R.J. Frankel is president of Southeast Asia Technology Co.Ltd. (SEATEC) in Bangkok, Thailand.

on time was always a problem in distant communities; without proper dosages the chemical coagulation-sedimentation portions of the plant operated ineffectively with the result that turbidity loads were almost entirely handled by the rapid sand filters; understanding of why or when to back-wash the rapid sand filter was generally lacking; mud bulls and short-circuiting were typical in many filters; proper sizing of sand was found lacking; and lack of sufficient operating funds often curtailed the use of chemicals and limited daily plant operation to 2-3 hours of discontinuous production. These difficulties left village leaders and villagers alike feeling cheated and deceived when what they received was seemingly an out-of-place and unworkable technology.

Thus, a proven technology in the developed world is not necessarily an easily exported commodity for the developing countries. The need for a new approach is evident in rural water supply, even in the seemingly well-proved areas of conventional water treatment. This study deals with a suggested approach for treatment of surface waters in Southeast Asia - an approach that has been under test for three years in Thailand.

The high levels of organic pollutants and colloidal particles found in surface waters of Southeast Asia contribute to the health hazards endemic in the region. The role of filtration in the total water use-reuse cycle is essential, so research was carried out to develop a simple, inexpensive filtering system, using local filter media, for the efficient removal of undesirable contaminants from water sources. The primary concern was for removal of colloidal and suspended particulates and microorganisms.

Various potential filter media were studied using local materials, including pea gravel, charcoal, coconut husks, bagasse ash, jute, ground corn husks, and rice husks. Sand was used as a control media for comparison. Single and series-filter systems were studied.

LABORATORY EXPERIMENTS AND DESIGN

A six-month preliminary study was carried out in search of efficient filtering material. Filter media was sought for both a primary, or roughing, filter and a secondary, or polishing, filter. Criteria for selection were abundance, ease of preparation and storage, filtering efficiency, service life, and low cost. Emphasis was placed on the natural state and size in which such materials were found locally in order to avoid complicated methods of preparation and to maintain simplicity in design and operation of the filters.

Use of the local filter materials as filter media for the treatment of surface waters was carried out in two stages: stage one at high-rate filtration and high turbidities in the range of 100 to 400 Jtu (Jackson turbidity units); and stage two at a slow rate of filtration with turbidities in the 15 to 40 Jtu range. Stage one was carried out to establish the effectiveness of the media as roughing filters - that is to remove a considerable portion of influent turbidity for a sustained period of time at low head-loss. Removal of other pollutants was considered secondary. In the testing of stage two, the prime objectives were to meet the drinking-water standard of clarity and to test the bacteriological removal efficiency of the filter media. When necessary, synthetic turbidity, in the form of kaolin clay, was added to the influent river or municipal water to insure a constant range of turbidity levels.

LABORATORY RESULTS

a) The primary (roughing) filter

The most successful filtering material found for the roughing filter proved to be shredded coconut husks. The raw husks are found throughout Southeast Asia and have little market value, except in areas where the husks are shredded for packing material and used in sugar making. Cost of obtaining the raw husk is generally transport cost only. In Bangkok shredded coconut husks were obtained from a coconut coir mill for US$ 0.125 per kg. In Figure 1(1) the efficiency of shredded coconut husk is shown for various depths of filter media. At the filtration rate of 1.25 $m^3/m^2/hr$, removal efficiencies of turbidity were consistently above 90 per cent. Duration of filter run increased directly with increasing depth of medium. Effluent quality was improved with greater depth of medium during the first hours of filter operation. Differences in effluent quality diminished however, as the duration of filter run increased. The filters were operated over a twelve-week period. Penetration of influent turbidity was substantially deeper than that of other media. Clogging started at the top layer, but the particulate matter penetrated deeper into the bed as the filter run progressed. The shredded-coconut-husk filter appeared to operate in a manner similar to an ion-exchange column. Once the exchange capacity of the upper portion of the bed had been exhausted, influent turbidity was removed in the lower portions of the bed. Head loss building was slow and was proportional to the depth of penetration of colloidal and suspended materials. An accumulated head loss of 1.2 m was used as a limiting criterion

---

1) Source: N. Jaksironont, Development of a Series Filtration Water Treatment Method for Small Communities of Asia, Thesis for the MS degree (mimeo), Asian Institute of Technology, Bangkok, 1972.

Figure 1

FILTER PERFORMANCE OF COCONUT HUSK FIBER AT DIFFERENT DEPTHS OF MEDIA

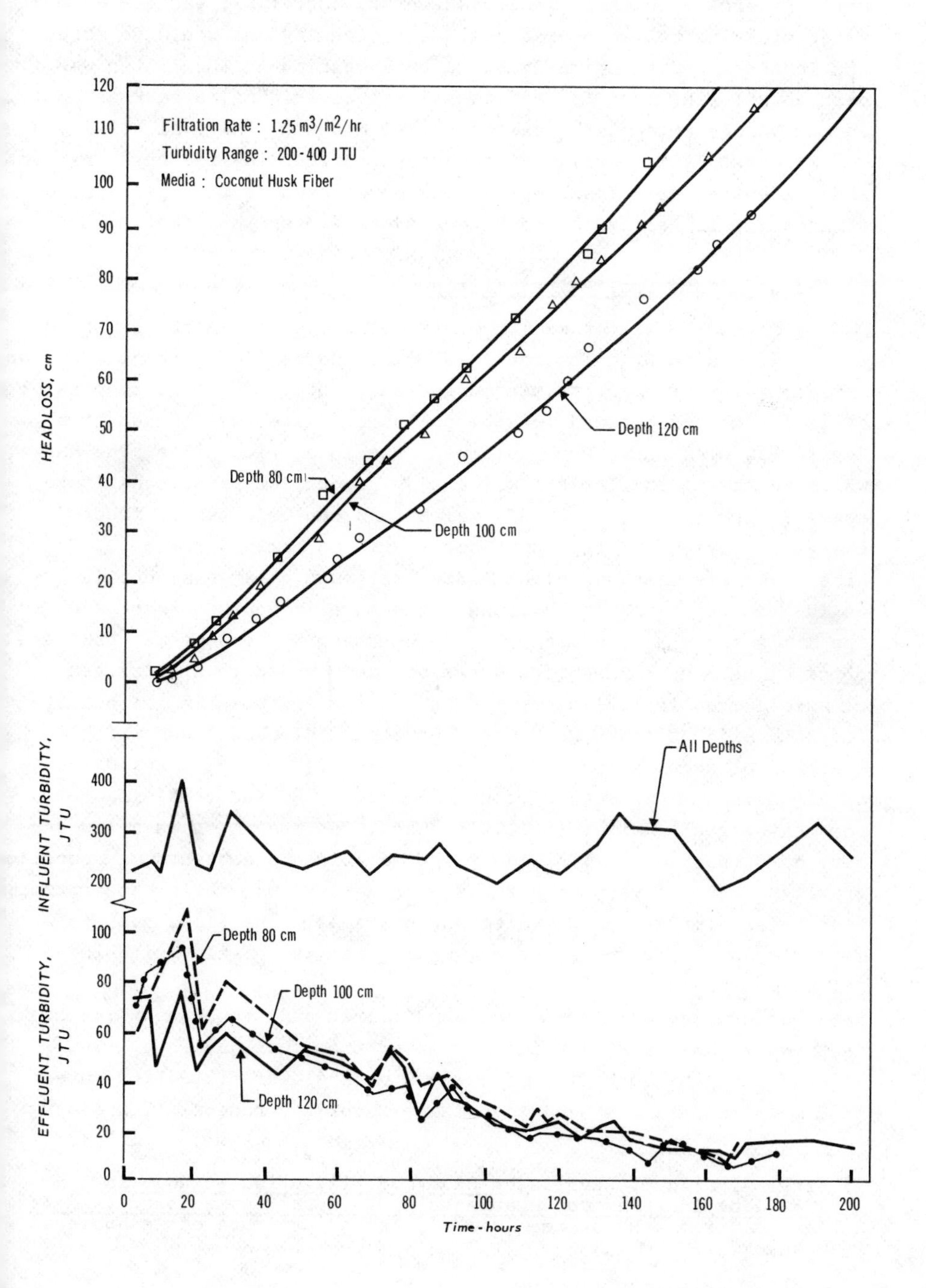

for duration of all runs. In no case did a deterioration in effluent quality occur prior to reaching the limiting head loss.

Operation revealed that the media were complicated to clean because of the deep penetration of particles into the filter bed. The coconut-husk fibres would require complex cleaning methods as well as large volumes of back-wash water. Therefore, wasting of media after clogging was more practical than cleaning and would be more applicable in rural operations. No back-washing of the filter was carried out. The need for additional valves, piping, a backwash pump, and a storage water tank was thereby eliminated. The availability and low cost of the medium also favoured discarding the husks rather than cleaning them. Ideal operations would require the operator to replace the filter media about once every 2-3 months.

b) The secondary (polishing) filter

A secondary filter medium was required to polish the water to the World Health Organisation's (WHO) standards (for turbidity colour and odour) and to improve microorganism removal. The most successful filtering material found for the polishing filter was burned rice husks. Raw rice husks were obtained from local rice mills. The husks, which represent the largest milling by-product of rice, constitute about 20 per cent of the paddy weight and can be obtained free of charge on payment of transport costs or at a nominal price ($1-2/ton) Rice husks are disposed of as waste, although a large amount is used as fuel by the mills in Thailand. Elsewhere in Southeast Asia, where the mills are run by diesel or oil-fed steam boilers, some husks are brick kilns. The burned rice husks are not burned to a white ash because combustion is incomplete at low temperatures in the boiler (less than 350°C). The husks are burnt to a blackened ash which consists of about 90 per cent silicon dioxide, 6-7 per cent oxides of magnesium, aluminium, calcium, and iron, and the remaining 3-4 per cent organic matter (mostly carbon). The medium shows a low density of compaction over a wide range of moisture contents, a specific gravity of 2.3, a very high surface-area-to-volume ratio, absorption properties similar to activated carbon, small pore size, high permeability and very low cohesion, making the ash highly suitable as filter material.

Performance of the control sand filter and the burned rice husk is shown in Figure 2(1) using raw water from the Chao Phya River with an influent turbidity of 25-85 Jtu. A slower filtration rate of 0.25 m^3/m^2/hr. was used to compare turbidity renewal efficiency

---

1) After Alberto S. Sevilla, A Study of Filtration Methods for Providing Inexpensive Potable Water to Rural Communities in Asia, Thesis for the MS degree (mimeo), Asian Institute of Technology, Bangkok, 1971.

Figure 2

COMPARISON OF SAND AND BURNT RICE HUSK AT THE SAME FILTRATION RATE AND INFLUENT TURBIDITY

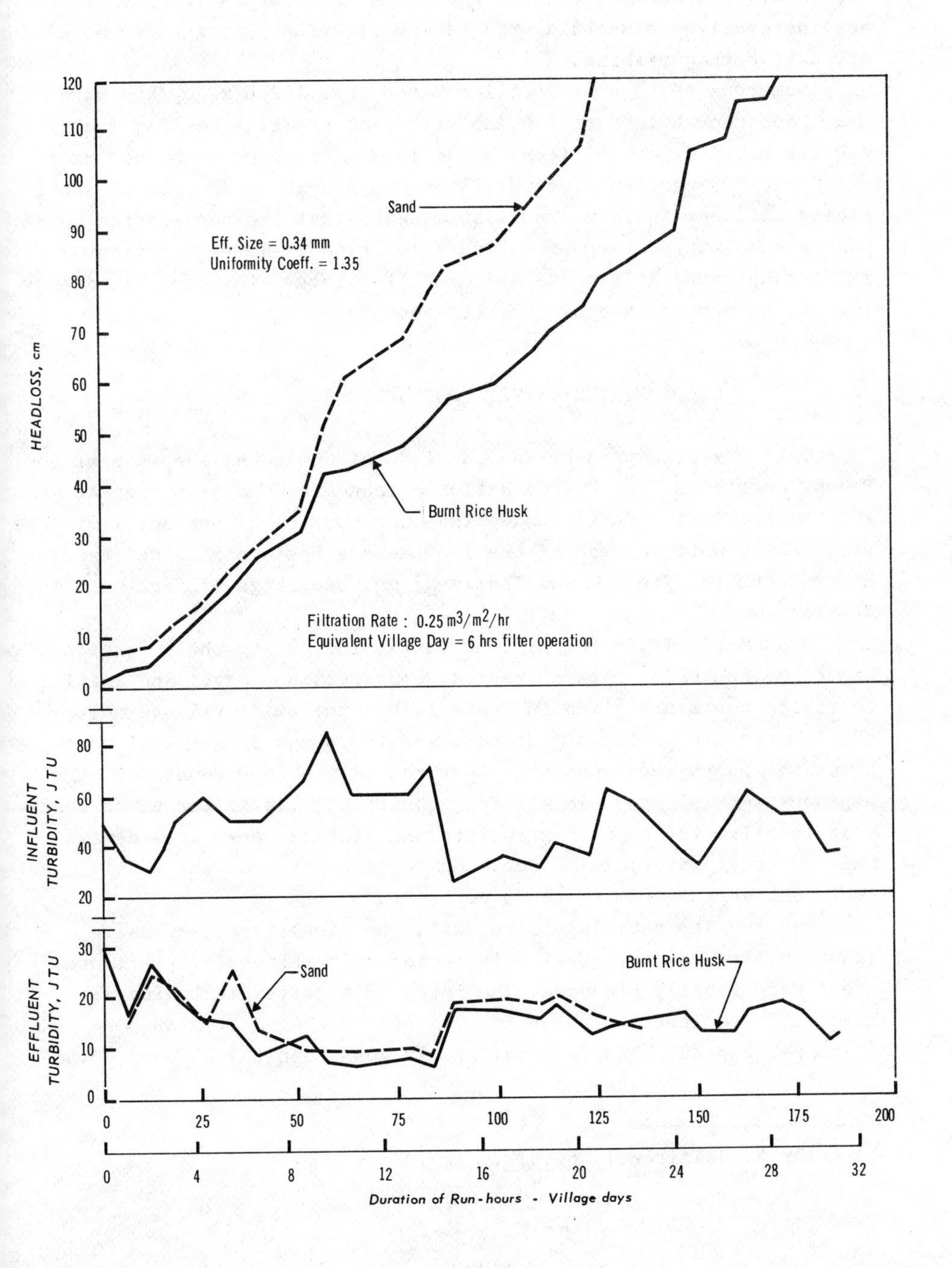

with that of a slow land filter design. Operation was stopped when the head loss reached 1.2 m. Throughout the length of run, effluent quality was excellent with a residual turbidity less than 0.5 Jtu. Penetration of the filter medium was superficial, since most of the turbidity was removed in the upper 2-3 cm. Thus, ideal operation would require scraping off the upper 5-10 cm of the medium about once every 2-3 months. Because the burned rice husks are plentiful and inexpensive, discarding the medium after use appears far more attractive than washing.

Numerous tests at filtration rates of 0.1-2.5 m^3/m^2/hr. were made comparing burned rice husks with sand as filter media. The results are plotted in Figure 3(1). In all tests using burned rice husks a 25-35 per cent longer filter run was achieved without sacrificing effluent quality. Thus, it appears that the burned rice husks can be substituted very effectively for sand in water-treatment filters operated at slow to medium filter rates, thereby eliminating the sizing problem associated with sand.

PILOT FILTER UNITS CONSTRUCTED AND OPERATED

This research project was carried out under the auspices of the Mekong Committee. The United Nation's Economic Commission for Asia and the Far East (ECAFE) funded the construction of one small village size filter unit in each of the four Mekong Basin countries (Laos, Khmer Republic, Vietnam and Thailand) and one large village unit in Thailand.

A summary of the pilot plant units, indicating their location, capacity, number of persons served, construction costs, and initial operation costs are given in Table 1. For the small village units, the largest single investment cost was the pump. In all four countries the pumps were imported; in three of the four countries, the pipework was imported. In all four countries, the filter boxes were made locally. The support structure and storage tanks were always made of local materials. Hence, the local portion of the total investment cost varied from 45 to 85 per cent.

For the large village size unit, the single most expensive imported item was the steel reinforcing bars. Essentially all other items were locally produced. The total unit costs (including all materials, equipment, labour and transport) ranged from \$0.30 - \$2.60 per capita. This is considerably less than the \$6.00 - \$8.50

1) After N. Jaksironot, op.cit.

Figure 3

COMPARISON OF SAND AND BURNT RICE HUSK AS FILTER MEDIA

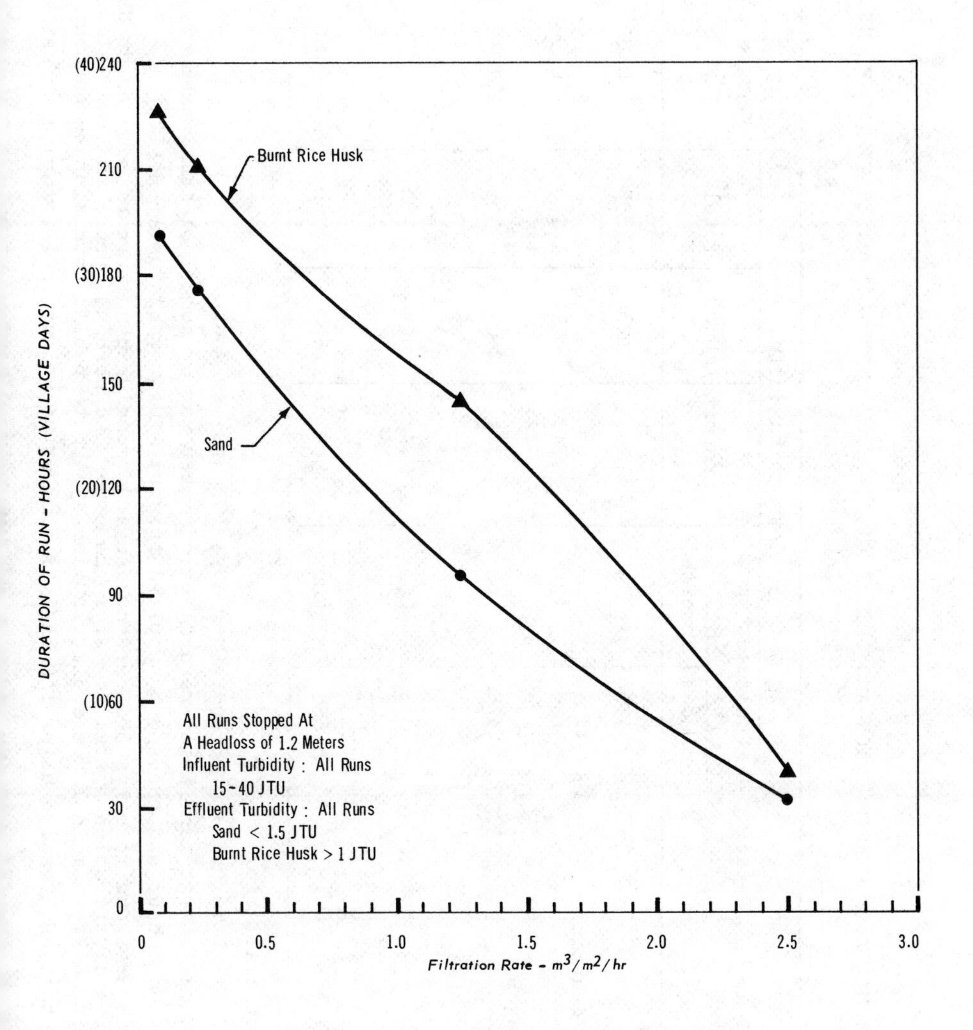

Table 1

SUMMARY OF CONSTRUCTION AND OPERATING COSTS OF THE PILOT WATER FILTER PLANTS
(in 1973 US dollars)

Location of Pilot Plant	Capacity m^3/hr	Construction Costs, $					Investment Cost per Capita	Village Operating Costs per month (1)	Operating Costs per Family per month (2)
		Equipment and Materials		Transport	Labour	Total			
		Local	Imported						
Ban Som, Thailand (830 persons)	1.25	330	74	105	126	635	0.80	26	0.30
Kambual, Khmer Republic (800 persons)	1.50	105	124	N.A. (3)	11 (3)	240	0.30	26	0.35
Hamlet Long Thong B, Vietnam (200 persons)	1.00-2.00	373	60	28	64	525	1.10-2.60 (4)	20	0.50
Nong Tha South, Laos (500 persons)	1.25	235	250	N.A.	N.A.	485	.95	N.A.	N.A.
Ban Nong Suang, Thailand (5,000 persons)	15.0	3,660	2,000	530	1,600	7,850	1.60	115 (5)	0.15

Notes: N.A. = not available.

1) Based on one full time operator 8 hours per day, 30 days per month.

2) Assuming on equal charge per family regardless of water use or number of family members.

3) All construction labour and transport supplied free by Ministry of Public Health.

4) Pilot unit was designed to serve 500 persons. An additional expenditure of only about US$30.00 was needed to complete the distribution system to serve the other 300 persons. Hence the two values of per capita cost. The pilot unit in Vietnam includes a distribution system to public fountains throughout the village.

5) Assuming "contact" filtration using 120 mg/l of alum as coagulant. Operating costs estimated.

per capita figure required to build conventional water treatment plants in Northeast Thailand(1).

Operating costs for all units remained low and varied between \$0.15 - \$0.50 per family per month. These costs included labour, fuel, filter media replacement and pump repair (and in the special case of Ban Nong Suang, chemical costs). Even at the low levels of income found in these communities, villagers indicated their willingness and ability to pay for the operation of the unit. In Ban Som, for example, the average water charge per family per month required to maintain the system was 4 Baht (US\$0.20). The average family income for comparison was 2,000 Baht (US\$100) per annum. When the village began to operate the system on its own and pay for total operating costs, the village headman set up a dual price system of 5 Baht per family per month after the rice crop was harvested and 1.5 Baht per family per month during the growing season when cash on hand was at a minimum. Insufficient time was available to observe how effectively the villagers would finance the system on their own. However, in comparison with other water systems operating in Northeast Thailand, the water charges needed to operate and maintain the water filter units were one fourth to one half the normal rates charged in similar villages(2). A village-wide survey made after four months of filter operation in Ban Som indicated that the average villager's willingness to pay was 10 Baht (US\$0.50) per family per month.

EXAMPLE RESULTS OF PILOT PLANT OPERATIONS

a) Ban Som, Thailand

Operation of the pilot filter unit at Ban Som was successful throughout the first year of operation. The media were changed first after 590 hours of filter operation (equivalent to about five months as the filters were operated only 2-5 hours per day). The coconut fibre was washed in the raw stream water and reused in the filter. The burnt rice husks were replaced with additional burnt husks obtained from the rice mill outside Khorat. The same fibre and husks were still working effectively when the media were changed for the second time some 418 hours of operation later. Water quality of the treated water was very good throughout the 9 months of operation and

---

1) C. Athikomrungsarit, Benefits and Costs of Providing Potable Water to Small Communities in Thailand, Master's thesis (mimeo), Asian Institute of Technology, Bangkok, 1971. The values relate to water treatment plant costs only, and would be considerably higher when adjusted to 1974 dollars.

2) Richard J. Frankel, Evaluation of the Effectiveness of the Community Potable Water Project in Northeast Thailand, Asian Institute of Technology, Bangkok, 1973 (Report submitted to the Environmental Health Division, Ministry of Public Health, Bangkok).

generally met the recommended WHO International Drinking Water Standards for clarity, colour, odour, and taste. The bacteriological content of the treated water varied from 0-72 coliforms per 100 ml as measured by the Millipore filter test. Removals of coliform micro-organisms varied between 25 and 100 per cent, with greater than 90 per cent being most typical. Unfortunately, many field test results were discarded because of poor sampling or incubation techniques in the field. More data are required before conclusions pertaining to bacteriological removal efficiency can be drawn.

Operational problems encountered included (a) insufficient treated water for the villages, the operator continuing to run the plant only 2-5 hours per day rather than increase the number of hours of plant operation; and (b) three pump breakdowns, totalling 16 days of plant shutdown, primarily due to the inability of the operator to repair the pump and his lack of tools to handle any such repairs (tools were later purchased). An additional 3 day shutdown was incurred when the operator went into Bangkok on personal business, and neglected to appoint a fellow villager to run the unit during his absence.

Acceptance of the water supply by the villagers was excellent throughout both the wet and dry seasons. In the dry season the filtered water was used exclusively for drinking and cooking. Villagers claimed they liked the taste of the water and had no complaints about taste or odour. The true test of acceptance was made during the rainy season to determine whether or not the villagers would use the filtered water as rain water equivalent. The results of a survey covering 100 families in the village are shown in Table 2. Eighty-three per cent of the families continued to use the filtered water for drinking and cooking, and were using it for washing and bathing, also knowing that sufficient rain water was available to meet all needs. Seventeen per cent of the families were not using the filtered water because they had to walk too far to fetch it (they lived close to the stream running by Ban Som) and had sufficient rain water storage to meet their drinking and cooking needs. In fact all families stated that during the rainy season they preferred rain water because they did not have to waste time in fetching the water (the villagers catch the rain water through roof gutters which convey the run-off to earthen or concrete storage jars). Interestingly enough, 75 per cent of all families stored both kinds of water. Fifty per cent stored rainwater and filtered water in separate jars; 25 per cent of the families stored the two waters together in the same jars. The other 25 per cent stored rainwater only.

b) <u>Hamlet Long Thong B, Vietnam</u>

The pilot filter unit at Hamlet Long Thong B worked well from

Table 2

RESULTS OF QUESTIONNAIRE ON FAMILY WATER USE IN BAN SOM
September 1973

Questions posed to some 100 families during the rainy season:

(1) How often do you use rain water for drinking?

- \- ... every day
- \- ... when it rains
- 100 ... until the supply is exhausted

(2) During the rainy season do you use filtered water from the water project for drinking?

- \- ... every day
- 30 ... sometimes
- 53 ... only when rain water is exhausted
- 17 ... do not use the filtered water because pilot filter project too far

(3) Do you store both rain water and filtered water at the house?

- 25 ... only rain water
- 52 ... both rain water and filtered water but in separate jars
- 23 ... store both waters in the same jar

(4) Apart from drinking, what else do you use the filtered water for?

- 83 ... cooking
- 83 ... bathing
- 83 ... washing dishes, clothes, house, etc.
- \- ... gardening
- 17 ... have sufficient rain water for all household uses

(5) Do you prefer drinking rain water over filtered water?

- \- ... No
- 100 ... Yes

Give reason why?

Because do not have to waste time in fetching the water. Also rain water has no colour or taste.

the time it was opened. Not one single day of shutdown was reported during the first 6 months of operations. Supervision of the unit and technical support for the operation by the Ministry of Public Health was excellent. The quality of filtered water met the recommended WHO International Drinking Water Standards for clarity, colour, taste, and odour. The data showed the dilution effects of the rainy season in the level of chlorides in the water, which are not removed by filtration. The villagers consumed water of 400-2000 mc/l chloride content during the dry season (the recommended WHO International Drinking Water Standard is 250 mc/l), and of only 10-40 mg/l chloride content in the wet season.

Turbidity levels during the high flow period of the river were higher than in the dry season months. The filter unit generally removed more than 90 per cent throughout the entire 6 months of operation. Bacteriological removal was erratic. Values of zero to more than 2,400 coliforms MPN/100 ml were reported in the filtered effluent The raw water coliforms count always exceeded 10,000 MPN/100 ml.

Acceptance of the water by the villagers was tested by an interview survey of the 42 families using the water. All 42 families responded that they used the water for drinking and cooking purposes, that they liked its taste and had no objections to its taste or odour. Nine families indicated that the filtered water had no taste.

CONCLUSIONS AND FUTURE RESEARCH NEEDS

The findings of the work to date may be summarised as follows:

a) Successful field testing of several different sizes and design of filter units have been accomplished in the Lower Mekong riparian countries. Two small village-size units, each serving several hundred persons in Thailand and in Vietnam, have operated continuously, almost trouble free, for a total testing time of 18 months. Units in the Khmer Republic and in Laos have not operated sufficiently long to judge operational effectiveness. A large village size unit has been constructed and operated in Thailand for approximately 5,000 persons.

b) Field data have indicated that the filters are capable of treating almost all surface waters successfully without the need for chemicals, except for those waters with abnormally high colloidal turbidities. However, these waters can be satisfactorily treated by the use of small dosages of coagulant chemicals. The required dosage is less than the standard coagulant demand, and that amount is sufficient to begin to induce particle agglomeration.

c) The physical quality of the treated waters was generally sparkling clear, with a turbidity of less than 5 Jtu, colourless, odourless, and of pleasant taste. Turbidity removals were generally greater than 80 per cent and as high as 97 per cent. Colour removal was similarly effective. Iron removal, where high iron containing waters existed, was sufficient to reduce all waters to within recommended limits. In general, treated water met WHO International Drinking Water Standards for clarity, colour, taste, and odour.

d) The bacteriological quality of the treated waters was generally 90 per cent or more improved over the untreated raw waters, a removal rate which compared very favourably with removal efficiencies obtained from conventional water treatment plants prior to disinfection. A small dosage of chlorine, probably less than 1 mg/l, would be sufficient to meet the recommended WHO International Drinking Water Standards. However, chlorination was not used because of the possible rejection of the water supply for drinking purposes by the villagers, and because of the questionable need for a 'pure' water supply, given the sanitation condition and water use habits typical of the villages.

e) Lengths of filter runs exceeded all expectations based on projections of laboratory findings. The variability of the turbidity loads and the greater range of particle sizes found in the field, as opposed to the more uniformly controlled laboratory raw water quality, allowed for longer duration times of filter media use prior to build-up of the design head loss without loss of filter efficiency or decrease in effluent quality. Thus filters were operated 4-5 months without a change of filter media, whereas in the laboratory the filter media had to be changed every month to six weeks.

f) Operational problems were minimal. The coconut fibre can be washed and reused. The upper portion of the burnt rice husks, some 5-10 cm depth, can be scraped off and discarded. Additional burnt rice husks can then be added to maintain the same depth of filter medium. Pump breakdown was the only significant operational problem, which accounted for some 15 days production loss during the pilot studies.

g) Considerable interest was generated, both in the villages near the pilot projects, and in several of the Ministries responsible for providing water supply to rural communities, to build and operate other filter plants of similar design.

Research Needs

Several unanswered questions present themselves as attractive research areas for future work as a result of the successful operation of the pilot plants. These include the techniques used to extend the service life of the filter media, to expand the range of waters that can be treated by the two-stage filter to include all types of turbidities of varying characteristics and particle size distributions, and to simplify design and lower capital as well as operating costs. These research areas include the following:

a) Development and field testing of the individual family size jar filters not previously tested in the pilot studies;

b) A systematic programme of research and development for studying the basic phenomena by which the coconut fibre and burnt rice husks achieve their removals of turbidity, colour, bacteria, and other impurities in water, with the objective of developing basic design criteria so that it will be possible, based on tests of a raw water, to design a filter system which will positively achieve the desired removals. The initial stages of this work would be limited to removal of clay turbidities of various types and particle size distribution. Further stages would investigate removals of bacteria, colour, oils, organic pollution, etc.

It is envisioned that the two-stage filter system, using both coconut fibre and burnt rice husks in series, with or without chemicals, represents an essentially new tool available to the engineer, competitive with the traditional rapid sand filter process, and which has a vast potential for application, both in water and waste processing, and in industrial filtration processes.

XII. THE SOLAR PUMP AND INTEGRATED RURAL DEVELOPMENT

by

J.P. Girardier and M. Vergnet*

The drought and acute famine in the countries of the Sahel have drawn attention to a very serious yet ancient problem: the lack or unsuitability of facilities for pumping water in desert areas. Professor Masson, Dean of the University of Dakar, and one of the authors of this paper (J.P. Girardier) developed a pump appropriate for these countries in 1964.

This pump runs on solar energy only. The plane solar collectors act as the hot source and the ground water as the cold source. Using this very small temperature difference, of the order of 20-30°C, an organic fluid working in a Rankine cycle transmits the energy to a gas expansion engine. This engine then drives the water pump and the auxiliaries necessary for the operation of the whole system.

The installation was designed to operate in arid areas, with minimal maintenance and supervision. This dictated certain basic parameters, particularly the use of plane collectors. These have considerable advantages: since they are stationary, they are not liable to break down, and they absorb solar radiation in its entirety. These features are highly attractive in areas where the atmosphere always contains sand in suspension.

The first solar pump was built in Dakar and was gradually improved and developed by J.P. Girardier, first in co-operation with the Dean of the University, and then within the Mengin Company and the Société française d'études thermiques et d'énergie solaire (SOFRETES).

In view of the essential requirements of reliability and endurance this equipment would have to meet, tests and experiments went on for a very long time. After the development and optimisation of the working cycle, all the components had to be tested under the most severe conditions.

---

* J.P. Girardier spent several years at the Dakar Institute of Meteorological Physics and is at present Chairman and Managing Director of the Mengin Company and of SOFRETES.
M. Vergnet, an engineer trained with the French Department of waterways and forests, worked for a number of years in Africa on problems of water systems and solar pumps.

During this development period, SOFRETES extended its co-operation to other solar energy research establishments in African countries: the National Solar Energy Board (ONERSOL) in Niamey (Niger), the Solar Energy Laboratory at Bamako (Mali), the Inter-State College for Rural Engineering (EIER) at Ouagodougou (Upper Volta) and the Faculty of Science at N'djamena (Chad).

This technical co-operation made it possible to determine the weak spots and improve those parts of the engine which were subject to the greatest amount of stress. By dint of considerable research in specific fields and particularly lubrication, this equipment is now operational under the most severe conditions.

The first practical outcome of this research was the pumping station at Chinguetti in Mauretania. This station is located some 700 km inside the Mauretanian desert and supplies water to the population of the oasis. Built in June 1973, the plant is an integrated unit which includes a school whose roof has been converted into a solar collector. This was made possible by the efforts of a group of architects who worked on the architectural integration of the buildings and designed collectors based on locally available components (e.g. cement asbestos panels of the 'canaleta' type).

THE BASIC TECHNICAL OPTIONS

Although a great deal of research and testing has been done on the conversion of solar energy into mechanical energy, industrial developments have been rather disappointing, if not practically non-existent. The energy policies pursued by the industrialised countries did not encourage the development of solar energy. In any case this substantial but low-density source of energy was considered inappropriate for countries with a very highly concentrated industrial structure. The true value of solar energy did not become clear until people became aware of the particular problems of the developing countries.

More recently, the energy crisis has considerably broadened the interest of the industrialised countries in solar power which until then had been considered marginal. In 1974 France spent 30 million francs on solar energy research and the United States has voted a substantial research budget to this effect(1).

A working party under the leadership of Professor Masson, supported by academics and architects, very soon became involved in the problems of applying this form of energy. The most appropriate region

---

1) For further details on research expenditures in solar energy, see Energy R & D - Problems and Perspectives, OECD, Paris, 1975. (Editor's note).

for the development of solar energy are clearly the desert and semi-desert areas, where sunshine is abundant and, above all, regular throughout the year. From the economic and human standpoint, these regions experience enormous problems in pumping their vital water. The transport of energy and spare parts is considerably hampered by the dispersion of sites and the difficulties of access, and costs rise accordingly.

The exploitation of solar energy is to a large extent a problem of storage. In the particular case of the solar pump, the solution is, if not elegant, at least simple: the water pumped from the aquifer is stored and can be used when required. The pumping and storage of water thus largely solves the problem of operating solar energy installations in desert regions. The design and construction options selected were therefore dictated by considerations of maximum suitability to the climatic and technical conditions in these regions.

The first choice had to be made between plane collectors and concentrators. The obvious advantage of the concentrator is that high temperatures are obtained and hence a fairly high potential Carnot efficiency (of the order of 30 per cent). However, such a concentrator uses only the direct radiation, which on the average is equivalent to 70 per cent of total radiation. The reason is that the sand suspended in the atmosphere diffuses about 30 per cent of the sun's energy on the average throughout the year. If direct radiation is to be used, the concentrator has to be reorientated either daily or at least on a seasonal basis. The turning mechanism must be extremely strong to withstand the frequent and violent winds in these areas. It must also be sandproof if it is to have a long life.

In order to ensure a high degree of concentration, the reflecting surface must be kept very clean and perfectly regular. These conditions are very difficult to fulfil in desert areas owing to the sand (cleanliness and surface erosion) and problems of transport and difficulties of construction (risks of damage).

The advantage of plane collectors is that they trap solar radiation in its entirety, an important feature when skies are frequently overcast. They operate well even in mediocre conditions of cleanliness. The collecting surfaces are made of easily assembled, prefabricated sections; transport is thus relatively easy and final erection can be carried out on site. These units are simple and robust, and as a result have a long working life.

On the other hand, the thermal characteristics of plane collectors give temperatures of the order of 70° only and the Carnot efficiency for this reason is no more than 12 per cent. Since the temperature difference between the hot source and the cold source (the water pumped from the aquifer) is very small, the use of plane collectors requires very careful study of the thermal cycle in order to make maximum use of the little energy available. In choosing the collecting system, the decisive factors were efficiency and long working life; for this reason plane collectors were chosen in preference to concentrators.

As far as the utilisation of the two sources of heat were concerned, the thermodynamic cycle was an obvious choice by virtue of its simplicity. An organic fluid moves between these two temperature sources and converts the available heat energy into mechanical or electrical energy.

TECHNICAL CHARACTERISTICS

a) Solar cells

A battery of static solar cells of a rudimentary design collects the solar radiation and converts it into heat. Each cell consists of a plate acting as a "black body". It is thermally insulated from the environment and placed under glass in order to benefit from the greenhouse effect. This plate gives up its heat through a heat transfer fluid, usually water, which circulates through the cells.

b) Thermal converter

The heat produced by the solar cells has to be put into a form suitable for conversion, in this case a thermo-dynamic potential. For this purpose, the heat transfer fluid collecting the heat from the cells gives up its energy to an organic liquid which is near its boiling point. The pressure resulting from the evaporation of this fluid at a suitable temperature operates an expansion-type thermo-mechanical converter (either a piston engine or a turbine). After expansion in the converter, the fluid is liquified in a condenser, and its heat is removed by the water.

c) Pumping

Water is drawn from a well or from a run-of-river inlet by a pump operated directly by the engine shaft in the case of the piston engine or by a centrifugal pump driven by an alternator keyed to the turbine.

d) Performance

On the average the piston engine can deliver 30 cubic metres of water a day at a head of 20 metres. This equipment can easily be adapted to other heads, as shown in the following table:

Heads	10m	15m	20m	25m	30m	35m	40m
Average Hourly Delivery	12	8	6	5	4	3.5	3
Average Daily Delivery(m^3)	60	40	30	25	20	17	15

The first turbo-alternator supplying electricity directly is currently in service in Mexico. It has an installed capacity of 25 kW, and provides water to the town of San Luis de la Paz. Solar power stations rated at 50, 100 and 150 kW will shortly be built in Africa and Latin America. The design of this equipment was also governed by a concern for reliability and simplicity of operation. All the components are derived from equipment which has operated satisfactorily over a long period.

FIELDS OF APPLICATION AND THE DIFFICULTIES OF INNOVATION

Setting up a rational water infrastructure is a basic feature in any policy for economic and social development. The use of solar-powered pumping equipment allows for a new approach to the delicate problems of supplying water for human needs, livestock and irrigation. These sealed systems are highly reliable and do not require maintenance which in any event would be difficult to provide without skilled staff; operating costs are low which is an advantage for a community whose resources are limited. These qualities are essential for equipment installed in the course of development programmes for arid regions.

a) Village water supplies

The solar pump provides a simple solution to many economic problems but its introduction does bring about fundamental social changes.

In the Sahel area, the job of getting water by traditional means usually occupies the children aged between 6 and 15 for a large part of the day. They and a few old people gather round the well where discussions take place in a kind of traditional school. Installing a pumping station in a village released the children from this job and breaks this traditional bond. It seemed important therefore to associate the installation of a pumping station with the creation of a school to replace the 'informal' school which developed around the well. The group's architects, J.M. and G. Alexandroff, studied the problem and succeeded in integrating the solar installation and the school, primarily by fitting solar cells to the roof of the newly-built school. (The roof is composed of rigid 'canaletas' panels). This integration has a twofold advantage: it cuts capital costs by using the roof as a collector, and it appreciably reduces the temperature inside the building, by 6-8°C at Chinguetti.

Another application of village pumping equipment is for improving bush dispensaries. These are integrated in the same way as the school; the system provides the basic sanitation needed for this type of building as well as keeping it cool. The possibility of introducing a small amount of refrigeration for improving the efficiency of these dispensaries is under study.

In Mexico, the installation of solar pumps is part of a vast programme for colonising the arid areas in the north of the country. Each new pump becomes a development pole around which the peasants rapidly gather and organise a community life.

The induced effects of a new solar pump, whether architectural or sociological, is attracting increasing attention from national authorities and is now being systematically investigated whenever regional development plans are being drawn up. Fairly advanced research is in hand in Senegal, Mexico and Brazil.

b) Water for pasture

The policy of establishing central water points for large grazing areas in the Sahel has been in most cases a failure and it appear preferable to establish a larger number of widely scattered small water holes. The fragile ecological equilibrium in these areas rapidl breaks down as a result of the excessive density of livestock around the water hole. The flora degenerates and in the present state of drought this phenomenon is virtually irreversible.

Pumping stations, which are completely self-contained and highl reliable, make it possible to build up a more balanced pattern of water holes, which give due regard to the constraints of epidemiolog and, particularly, agrostology. In addition to pastures deserted through lack of water, these stations can also serve as supply point for goods traffic or transhumance.

c) Water supplies for agriculture

An essential problem of development in the Sahel countries is to increase agricultural production. Soil studies have revealed an unlimited water potential which can only be exploited through pumping. Solar powered pumping stations are particularly well suited both to large settlement areas, which require capacities of 50, 100 or 150 kW, and to small village-type development projects.

OBJECTIVES

The first stage of developing pumping installations based on solar energy is coming to an end with some 30 stations currently completed or under construction in Africa and Latin America (Upper Volta, Mauritania, Chad, Niger, North Cameroon, Mexico and Brazil).

The feasibility of these systems is now proven and two new approaches have been adopted. The first is to improve them and to produce them on an industrial scale. This task is being undertaken within SOFRETES, enriched by new and important partners: the Régie Renault through its subsidiary Renault Moteurs Développement, with its marketing expertise and its industrialisation know-how; the Commissariat à l'Energie Atomique with its important interests in heat exchange and conversion; and finally the Total group which will work in particular with the Technigaz Company on aid conditioning, refrigeration and heating.

The second approach is that of the transfer of technology. The design options which were chosen (plane cells, thermodynamic cycles of the steam engine type) are paving the way to the gradual development of local manufacturing. Senegal and Mexico are already considering building 50 per cent of these systems locally.

The research co-operation with the different countries will have to be extended to cover further modifications and improvements and to use existing industrial facilities or those which are in the process of being set up.

CONCLUSION

Life is gradually disappearing from these immense regions despite the fact that they have an inexhaustible source of energy at their disposal. Underground, they have in abundance the water which is lacking on the surface.

They do, however, lack the appropriate means to extract water and the industrialised countries have not for the most part tried to meet this need. The development of these regions could be greatly facilitated by new installations which have been designed and built specially for them.

XIII. GARI MECHANISATION IN NIGERIA: THE COMPETITION BETWEEN INTERMEDIATE AND MODERN TECHNOLOGY

by

P.O. Ngoddy*

INTRODUCTION

Gari is a dehydrated food product made from cassava. Of the traditional staple foods of West Africa, it is certainly one of the most widely known and it is eaten particularly among the communities along the coastal belt, where it is known to contribute as high as 60 per cent of the total calorie intake(1). The present pattern of production in Nigeria has evolved over the past 75 years and a reasonably stable equilibrium in the division of labour and the structure of rural-urban trade relations has been established. However, two significant sets of factors threaten to disrupt this state of affairs. The first is a rapid increase in the demand for gari in the country in the face of what appears to be a static, or even a contracting supply. The second is the imminent likelihood of a partial or total mechanisation of the production process. These two developments are interlinked, with the first generating the second.

In a free economy, and in the absence of any form of governmental intervention, the potential for innovation is related to the opportunity for earning a considerable private profit in mechanising what is at present a labour-intensive method with a low productivity. Given the wide divergencies which exist between net private profit and social benefit, innovation in this case assumes a great importance not only because of the pre-eminence of gari in the socio-economic framework of the country, but also because gari is the first staple food in the processing of which mechanisation is being introduced on a significant scale. What happens in gari mechanisation could set important precedents in the processing of other basic foodstuffs.

---

* Dr. Ngoddy is Senior Lecturer in Food Engineering and Processing at the University of Ife in Nigeria.

1) W.O. Jones, Manioc in Africa, Stanford University Press, Stanford, 1959.

We are faced here with a classic dilemma of contemporary growth in a developing economy. A technological innovation is generated in response to a genuine national need. The introduction of such an innovation is accepted as one instrument for accelerating economic growth while achieving a measure of modernisation in the local society. However, each step in the innovation process implies a series of trade-offs which must be made in social terms, and the challenge is to maximise the overall benefits of the innovation while minimising its ill-effects.

It is in quest of an answer to this dilemma in one specific case that an interdisciplinary research project was initiated at the University of Ife in 1972 to examine the impact of innovation in gari production in a number of spheres of the Nigerian economy. This project was undertaken in co-operation with the Science Policy Research Unit of the University of Sussex and is receiving general support from the Canadian International Development Research Centre (IDRC). When completed it is hoped that a set of policy proposals will be produced for the Federal and State Governments of Nigeria. At this stage, only a preliminary phase of the study has been completed and the present paper will therefore focus only on those aspects of the main issues which have been covered so far.

CASSAVA IN TROPICAL AGRICULTURE

Cassava, also known as manioc, is a widely grown tuber of the tropical world. It grows best between latitudes 20° North and 20° South but can also be found as far North as 30° and as far South as 30°. Precise statistical data do not exist at the moment on the total world or regional crop. FAO estimates have placed the 1969 crop at a world total of 91 million tons, grown on 9.7 million hectares of land(1). Africa with 36 million tons and South America with 35 million tons were reckoned in that year to be the major producing areas. The other big producing centres are South-East Asia and India.

Cassava is attractive to producers in these regions for a number of reasons(2):

a) It is a hardy crop, able to resist extreme drought, and it is attacked only by leafy mosaic. It is grown in many of these areas as a reserve foodcrop which farmers harvest when other more attractive crops (e.g. yams) are out of season or destroyed by drought.

---

1) FAO, Production Yearbook 1970, Rome, 1970.

2) See T.P. Phillips, Cassava Utilization and Potential Markets, Research Publication No.020, International Development Research Centre, Ottawa, 1974.

b) It is easy to grow. Small mounds are prepared, and a cutting of the cassava stem is planted. It is weeded once or twice and subsequently left unattended until it is harvested anywhere from twelve to twenty-four months later.
c) Its yield is high, ranking only second to plantain banana as the highest source of carbohydrates per acre of any tropical crop.
d) It has a wide range of uses, ranging from industrial starches and glues to animal feeds and processed foodstuffs. As food, it is used in a variety of forms - flour, fermented mash, gari or simply boiled.

GARI AS A FOODSTUFF

In West Africa, and in particular in Nigeria, the most commonly eaten form of cassava is gari. There are no precise statistical data on the production and consumption of gari in Nigeria. However, variou sources have estimated that the national consumption is between 1 to 1.5 million tons per year(1). This would indicate that more than 70 per cent of the total cassava crop goes into gari manufacturing. Market turnover is estimated to be somewhere between 100 and 120 million naira per year (i.e. $150-180 million)(2). These figures suggest that gari ranks among the top two most important staple foods in the country. It is without question the most important food in the South where it is often eaten once and sometimes more than once daily among the low-income, and sometimes the middle-income family groups.

There is strong evidence that the spread of cassava has depended to a large extent on the diffusion of technologies for processing it into safe edible products(3). Historical documents show that cassava was introduced to West Africa in the 16th century by Portuguese slave merchants, but it was not accepted widely as a food until three centuries later when freed slaves immigrating to this region from Brazil introduced a method of processing it into gari. With the advent of gari came a steady expansion in the production and consumption of

---

1) See I.A. Akinrele, "Nutrient Enrichment of Gari", West African Journal of Biology and Applied Chemistry, Vol.10 (1), 19, 1967; I.A. Akinrele, H.M. Joseph, R.O. Okotore and F.O. Olantunji, "Perspectives of the Protein Enrichment of Industrially Prepared Gari", paper presented at the High Protein Foods Symposium, University of Ife, 1971; and Federal Institute of Statistics, Food Section, Production of Major and Some Minor Crops 1960-1969, Lagos, 1970.
2) I A. Akinrele, M.I.O. Ero and F.O. Olatunji, "Industrial Specifications for Mechanised Processing of Cassava Into Gari", FIIR Technical Memo No.26, 1971.
3) D.G. Coursey, "Cassava as Food: Toxicity and Technology in B.Nestel and R. McIntyre (eds.), Chronic Cassava Toxicity, IDRC Research Publication No.010, Ottawa, 1973.

cassava throughout West Africa. The basic formula for converting cassava into 'farinha de mandioca' (a gari analogue still widely eaten in South America) appears to have originated from Indian ethnic groups of tropical America. It was one variant of a range of technologies devised by Amerindians to detoxicate the more poisonous varieties of cassava.

The traditional processing method consists of the following stages. Both the corky outer peel and the thick cortex of the cassava root are removed manually with the aid of a kitchen knife and the root is then grated by hand on home-made raspers into a pulp. The grated mash is dewatered on a primitive press consisting of weighted cloth bags in which the pulp is left for 3 or 4 days. During this time fermentation and de-watering take place simultaneously. Home-woven vegetable fabrics are used as a sieve to sift out the fibres and the ungrated ends and stumps. The dewatered pulp is then fried in a large cast iron pan over an open fire, with or without the addition of palm oil. The finished product is a dried granular mass with a moisture content of between 10 to 15 per cent. Besides its characteristic sour taste, one of its critical features is its ability to swell in water. Gari is usually consumed in two forms: either reconstituted in cold water and eaten more or less like a porridge or reconstituted in boiling water and worked physically to a soft dough which is eaten with vegetable soup and meat. In both forms it serves as a precooked convenience food, easy to keep and easy to prepare.

The demand for gari has increased steadily since the turn of the century and particularly since World War II. This growth in demand is attributable to the rapid rate of urbanisation and in particular to the growth in the cities of a new class of landless poor who consume income-inelastic foods. Since the Nigerian Civil War, all these trends have become more pronounced and have been reflected in the price structure. In Lagos for instance the price of gari has risen from about ₦100 per ton in 1968 ($150) to above ₦200 ($300) in 1971. This rapid rise is far higher than that of the general price index (from 123 in 1968 to 165 in January 1971) or that of the food-price index (from 117 in 1968 to 186 in January 1971(1). This disproportionate rise is directly related to the inelasticity of supply which in turn results from the place of cassava in traditional agriculture and from the relationship between agriculture and other sectors of the economy.

---

1) Report of the Federal Office of Statistics, Lagos, 1972.

THE LINK BETWEEN TECHNOLOGICAL INNOVATION AND MARKET NEEDS

For a long time following its initial introduction from Brazil, gari production remained primarily a household operation characterised by low daily output and high labour requirements. As long as the purpose was to serve the immediate requirements of the farmer's family, there was no compelling pressure for far-reaching innovations.

Things began to change with the development of the cities - a trend which can correctly be traced to the establishment of a permanent colonial administrative capital supplemented by a number of regional headquarters. These 'townships', as they were then called, constituted new centres of gravity, attracting people to new opportunities for jobs and training. However the full momentum toward urbanisation was not realised until after World War II. Once military conscription got under way in the war, the low cost and ease of preparation of gari combined to make it the most important component of the ration in army barracks. A scale of trade in gari completely unknown prior to the war got underway in an effort to supply the large requirements of the army. In order to earn some cash, small producers (village housewives in most instances) started to make more gari than was absolutely necessary to meet the family's daily requirements.

The first recorded attempts at innovation in the manufacturing of gari appears to have occurred in Francophone West Africa(1). Sometime during World War II, the French assisted the Togolese in building small 'factories' for manufacturing gari and tapioca. These were a considerable advance over the usual manual method. Each factory consisted of a small hand-fed rotary grater driven by a gasoline engine, batteries of concrete tubs for starch sedimentation, and a row of low wood-fired cookers for drying the product. The women in the villages used this improved equipment to prepare their own gari or tapioca.

The engine-driven grater is believed to have found its way into Nigeria in the 1940's. This improved technology was welcomed in a handful of areas where its superior output brought demonstrable economic advantages to the producers supplying gari to the towns.

LARGE-SCALE MECHANISATION EFFORTS AND THEIR SPILLOVER

Innovation in gari production was confined to the mechanisation of the grating stage until the early 1950's when the newly established Federal Institute of Industrial Research (FIIR) in Lagos initiated a

1) See W.O. Jones, <u>op.cit</u>.

major effort aimed at modernising the technology for gari processing. Evidence shows that full mechanisation was the goal from the outset(1). The justification for the project was to alleviate the drudgery of the traditional method, improve both hygiene and quality in the process and attain large-scale production of a basic food required to feed the rapidly growing populations of the urban centres. The aim was also to assist industrialisation in whatever way possible by generating attractive commercial opportunities which either private investors or the government could exploit.

The earliest effort was directed at the development of a mechanical fryer. Failure to produce a product acceptable to consumers as genuine gari led the Institute to the conclusion that its understanding of the chemistry of the process was incomplete. Fundamental research was henceforth undertaken to study the biochemical changes which accompany the conversion of cassava pulp into gari. Co-operative work between FIIR and Professor Collard of the Department of Bacteriology at the University of Ibadan was initiated and subsequently elucidated the full sequence of changes in the fermentation of cassava pulp(2). While this fundamental work was in progress, engineers at FIIR, somewhat impatient with the slow pace with which results evolved on the biological side of things, completed detailed plans for a fully mechanised pilot plant around 1954. However, owing to lack of funds, the implementation of these plans were abandoned till several years later. It was apparently in lieu of this mechanised pilot plant that a decision was taken to develop a completely hand-operated process which would be superior to the existing village method. A comprehensive report giving the engineering details of the hand-operated system was published in 1958(3).

Two major contributions were made by the research effort embodied in this report. The first was a detailed and simplified guide, including working drawings, for the local construction of mechanical graters. This development was quickly popularised in the gari producing communities of Southern Nigeria through the active promotion of government agencies concerned with rural development(4). Rural carpenters and blacksmiths were soon manufacturing mechanical graters and the

---

1) Report of the Department of Commerce and Industries, Lagos, 1950-51.

2) P. Collard and S.S. Levi, "A Two-stage Fermentation of Cassava", Nature (London), Vol.183, pp. 620-621 (1959).

3) S.S. Levi and C.B. Oruche, Some Inexpensive Improvements in Village-Scale Gari Making, FIIR Research Report No.2, Lagos, 1959.

4) M.Th. Zwankhuizen, The Improvement and Utilization of Copra, Cassava (Gari), Rice and Cashew Nuts Suitable for Adoption in Rural Industries, FAO, Report No.1529 to the Government of Nigeria (Eastern Region), Lagos, 1962.

motorised unit capable of grating several hundredweight of cassava per hour rapidly found its way into village co-operatives. Rural entrepreneurs were encouraged by parallel developments in the milling of maize and 'lafun' (a dehydrated piece-form cassava) where private owner-operators of disc-attrition mills were doing a profitable business operating on a 'mill-and-charge' basis.

This type of operation is now commonplace in all rural communities where gari is produced for commercial purposes. Large-scale co-operative producers have their own mechanical graters. Alternatively, privately-owned graters mill the cassava delivered by customers in the neighbourhood. Privately-owned mills of this nature, which are used in the grinding of a wide variety of food items (beans, cereal grains, yams, pepper, tomatoes, onions, cassava, etc.) must rightly be seen as the earliest and still the most important form of grass-roots food processing in many parts of West Africa(1). New arrangements to encourage the use of mechanical graters are continuing to be evolved by private owners. One interesting development which has gained increasing importance in recent years is the introduction of an engine-driven mobile grater mounted on a 4-wheel or 2-wheel trailer which is wheeled from one place to another to serve the customers.

High user satisfaction has sustained a steady climb in demand for grating equipment. As a result, a wide variety of indigenously manufactured graters are now available. The Western Nigerian Technical Company in Ibadan is the biggest manufacturer with an estimated annual output of 300 to 400 machines. There are other small companies making 20 to 50 units annually. The overall development in cassava grating thus appears to be a balanced and effective one, not only from the standpoint of economic success but particularly when viewed from the standpoint of the potential linkages with indigenous technological capabilities.

The second significant contribution made by the FIIR work was the development of a successful multiple frying range for gari. The range consisted of four frying pans in a row. In the process of frying, gari is transferred successively at 6-minute intervals from the coldest pan next to the chimney to the hottest one next to a wood-fed fire-box. Sanitation and general working conditions were improved. Two women operating the range achieved an hourly output slightly more than double the traditional fryer. It is not clear whether the multiple-range frying technique was promoted as vigorously as the mechical grater. For if in fact it was actively promoted, it remains unexplained why gari producers, and in particular the co-operative

---

1) Francis Aylward, Report to the Government of Ghana on Foods and Nutrition, ETAP Report No.1449, Accra, 1961.

operators have not taken advantage of its superior productive capacity(1). The situation is rather interesting: here is a case of two concurrent innovations with demonstrable labour productivity advantages, originating from the same source, in more or less identical circumstances. One of them makes a tangible impact, while the other turns out to be a failure.

Around 1957 the collaborative work with Professor Collard began to yield definitive results and the two-stage fermentation of gari was elucidated. Armed with this knowledge, FIIR proceeded to develop a seeding technique suitable for use in large-scale mechanised production. These first significant successes gave a much needed impetus to the project. The 1954 proposals were revised, equipment ordered and trials on a mechanised pilot plant initiated in 1960. The engineering problems proved no less difficult and protracted than the earlier fundamental research. By 1964 the hardware had undergone a series of redesigns, culminating in a pilot plant which functioned more or less as follows. The cassava roots were placed in a water-filled concrete mixer adapted with wooden mixing arms. When the arms were set in motion, the rubbing of the roots against one another and against the side wall removed the outer corky tissue. The roots were grated in an impact mill and the resultant mash was put into aluminium tanks seeded with 4-day old cassava liquor and left to ferment for 24 hours. Next, the mash was centrifuged to reduce its moisture content to about 50 per cent and then granulated and sifted to separate the particles and remove as much of the fibre as possible. The particles were 'garified' - a process involving partial gelatinisation and drying to between 6-8 per cent moisture content in a cascading louvre dryer equipped with a patented garifying section.

In 1965 this 1-ton/day pilot plant was loaned to a group of entrepreneurs in Ijebu-Ode(2). Beset with rising cassava prices, difficulties in the frying process and frequent breakdowns, the plant was returned to FIIR in 1966. The Institute, which had made cost projections on a 10-ton/day plant, continued development work at a rather low profile on the frying unit. Progress is believed to have been significantly slowed down by staff dislocations and other difficulties associated with the civil war. Consequently it was not until the very end of the 1960's that the first major redesigned frying hardware, embodying a separate garifier and a separate dryer, was supplied by Newell Dunford Engineers, a British firm with a reputation

---

1) According to officials of the Eastern Nigeria Development Corporation, efforts to promote the adoption of the multiple range fryer did not yield much results. It was believed that the equipment did not work satisfactorily. See P. Kilby, Industrialisation in an Open Economy - Nigeria 1945-1966, Cambridge University Press, Cambridge, 1969.

2) See P. Kilby, op.cit.

in dryer manufacturing. This firm became increasingly involved in the hardware construction aspects of the project since it had supplied the first louvred drying kiln in 1960. That kiln had by now undergone a series of redesigns(1). Further modifications were to follow, leading in the 1970-71 period to a 3-ton/day capacity plant still in operation at FIIR and giving quite satisfactory results(2).

At the beginning of 1969, a Lebanese entrepreneur from Gambia, Mr. Masri, approached the Tropical Products Institute (TPI) in London about setting up a starch factory(3). The TPI suggested that he might consider gari production and put him in touch with Newell Dunford, who supplied Edgar Masri and Co.Ltd. with a 10-ton/day plant based on the FIIR design. The performance of this plant was faulty, suggesting that the developmental work on the hardware (the garifier in particular) was incomplete. After a dispute between the two parties Newell Dunford undertook to bring the machine up to the specifications of the contract. This was achieved by 1972 with the active collaboration of FIIR.

A number of high level technical staff at the Institute visited Gambia during this time on the invitation of Newell Dunford to assist the intensive R and D effort in progress there. The end product of this co-operative effort was the Mark III Newell Dunford gari plant. Its estimated cost in 1972 was £55,000 sterling - a considerable mark-up from the original £13,000 which Edgar Masri and Co.Ltd. had paid for the Mark I plant in 1969. The present situation is that the Mark III plant is functionally successful although its operation is hampered by the lack of adequate supplies of cassava. There is also the problem of developing a local market for the product in Gambia.

The fully mechanised technique as embodied in the Mark III hardware is in the process of being introduced in Nigeria for the first time(4). The Ido (Ibadan) Co-operative Farming and Produce Marketing Society Ltd. with a membership of 132 farmers has completed plans for a vertically integrated programme involving 5,000 acres of plantation which will supply all the cassava requirements of a fully

---

1) See O. Adeyinka and C.D. Akran, Improvements on the 1 Ton/Day Gari Plant. FIIR Research Report No.29, Lagos, 1964.

2) I.A. Akinrele, M.I.O. Ero and F.L. Olatunji, Industrial Specifications for Mechanised Processing of Cassava into Gari, FIIR Technical Memorandum No.26, Lagos, 1971.

3) A visit to Gambia by two members of the gari project team from the University of Ife enabled us to interview the owners of the Gambian gari plant.

4) Chief Obafemi Awolowo gave a press conference on the 9th of November 1973 at the launching ceremony of the co-operative. The statement was later carried in several leading national dailies.

mechanised gari processing plant. Planting on nearly 2,000 acres of this land was completed in 1974 and the installation of the Mark III plant was in progress in 1975.

THE ALTERNATIVE APPROACH: INTERMEDIATE TECHNOLOGY

Because of the overriding commitment to full mechanisation from the start, it is not surprising that the logical progression from a completely hand-operated process through a range of intermediate developments to the fully mechanised process did not materialise either at the FIIR or collaterally with its work elsewhere. Instead, intermediate technology appeared on the gari scene as a one-sided reaction to the large scale, high cost and considerable sophistication of the technology resulting from nearly 20 years of research effort at the Institute.

The first sustained effort to develop an intermediate technology for gari processing has its roots in the Nigerian Civil War. Early disturbances in 1966 had resulted in large-scale repatriation of people of Eastern Nigerian origin from other parts of the Federation. Among these returnees were many scientists and technical people, some of whom had been involved in early developments of gari mechanisation at FIIR. Several of these scientists became part of a 'think-tank' set up by the then Eastern Nigerian Government. For much the same reasons that make-shift refineries and distilleries were developed by this group, preliminary designs on a low-cost mechanical gari fryer were initiated. Under the circumstances of the war it would have been logical not to give priority to the development of a technology for which a labour-intensive but otherwise effective alternative existed. The development was in fact postponed, but after the disturbances, the East Central State re-established this activity in a unit called PRODA (Products Development Agency). Since 1970, the work at the Agency has progressed through three prototypes.

Some of the technicians of the 'think-tank' drifted back to the Mid-West State, where in collaboration with others they became involved in the mechanisation of gari. They formed a firm called Fabrication Engineering and Production Company (FABRICO) whose primary activities were in metalworking and furniture manufacturing. FABRICO then developed a motorised gari fryer which has been in commercial operation at Issele-Uku (55 miles East of Benin) since 1971. The first sale of the gari fryer to a government-owned company was made in the East Central State(1) and FABRICO is currently developing three other machines for co-operatives and private operators.

---

1) The customer was the Agricultural Development Authority (ADA) of the East Central State Government.

Both the PRODA and FABRICO machines are of similar design. They are simple and low-cost when compared to the FIIR design (which served as a catalyzer in their development). Both machines represent a considerable advance over the traditional method: in each case the fryer can handle one ton of finished gari in an 8-hour shift. The traditional technique, by contrast, has remained a manual operation, and the only significant innovation in the last 75 years has been the introduction of mechanised grating. For the rest it has remained totally insulated from modern scientific and technological improvements.

The heart of the FABRICO machine, and its most significant innovation, is the frying unit. The man-hour output of 150 lbs is clearly phenomenal when compared with the 5 lb output of the traditional frying process. This fryer can be described as a modified horizontal feed mixer. The original inspiration for its design had dawned on one of its developers while watching a horizontal feed mixer in operation. The machine is remarkably simple. All its components are locally fabricated. It is fired with wood and housed in an inexpensive open structure.

The contrast with the ND machine in Gambia is striking. The ND machine is characterised by its relatively sophisticated technological inputs stemming from Newell Dunford's long history of industrial expertise in the drying field. The ND-MK III plant is a package of 20th century technology housed in a well built modern factory. The garifying unit is a patented innovation, but the rest of its hardware is available for purchase as standard equipment in the international process engineering market. The main thrust for its development has been of the integrative engineering type. Although most of the basic research and much of the development work took place at FIIR, the hardware as it now stands is basically an imported technology.

THE ECONOMICS OF INNOVATION(1)

As could be expected, the ND-MK III plant costs a great deal more than the FABRICO plant. There is some difficulty in valuing each one since the commercial market has not yet materialised. However, on the basis of conversations with the respective parties, we have valued the installed Gambian machine at ₦120,000(2) and the

1) This section is based on a detailed analysis prepared by Raphael Kaplinsky of the Institute of Development Studies of the University of Sussex with the assistance of the author. The full study will in due course be published elsewhere.

2) Or approximately $180,000. Newell Dunford published a feasibility projection on the MK III plant in 1971. However the Ido (Ibadan) Co-operative Project has estimated the cost at a much higher level. See Chief Awawolo's press release.

FABRICO equipment at ₦10,000(1). If anything, these figures overstate the cost of the local intermediate technology and understate that of the imported fully-mechanised technology. When operating at full capacity (three shifts per day), the FABRICO plant produces 3 tons of finished gari a day and requires 41 workers; the ND plant produces 10 tons of gari with 83 people.

Both techniques are basically efficient: the FABRICO machine has a somewhat higher productivity of capital, but this is counterbalanced by the higher productivity of labour in the ND machine, and all other things being equal, one technology is as good as the other. This first conclusion must be qualified. If one assumes that the two machines are operating not at 100 per cent capacity, but at 33 per cent (a more realistic assumption), the production costs per ton of gari are relatively higher with the ND machine(2). A number of factors account for this: the FABRICO machine's comparatively high efficiency in terms of capital productivity, the smallness of ND's advantage in labour productivity, the efficient use of raw materials in the FABRICO process, and the relatively high wages paid to workers of the ND plant.

An entrepreneur planning to set up a gari factory can choose either one of the two technologies, but the decision is far from simple. Viewed in the perspective of private optimality, his choice will depend on a number of factors, and in particular on the cost of labour and the interest rates on borrowed capital. If the plant is to be set up in a rural area, where wages are low, the FABRICO machine is clearly more profitable. In urban areas, where salaries are higher but a qualified labour force more easily available, the ND machine is a better choice. The higher the prevailing interest rates, the greater the comparative advantage of FABRICO's intermediate technology. Taking these two constraints into account (wages and interest rates), one can draw a boundary line between the two techniques and show at what level of salaries or interest rates a switch-over from one technique to the other becomes inevitable.

The choice between the two techniques can also be made on the basis of social optimality. In this case, the criterion is not the profit made by the individual entrepreneur, but the benefit accruing to the country as a whole. Measurements of social optimality tend to be rather shaky, for they are based not on market prices (for wages,

1) Or approximately \$15,000. The ADA paid ₦3,500 (\$5,250) for the FABRICO machine in 1973. Two additional FABRICO units were sold to a private owner for ₦11,000 (\$16,500).

2) At full capacity, the production cost of one ton of gari is ₦97.5 for the ND machine and ₦74.6 for the FABRICO machine. If capacity utilisation falls to 33 per cent, the costs are ₦112.7 and ₦81.8 respectively. In the first case, the ND machine's production costs are 31 per cent higher, and in the second case 38 per cent higher.

raw materials, etc.) but on 'shadow' prices and 'opportunity' costs. Furthermore they introduce a certain number of qualitative factors - for instance the social value of indigenous innovations or the need to penalise imports - which can be quantified only in a very rough way. Despite their shortcomings, such calculations are important both for the policy-maker at the national level and for the managers of a country's R and D laboratories, for they try to take into account what might be called the national interest. They also show that the social utility of an innovation is not necessarily the same thing as its private utility.

In the case of the ND and FABRICO machines, these calculations of social optimality - based on shadow wage rates, foreign exchange conservation, the value of indigenous innovation and the effects of innovation on the price of gari - all tend to show that the FABRICO machine is more attractive than the imported ND machine. In other terms, the conclusions yielded by a social optimality analysis happen to go in the same direction as those suggested by the earlier private optimality analysis. This of course does not mean that private entrepreneurs will necessarily all choose to use FABRICO machines in their gari plants: in a free market economy, it is very difficult to integrate social optimality into the price system, and business decisions are not based entirely on a rational analysis of the available economic data.

OBSTACLES AND INCENTIVES TO INNOVATION(1)

The fact that there have been some innovations in gari production in Nigeria over the past 40 years confirms that there are profitable opportunities for innovation. However, there are also a number of institutional, organisational and managerial obstacles. The FIIR suffers from the traditional weaknesses of scientific institutions in developing countries. Poor financing, red tape, excessive staff mobility and the lack of internal co-ordination are among the main obstacles to the application of science and technology. What is perhaps even more important is the fact that these weaknesses effectively prevent or slow down the development of what might be termed a "realised demand" for scientific and technical knowledge.

Of equal interest to our current concern are some of the more subtle and largely unexplored aspects of underdevelopment. There is evidence to suggest that in the case of gari, the basic simplicity of the necessary innovation may have been obscured by the professional

1) Many of the ideas discussed in this section have resulted from interaction with various members of the Institute of Development Studies, University of Sussex, England.

complexity of the work done at FIIR and Newell Dunford. The high degree of scientific and professional attainment of FIIR staff is at the basis of the high level of scientific inputs in the ND technique. The preoccupation with full mechanisation from the start is partly a reflection of the dominant national mentality at the time, but also partly a reflection of the fact that research officers confronted with a problem have delved into their vast stock of technological knowledge for an answer. Armed with a comprehensive knowledge of a wide assortment of processing operations, they were inexorably propelled in the same direction. The outcome is a technology package which appears to have given very scant attention to the environmental conditions in which it is to be used.

The obvious discrepancy between the resource-mix for which such modern technology was designed, and the actual resource-mix which exists in a developing country like Nigeria places this technology at a fundamental disadvantage. This approach to design also presupposes the ready availability of complementary production factors like workshops, materials and highly qualified manpower. Because these factors were not available, Newell Dunford Engineers became involved as co-developers. The outcome of this relationship raises a nagging question: to what extent have FIIR in particular, and Nigeria in general, gained from their share in the inventive and innovative process? The term 'gain' referred to here does not pertain primarily to property rights, although this is important, but concerns the development of an indigenous technological capacity and the ability to make it a cumulative and self-sustaining process. At the moment, the answer to this question remains at best uncertain.

In contrast to the above picture, and precisely because it lacked high-powered scientific and engineering skills, FABRICO was able to see through the 'skeleton' of the problem. As a consequence, the simplicity of the FABRICO hardware blends it naturally into the rural environment in a way that the ND technique cannot. Qualitatively, it represents a purposive development in the historical process of innovation but it does not conform to the slow pace of change which is generally characteristic of traditional technology. It has been aided from abroad in its conception but retains its indigenous character. In this sense a positive transfer of technology has taken place and there have been indisputable gains in indigenous technological capacity at the grass-roots level.

The weight and orientation of science and technology in the advanced countries might have a decidedly negative impact on the development of technologies which are appropriate to the specific requirements of the developing countries. In the gari case, there has been a double-edged 'blinding' effect which derives directly from modern engineering approaches to process design. The spectrum of available high-cost components from which sub-units were selected by FIIR and ND for the integrative design effectively precluded a consideration of low-cost, labour-intensive alternatives. In the FABRICO case, the lack of conventional engineering knowledge, particularly of modern techniques of design optimisation, resulted in a relatively efficient, low-cost and labour-intensive machine whose functional potentialities, we strongly believe, has not been fully exposed. We cannot, however, condone the ignorance implicit in FABRICO circumstances as a means of 'setting the mind free' to innovate without the sometimes stifling effects of formal technical education. The continued lack of these basic skills will, in the long run, impede the development of the innovative potential of organisations such as FABRICO.

Under present conditions everything appears to conspire to prevent developing countries from selecting, even from the existing spectrum, the capital-saving and employment-generating technologies which may be optimal. The consulting firms which they employ in preparing projects, making feasibility studies, drawing up specifications and examining tenders are all steeped in the technological outlook and traditions of the richer countries. The same is true also where local professionals are employed since their basic outlook derives from the same dominant monolithic orientation in science and technology. What is needed among other things is to politicise scientists and engineers in research institutions like FIIR to a point where social optimality takes on a new importance in their design considerations.

THE SOCIAL IMPACT OF INNOVATION

We can now attempt to evaluate the likely impact of innovation in cassava and gari production. At the one extreme are the existing traditional techniques for producing cassava and gari. The processes are very labour-intensive but there is no clear idea of the opportunity cost of this labour. The distribution of income is largely a reflection of the division of labour between the sexes (men produce cassava, women make and sell gari). The concentration of income is not great, but there are indications that distributive margins have increased in real terms, thus transferring income from producers to marketers. There is no clear idea of the saving propensities of these two groups but it is unlikely that a significant proportion of this money is reinvested. At any rate, investment is likely to be located within the immediate areas, and most of it probably goes to expanding marketing activities. Innovation is neither rapid nor widespread and the pattern of social relations, while by no means static, has evolve over the past 100 years in such a way that the production of cassava and gari has become an integral part of the social system.

At the other extreme, the fully mechanised plant is likely to have wide-ranging consequences. The plant producing 3,000 tons of gari per year requires only 83 workers and will displace a large number of people. There will also be a substantial displacement as a result of mechanisation in cassava production although the extent of this is not known. The effect of this upon other crops depends on the seasonal availability of agricultural labour. Income from the production of gari will be redistributed from women producers to the owners of the plant and its workers, and only a small number of the latter could be the same women.

In the production of cassava, income will be redistributed from peasant farmers to the owners of the plantations, the agricultural labourers (some of whom could be the same farmers) and the producers of the equipment, who are predominantly foreign. The concentration of income will increase significantly, since it is very likely that both mechanised cassava production and mechanised gari manufacturing will be in the hands of the same entrepreneurs. The growth of employment and output will depend on the investment propensity of this innovating group. The total investment will at any rate be higher than with the existing producers.

The location of new investment will depend on profit opportunities, but it is unlikely to be as locally-centred as it would be with small-scale producers. The trickle-down effect of innovation on peasant producers will be virtually zero. However, the success of this innovation will influence future innovations by the same entrepreneurs or others organised in an identical fashion. The displacement of peasant farmers will increase the speed and extent of class formation in the rural areas. If the fully mechanised plant is owned by a co-operative, it will have less marked effect on the concentration of income and the extent and location of new investment. Co-operative ownership will slow down and change the pattern of rural class formation but is unlikely to result in significant differences with regard to other factors. The organisation of successful co-operatives must be recognised as one of the main difficulties here.

Intermediate technology innovation in cassava production can take either one of two institutional forms - peasant co-operatives or core-plantations complemented by peasant producers. In each case, there is no major imperative for mechanisation, although the private profit inherent in mechanisation may make it attractive to individual innovators. Growth of output and employment will be affected by the investment propensity of the groups involved, but the core-plantation arrangement is more likely to result in investment outside the locality. Both forms of organisation will affect social relations and the process of rural class formation, but the co-operative system should be less disruptive.

XIV. APPROPRIATE TECHNOLOGY IN GHANA: THE EXPERIENCE OF KUMASI UNIVERSITY'S TECHNOLOGY CONSULTANCY CENTRE

by

B.A. Ntim and J.W. Powell*

DEVELOPMENT GOALS AND THE PREREQUISITES FOR DEVELOPMENT

In a recent conference on international co-operation in rural development in Africa(1), it was generally agreed that the basic aim of development was to raise the social and economic level of the rural populations and that in order to achieve this, there was a need for better infrastructures in fields like education, health, shelter, water supplies and employment. The conference also stressed the following elements of an integrated rural development strategy:

a) At the local level, there is a major need for detailed surveys of the economic, social and technical resources of the community. Such surveys which must be made as far as possible with full participation of the local population, are the basic way to identify the most urgent needs and to help in the implementation of new development projects.

b) Since the main occupation in developing countries is agriculture, the starting point for development lies in the diffusion of improved production techniques, the supply of more appropriate agricultural implements, a wider use of fertilisers, better animal husbandry and the provision of services such as storage facilities and supplies of improved seeds.

c) Additionally, it is essential to encourage the establishment of small-scale industries in fields like metalworking, carpentry, and food processing to support the agricultural sector and to meet the basic needs of the local population.

---

* Dr. J. W. Powell is the director, and Dr. B.A. Ntim the deputy director of the Technology Consultancy Centre of the University of Science and Technology in Kumasi, Ghana.

1) Report of the Third Sub-Regional Workshop on International Co-operation in Rural Development in Africa, sponsored by the Economic Commission for Africa (Report No. E/CN.14/SWCD/66), November 1974.

d) Continued intensive multi-disciplinary research programmes on rural development should be set up to identify the major innovation needs. The objectives, scope and nature of such programmes must in the first instance be explained to its potential beneficiaries.

e) United Nations agencies, national and international voluntary agencies could, whilst co-operating with governments, play a useful role in the promotion of rural development by undertaking and financing the feasibility studies for such projects, providing technical experts to help in the promotion of rural development, and assisting rural training centres and national information services in the collection and dissemination of information on rural development.

Essentially the conference called for an integrated approach to the development of rural areas and emphasised the fact that the choice of technology to be applied in these areas must take into account the nature of the production process, the availability of raw materials and manpower, the size of the market and other economic, social and cultural factors.

During the early years of technology transfers from industrialised nations to developing countries, much frustration was suffered by the latter group because whole industries were imported unmodified without due regard to the local economic and technical resources. The case of agricultural machinery is a typical example. After independence, a great many countries increased their imports of agricultural equipment for supposedly boosting up agricultural production. This machinery could not be used and maintained for any length of time, due mainly to the lack of technical know-how and to insufficient maintenance, not to mention the fact that most of the machinery had nearly reached the end of its useful working life. Furthermore, agricultural pilot projects which were financed by industrialised countries and directed by foreign experts used equipment and methods which were completely unknown in the project area and which met with the disapproval of the local population. Failure was the inevitable end result.

It is not surprising therefore that a call was made at the United Nations for UNCTAD to lay down some conditions for the transfer of technology to developing countries and in particular for the inclusion in any transfer of those elements of technical know-how which are normally required in setting up and operating new production facilities.

MODERN TECHNOLOGY AND ECONOMIC DUALISM

We are all aware of the growing gap between rich and poor nation and, within developing countries, of the gap between the towns and the rural areas. High rates of unemployment, massive migrations to the cities and excessive dependence on imports from the industrialise countries are common to almost all developing countries. These proble are among the many well known manifestations of what has rightly been called a dualistic economy. What is perhaps less obvious is that this dualism has a lot to do with technology, and in particular with the import of capital-intensive labour-saving technology which requires a high level of skills, a sophisticated educational infrastructure and a well developed maintenance and support system.

It is obvious that developing countries need certain large-scale industries based on modern technology, were it only to produce the goods which would otherwise have to be imported. However these industries do create problems, and their promotion must be analysed in a more critical way than heretofore. This is particularly true in the case of a country which, like Ghana, suffers from a chronic shortage of foreign currency. In that country, the cement, sugar, beer, tobacc and soap industries, which rely either on foreign raw materials supplies or on foreign maintenance support, are all operating today at less than 80 per cent of their capacity, and in some cases even much less. After several years of operation, the production level of the sugar industry for instance is still at 10 per cent of its maximum. Cane production has not managed to keep up with the demands of the cane crushers and it is doubtful if the full capacity of the machiner will ever be used before the best part of it is scrapped.

The use of highly productive modern technologies in situations where their products compete with those of indigenous crafts (e.g. textiles produced by weavers or pottery made by local potters) result in the destruction of these crafts and reduces the number of people required to satisfy the need concerned. This is often without any overall economic gain to society. Modern factories have to be located in urban centres; the little employment they offer is created in the metropolitan areas whilst their establishment causes unemployment mainly in the countryside.

The widespread adoption of highly advanced technology as a means of development in all sectors of the economy is not possible because of the constraints imposed by the lack of capital, the shortage of technical and managerial manpower, and the cost of processed raw materials.

The problems of modern technology may be met by adopting a strategy based on appropriate or low-cost technologies which involve low capital costs per work place, per unit of output and per machine, and a heavy reliance on local inputs, both in manpower and materials. The experience of the Technology Consultancy Centre in Kumasi, Ghana, shows how such a strategy might be implemented and brings to light some of the problems of diffusion of low-cost technology.

OFF-CAMPUS BUSINESS GENERATED WITH THE DIRECT ASSISTANCE OF THE TECHNOLOGY CONSULTANCY CENTRE

The Centre was set up at the University of Science and Technology in January 1972 to make available to the public the expertise and resources of the University and to promote the industrial development of Ghana. The Centre has concentrated much of its effort on the development of small-scale industries, as can be seen from the following case studies.

a) The manufacture of spider glue - Mr. S.K. Baffoe sought the help of the Centre in 1972 for the development of a formula to manufacture paper glue from cassava starch and alkali from plantain peel - two raw materials which are in abundant supply on the local market. The Centre provided the technical know-how, built a production plant capable of producing 40 gallons of glue a day and assisted the entrepreneur in obtaining a financial loan. Within a year, the entrepreneur was supplying the best part of Ghana's needs for glue, thereby saving substantial amounts of foreign currency. Additionally, this project is providing employment to some 25 rural dwellers who would otherwise be unemployed. The currently envisaged expansion of this project is likely to bring an end to imports of glue and the prospects for exports are very bright. In fact, the entrepreneur has concluded an agreement to manufacture the product in the Ivory Coast, and Senegal has placed a token order to sample the local market. This project is likely to lead to the development and adoption of other technologies. For example, there are arrangements to manufacture locally the plastic containers which have to be imported and which often run short on the local market.

The glue project is outstanding in that after the development of the glue formula, most of the remaining work (production, know-how and market research) was carried out by the entrepreneur who kept a close contact with the Centre for advice on industrial production methods and management.

b) The Schools Scientific Import Substitution Enterprises (SSISE) - In October 1973 the Centre imported two machine tools for SSISE. The first was an 8 inch centre lathe and the second a high-speed wooden handle turning machine. The latter is capable of producing wooden handles for chisels, files and screwdrivers at a rate of 200 per hour. Some instruction has been given to SSISE employees in the production of handles and the fitting to chisel blades. The use of good quality tropical hardwoods should enable the handles to be exported to Britain which has already placed a large order. The only problem now is to produce these items at competitive export prices. SSISE have also contributed enormously to the building of soap plants originally designed in the Centre and which are being installed in most parts of the country. They have manufactured all the platforms, soap moulds, cutting tables, lockers, duck boards and tressels. Another project in which SSISE has been involved is the manufacture of the broadlooms for village weavers. More than 45 of these looms, which cost about $100 a piece, have been produced by SSISE since the middle of 1973.

c) Other projects - The Centre has given assistance in various forms to numerous other people. In many cases, there was no feedback concerning the effect of technical advice or other assistance rendered. In several cases, products were submitted to the Centre for analysis. This work was undertaken in the Faculty of Science and the client was advised if the quality conformed to the National Standard and, if not, how an improvement might be made.

The chemical starch production for laundries was a typical case of the above request. A local entrepreneur presented to the Centre a sample of laundry starch prepared from cassava starch and the alkali from a wild plant. The sample was tested and found successful. The Centre then proceeded to advise on a process plant for commercial production, and market research was conducted in the commercial centres of the country. Presently sufficient orders have been received for purchase of the chemical starch and arrangements have been finalised to secure a bank loan to produce the starch on a permanent commercial basis.

Some clients ask for help to improve their existing skills, others for technical assistance to set up import-substitution production units and still others want to know where they can obtain production equipment or packaging materials. Sometimes these questions are quickly answered but in other cases they entail lengthy enquiries. Often, no satisfactory answer is available because of the difficulty of importing essential equipment or materials.

ON CAMPUS PRODUCTION UNITS AS A MEANS OF PROMOTING SMALL-SCALE INDUSTRIES

When the need for a new product is identified, but local industry is reluctant to take it up for various reasons, the Centre sometimes chooses to establish a small-scale production unit. Basically, these units have the following aims in mind:

- to train craftsmen and managers in the skills of the new industry;

- to complete product development under production conditions;
- to test the market for the product; and
- to demonstrate to entrepreneurs the viability of a new industrial activity.

So far there are three of such units in the Centre for the manufacture of nuts and bolts, pale bar soap and weaving items.

a) Nuts and bolts

In the case of nuts and bolts, as much as half a million dollars worth are imported into the country each year, and until the Centre started up production, there was no establishment in the country which produced them in any quantity on a regular basis. The little amount which was produced by local blacksmiths was usually of very poor quality. With the establishment on the campus of this production unit, which uses second hand imported machines, all the production and commercial problems associated with the enterprise have been studied, analysed and documented(1). The quality of the products is as good as the imported item, and marketing is the least problem. What remains at this stage is the transfer of the technology from the campus. Several enquiries have been received from interested entrepreneurs who, at a glance, perceive the advantages of pilot production units. The difficulty with the transfer of this technology lies with the importation of the necessary machinery. However, efforts are being made to secure import licences for entrepreneurs to bring into the country machines similar to those used in the pilot production unit.

b) Soap production

The soap market in Ghana is quite erratic. The main producer of soap in the country is Lever Brothers Ltd. who use mainly imported raw materials (oils, caustic soda and additives). At the present world-wide high inflation levels, these imported materials are rising in costs. However, the price of soap products is state controlled and consequently there are frequently times when these products cannot be produced at the controlled prices. As a result there are shortages of soap, especially in the rural areas.

The Centre therefore embarked upon a project to develop a simple plant for making soap in bars from locally available raw materials. Its plant uses caustic soda produced from local slaked lime and sodium carbonate. The former is a waste product from a local firm whereas the latter is imported, but it is easily available at prices

---

1) Technology Consultancy Centre, Second Annual Report on the Steel Bolt Production Unit, Kumasi, April 1974.

much lower than those of imported caustic soda. A pilot soap plant has been built on the campus and it is being used to instruct potential soap makers who are planning to install plants in different parts of the country. Additionally, a soap factory has been built by the Centre to house a soap co-operative society in a village five miles from the campus. A follow-up of this project at the village level is the establishment of an oil palm plantation and an oil mill to feed the factory. The successful implementation of the plantation and the mill should completely eliminate unemployment in this village.

c) The Broadloom Weaving Project

The traditional loom of the Ghanaian weaver is capable of weaving a width of 4 inches only. The development of the 40 inch broadloom on the campus arose from the need to increase the productivity of the village weaver. The production unit on the campus has trained a number of weavers on the broadloom who have returned to their villages. The cost of the broadloom with all the accessories is only about $150 and consequently the rate of transfer of this technology has been much more rapid than it has been for nuts and bolts and soap production. However, the high cost of yarn makes the weaving venture rather unprofitable.

ECONOMIC ANALYSIS OF THE CENTRE'S PROJECTS AND THE NEED FOR TECHNOLOGY PACKAGE

The following table shows some of the basic data on the projects undertaken by the Technology Consultancy Centre or with its direct assistance. The glue and nuts and bolts are manufactured in existing buildings and the figures have been adjusted to make provision for housing these projects in a factory, as is the case of the soap manufacturing co-operative.

The small soap plants and the glue and weaving projects are typical cottage industries, as may be deduced from the figures of this table. They all need less than $1,000 of investment in plant and machinery per worker and a total of $5,000, enough to establish a small plant. They all rely on local supplies of raw materials and consequently the horizontal transfer of the technology is easy and rapid. Another point is that the profitability of the glue and soap manufacture is relatively high and offers a useful incentive for a quick transfer of the technology. On the other hand, because of the high cost of yarn, the profitability of the weaving project is quite low.

The nuts and bolts manufacture is quite clearly a capital intensive venture, relying, as it does, on the importation of machine

BASIC DATA ON THE TECHNOLOGY CONSULTANCY CENTRE'S INDUSTRIAL PROJECTS

	Glue (1)	Nuts & Bolts (1) (2)	Soap Factory	Small Soap Plants (3)	Weaving Project (3)
Total Investment Required (Land, Buildings, Plants & Machinery)	50,000	150,000	50,000	4,000	4,000
Investment in Plant & Machinery	6,000	100,000	8,000	3,000	3,000
Total Employment (Number of workers)	50	45	15	4	24
Total Investment per worker	1,000	3,300	3,300	1,000	167
Investment in Plant & Machinery per Worker	120	2,200	530	750	125
Annual Turnover	300,000	144,000	200,000	50,000	24,000
Turnover per worker	6,000	3,200	13,000	12,500	1,000
Profitability	Good	Fair	Good	Fair	Poor

All figures are given in Cedis; 1 Cedi (¢) = $ 1.15

1) Additional capital costs in building have been added to the present level of investment.
2) This level of production, which will provide 30 per cent of the country's requirements, is about ten times the present level of production.
3) Production is at the cottage industry level, and only needs a shed as shelter.

tools, even though the raw material inputs are locally available. There have been several inquiries from entrepreneurs who want to go into this venture but until now the technology has not been transferred. This is clearly due to the large initial capital requirements, and in particular to the need for foreign exchange. Important efforts are nevertheless being made to establish this industry outside the campus.

Three guidelines have helped in the rapid transfer of technology from the University Campus to the public. The first one is that the request for a study into a particular technology should come from an entrepreneur who must keep in regular contact with the agent of change throughout the development of the technology. The second is the requirement that all inputs (i.e. raw materials, financing, etc.) must as far as possible be available locally. The third is that the

Centre should always be ready, when the need arises, to extend its assistance on production and management.

The above factors came into play in the successful establishment of the glue project which now has a production level of $20,000 per month and supplies about 80 per cent of the country's requirements. In this case, the Centre helped in the development of the glue formula, built a process plant, assisted the entrepreneur in securing a banking loan, assisted in costing the commercial product and finding a market. Many projects, however, have failed because of the disregard of the above guidelines, and in particular because of the lack of regular contact between the entrepreneur and the agent of change.

INTERNATIONAL CO-OPERATION IN RURAL INDUSTRIAL PROMOTION

International co-operation is indeed not new to developing countries. In the case of the Centre, much help has been obtained from external sources in the United Kingdom and the United States through the Intermediate Technology Development Group, VITA, voluntary agencies and universities. In the specific field of development of low-cost technology, four areas for co-operation may be identified:

a) Research

International co-operation could be improved in a number of ways. The research laboratories in advanced countries could accept for instance, as one of their tasks, to undertake research work directed to the development of technologies for rural areas in developing countries. Developed countries could also assist in the financing and establishment of research and development activities in developing countries directed to the development of specific technologies, with suitable arrangements for secondment of experts and counterpart training.

b) Manpower Training

In the developing countries there is a shortage of technical manpower capable of developing appropriate technologies to suit local needs. International co-operation could be improved if industrialised countries were to accept an expansion and improvement of their facilities for educating and training technical people from developing nations.

c) Regional Co-operation

Technical contacts between individual developing countries and industrialised countries are widespread. However, there is little contact between neighbouring developing countries and consequently there is much duplication of activities in development work. Agencies for the promotion of industrialisation in rural areas should place much more emphasis on the development of closer working relations with their counterparts in neighbouring countries.

d) Forms of Technology Transfer

Attempts should be made in developing countries to specify in detail the areas in which technology transfers from the industrialised countries are most needed, as well as the mode of transfer (i.e. whether in the form of capital goods or in the form of direct investments). In the case of rural development, it is expected that little capital machinery would be involved in the promotion of labour-intensive projects. Developed countries must appreciate this and offer direct investments opportunities.

CONCLUSION

Technology practitioners and development planners should bear in mind the fact that the promotion of industry in rural areas must take into account the type of production, the availability of manpower, the size of the market and all the other social, economic and cultural factors. In particular, initial studies of appropriate technologies must as far as possible involve the participation of the indigenous entrepreneurs.

The experience of the Consultancy Centre, limited as it may be, has evidently indicated that such centres have a major role to play in the industrialisation of developing countries. In particular, the experience has demonstrated the need for reviewing and adapting the traditional university curricula to suit the needs of developing countries. Such establishments can also contribute to a closer co-operation between industrialised and developing countries and accelerate the diffusion of low-cost technology between neighbouring countries.

It is, however, doubtful if one can defend the establishment of production units on university campuses in the context of the traditional university curricula. But it is equally true that teaching and research in institutions of higher learning are meaningless if they are not linked with the industrial and economic development of the areas they serve. The purpose of such production units is to show to potential entrepreneurs what can be done and to demonstrate how it can be done. The campus production units, when proved technically and financially viable, will hopefully fill a vacuum in the rural industrialisation policies of developing countries.

XV. SMALL-SCALE DISTILLATION OF POTABLE SPIRITS FROM PALM WINE

by

I.A. Akinrele*

INTRODUCTION

The development of the traditional art of distilling alcoholic spirits in Nigeria has largely been influenced by the ecological and socio-economic conditions prevailing in the local communities. Until recently when molasses became available from sugar mills, the traditional raw material for distillation was raphia wine derived from the sap of *Raphia hookeri* and palm wine from the oil palm *Elaeis guineensis*. The raphia plant grows extensively in the riverine areas and delta creeks of the country. The communities in these regions are composed mostly of fishermen, and both the men and women relish a drink of 'ogogoro' (raphia wine spirit) on social occasions. They also use it for medicinal and ritualistic purposes.

The traditional distillation technology is very primitive and leads to serious contamination of the product with organic and metallic toxicants. Because of this, the sale of the product was declared illegal by the government. The market as a result was driven underground and distillation operations were transferred to the backwoods to avoid detection by the police. The processing technique thus became very secretive and remains totally insulated from modern improvements.

Another factor which has restricted the development of the industry is the fact that the terrain over which the raphia plant grows is swampy. Transportation of raphia sap or wine is difficult, and this accounts for the fragmentation of the distillation units. Similarly on dry land where the oil palm thrives, the wide scattering of the trees and their low yield (about one litre of sap per tree per day) make collection on a large scale uneconomic.

It is estimated that in the 1,400 square miles of fresh water swamps in the Niger Delta area there are some 300 million raphia

* The author is Director of Research of the Federal Institute of Industrial Research, Oshodi, Nigeria.

palm trees, with a potential annual production of more than two billion litres (450 million gallons) of palm wine. It is also believed that about one tenth of this quantity (200 million litres or 45 million gallons) could be obtained from oil palms(1). If this wine were distilled and converted to ethyl alcohol, some 91 million litres (20 million gallons) of potable spirit with a value in excess of ₦ 20 million ($33 million) per annum before excise duty would be available for manufacturing various alcoholic beverages and for industrial uses.

The exploitation of this vast potential resource of potable spirit and the development of the associated native industry are being tackled in the following way:

a) A legalisation of rural distillation operations through licensing arrangements;
b) The design and manufacturing of a cheap and sturdy still that can be used in a rural environment by unskilled labour after a short period of training;
c) The granting of a concessionary excise duty to enable the industry to survive economically against imported alcohol concentrates;
d) The permission to market less refined, but nevertheless potable, spirit beverage at a price low enough to meet the demand of the low-income rural market;
e) Ensuring that the main economic benefit returns to the producers and is not captured by a new class of middlemen or entrepreneurs.

For a number of years, the Federal Institute of Industrial Research (FIIR) in Oshodi has been studying the ways of improving the native art of distilling locally prepared worts for spirits. The investigations carried out involved a survey and evaluation of the economic potential of current illicit distillation; an investigation of the microbiology of palm wine fermentation; the design and development of a distillation method based on modern scientific and chemical engineering principles; the biochemical screening of the distillation product for toxic residues that can affect the health of the consumers; a sensory evaluation analysis of the product and finally market research on the acceptability of the final products to the consumer.

Field tests with a commercially manufactured prototype of the still were carried out under a co-operative arrangement by the Federal Institute of Industrial Research, the Nigerian Brewers and Distillers Co-operative Society and the Addis Engineering Company Limited.

---

1) Source of data: private communication from the Secretary of the Niger Development Board.

THE DEMAND FOR ALCOHOL PRODUCTS

Nigeria, with its population of 80 million, currently spends some ₦2 million a year ($3.3 million) on spirits and alcoholic beverages. The available statistics indicate that the legal imports of ethyl alcohol at 80 per cent per volume amount to more than 1 million litres of concentrates, with an average yearly value of ₦200,000 ($330,000) while the corresponding value for alcoholic beverages is around ₦875,000 ($1.4 million) for some 830,000 litres(1). It has been reliably claimed that about the same quantities are illegally smuggled into the country, and that equivalent amounts are supplied by native rural distilleries. A conservative estimate thus suggests that the total market in Nigeria for alcoholic beverages is around 2.5 million litres (550,000 gallons) per year.

The average yearly growth rate for the legally imported alcoholic beverages is 18 per cent. However, the statistics must be interpreted with caution since they grossly underestimate the size of the market. In the low-income rural areas, the consumption of imported products is very much restricted, but there is nevertheless a large potential demand for potable spirits marketed at low cost. For this reason the FIIR project was aimed at making inexpensive potable spirit available to the large majority of the rural consumers. Furthermore, with the expansion of the industrial sector, it is estimated that in the next five years the average annual growth rate of demand for industrial alcohol will not fall below 27 per cent - the rate of growth achieved by the industrial sector during the first year of the 1970-74 Second National Development Plan(2).

THE BACITA FACTORY

A new factory, the Nigerian Yeast and Alcohol Manufacturing Company Limited (NIYAMCO) has recently been established at Bacita, in the state of Kwara. This company, which is owned by the Federal Government in partnership with other institutions, is designed to produce industrial and potable alcohol, spirits and yeasts, and animal feeds. NIYAMCO will use annually some 9.5 million litres of molasses, which as by-products of the Nigerian Sugar Company, have until now been wasted. Its capital cost was 1.7 million ($2.8 million) and its initial production capacity of 2.5 million litres of ethyl

1) Federal Office of Statistics, Trade Tabulations, December 1972 and December 1973, Lagos, 1972-73.

2) Federal Ministry of Economic Development and Reconstruction, Second National Development Plan 1970-1974, First Progress Report, Lagos, 1972.

alcohol should rise to a maximum of 3,375,000 litres a year. The NIYAMCO plant was designed for the urban domestic market and for export, and the FIIR plant for the rural/urban domestic market. However, the price of alcohol produced by the FIIR plant is expected to compete favourably with that of NIYAMCO.

THE TECHNOLOGY OF DISTILLATION

a) The traditional distillation technology

In large areas of Nigeria, people illegally distil palm wine into a crude spirit containing between 30 to 40 per cent of alcohol by volume(1). The equipment for doing this consists of a steel distillation drum connected to a copper pipe which passes through a wooden cooler containing cold water. The distilled crude spirit collects in a clay pot. The drum containing the palm wine is heated with a wood fire, the alcohol vapour passes through the copper pipe, condenses in the wooden cooler, and drops from the open end of the copper pipe into the clay pot.

The final product is usually water-white in colour, sometimes straw-yellow. Solid contamination is usually visible. The flavour is characteristically strong. Heavy losses of alcohol occur during processing due to the lack of flowing water and to the evaporation of the hot distillate from the open receiver. The first distillate is usually redistilled.

b) The intermediate technology designed by FIIR

The FIIR still has been designed to improve on this traditional village technique and to give an optimal product yield and a potable quality. The main design feature of the still is a column with predetermined packing height, so as to separate the volatile components present in the fermenting liquor. It is packed with charred palm kernel shells, but other packing materials can also be used (glass beads or rings, nut shells, wooden fibres or pebbles). The column is well lagged so that the temperature profile in the packed column is due to the pressure drop of the vapours across the packing. A thermometer is provided at the top of the column to read the temperature of the vapours emerging from the column prior to condensation. A soft wooden disc, highly polished on one side, is attached to the middle of the column; the polished side facing the pot minimises the

---

1) Fresh palm juice, which has a sucrose concentration of 13g/100ml, ferments very rapidly. Within 24 hours, the process of fermentation is usually complete, and the resulting liquor contains about 7 per cent of alcohol by volume.

heat transfer by convection and radiation from the pot to the condenser.

A 25 gallon charge of fermented liquor (114 litres) containing 5 per cent alcohol by volume gives 3 gallons (13.65 litres) of alcohol at 35 per cent on first distillation in two and a half hours. The recovery rate is about 84 per cent. On second distillation a charge of 2½ gallons (11.4 litres) of alcohol at 30 per cent gives about 0.8 gallons (3.64 litres) of alcohol at 80 per cent. The recovery rate is 85.5 per cent.

The final stage of processing consists in blending and flavouring the concentrates. Most of the equipment recommended by FIIR for this stage is operated manually or semi-automatically. The rationale for this is that since the factory is to be sited in rural or suburban areas the non-availability of electricity should not diminish its overall operational efficiency. Maturation of the alcohol concentrate is allowed to occur in wooden casks for periods ranging from three to six months. Demineralised or deionised water is then used to dilute it to potable level, and the blend is clarified by using activated carbon or other suitable clarifiers. To the diluted spirit, the appropriate flavour is added, usually about 0.5 per cent by volume. The final product is filled into standard glass bottles of 1.2 litres and capped.

PRODUCTION ECONOMICS

Based upon an estimated average yearly demand of 234,000 gallons (1,064,700 litres) of concentrates and 183,000 gallons (832,650 litres) of potable spirits, the recommended plant capacity is 500 bottles per day (26.6 fluid ounce bottles), i.e. 25,000 gallons (113,750 litres) per year. A factory with this capacity will use the distillates of sixteen FIIR stills, each of which produces about 792 gallons (3,604 litres) of alcohol at 80 per cent per volume annually. It is estimated that about 647 FIIR stills will be required to supply the entire Nigerian market for alcohol. Table 1 summarises the capital costs and capital structure for this small-size processing plant and Table 2 its production costs and profitability.

The cost estimates for this project are based on the following assumptions:

- The palm wine used for distillation in the rural area will be undiluted and hence should contain maximum alcoholic content;
- The whole project will be integrated, i.e. the village distilleries will be linked with the blending, flavouring and bottling centres.

Table 1

CAPITAL COSTS AND CAPITAL STRUCTURE FOR A SMALL-SIZE PRODUCTION PLANT(1)

Fixed Capital		
Land and building: (2800 sq.ft. warehouse at ₦2 per sq.ft.)	₦	5,600
Furniture and fixtures		400
Equipment and plant:		9,930
- 2 holding tanks of 2275 litres for alcohol	1,200	
- 1 holding tank of 2275 litres for water	80	
- 90 maturing wood casks	6,000	
- 1 blending vessel with stirring unit	1,000	
- 1 filling machine (capacity: 800 bottles per hour)	500	
- 1 capping machine (capacity: 480 bottles per hour)	150	
- 1 water purifier and 1 water de-ioniser	400	
- 1 filter press	600	
Total fixed capital costs	₦	15,930
Working Capital		
Two month supply of alcohol at 80% per volume	29,100	
Six month supply of flavour	8,100	
Six month supply of packaging materials	16,400	
Three months of labour costs (1 manager at ₦100 per month, 1 supervisor at ₦50 and 10 workers at ₦26)	1,230	
One month's value of excise duty	2,500	
Accounts receivable	3,000	
Total fixed capital costs	₦	60,330
Capital Structure, Interest and Depreciation		
Total initial investment required (Fixed capital plus working capital):		76,260
Source of fixed capital: 40% from share capital (₦6372)		
60% from loan capital (₦9558)		
Source of working capital: 100% from short term bank loans		
Interest charges 8% p.a. on loan capital, i.e.	765	
12% p.a. on working capital for 6 months (2), i.e.	3,620	
Depreciation charges: 4% p.a. on the building, i.e.		
10% p.a. on plant & equipment	993	

Notes: (1) All figures are given in Naira (₦). approximate rate of exchange: ₦1 = US$1.6375.

(2) This working capital will be raised from commercial institutions and should be paid back within a period of six months.

Table 2

ANNUAL PRODUCTION COSTS AND PROFITABILITY FOR A MEDIUM-SIZE DISTILLERY

Production Costs	
Raw materials:	
12500 gallons (56,875 litres) of 80 per cent alcohol per volume at ₦14 per gallon	₦ 175,000
125 gallons (625 kg) of flavour at ₦26 per kg	16,250
Packaging materials: 150,000 bottles with bungs and labels at ₦0.22 per bottle	33,000
Labour	4,920
Supplies	220
Excise duty (₦1.20 per gallon, or ₦0.26 per litre)	30,000
Electricity	210
Rent and insurance	318
Interest on loan capital	765
Interest on short term working capital loan	3,620
Depreciation	1,233
Total production costs	₦ 265,536
Production cost per bottle: 1.77	

Profitability	
Total revenue: 150,000 bottles at ₦2.50 each	₦ 375,000
Total production costs	265,536
Profit before tax	₦ 109,464
Profit before tax as a percentage of total investment:	143.5%
Profit before tax as a percentage of sales:	29.2%

As can be seen from the figures below, profitability varies according to the price at which each bottle is sold:

Price per Bottle	Total Production Cost	Total Revenue	Profit Before Tax	Profitability as a percentage of sales	Total Investment	Yield on Capital %
2.00	265,536	300,000	34.464	11.5	76,260	45.2
2.20	265,536	330,000	64,464	19.5	76,260	84.5
2.30	265,536	345,000	79,464	23.0	76,260	104.2
2.50	265,536	375,000	109,464	29.2	76,260	143.5

- The FIIR stills will be located in rural communities in order to minimise the cost of transporting palm wine;
- The factory wall is built of concrete blocks strong enough to offer security to the potable spirits, but fitted with the type of inexpensive materials used in low-income buildings so as to minimise cost;
- The project is labour intensive and although Nigerian labour is relatively costly, family labour should be used when available;
- The entrepreneur will organise the factory to receive distillates from sixteen FIIR stills located close to the factory;

- The factory avoids the use of middlemen in procuring the concentrates from the rural distilleries;
- Grid supply of electricity is not necessary since most of the equipment can be operated manually or with batteries;
- The production costs were based on the FIIR Pilot Plant operations, after applying the correction factors, due to the inherent imbalances in pilot plant operations. Measures have been taken to accommodate the commercialisation of the product by private entrepreneurs. Considerations were also given to the effect of government excise duty and its effects in terms of favourable competition with imports.

The seemingly high cost of local production is explained by the high cost of palm wine. At the distillate level, the palm wine constitutes about 90 per cent of the production cost. This means that any change in the cost of palm wine will significantly affect the profitability of the entire project. The incidence of the cost of palm wine on the price of the alcohol supplied by the FIIR stills to the processing can clearly be gauged from the data presented in Table 3.

Table 3

ESTIMATED ANNUAL PRODUCTION COSTS FOR ONE FIIR STILL (IN ₦)

Cost of Palm Wine per Gallon (4.55 litres)	Cost of 75,075 litres (16,500 gallons) of Palm Wine	Other Production Costs	Total Cost	Average Cost per Gallon of alcohol at 80 per cent
0.50	8,250	850	9,100	11.49
0.45	7,425	850	8,275	10.45
0.40	6,600	850	7,450	9.41
0.35	5,775	850	6,625	8.36
0.30	4,950	850	5,800	7.32
0.25	4,125	850	4,975	6.28
0.20	3,300	850	4,150	5.24
0.18	2,970	850	3,820	4.82
0.17	2,805	850	3,655	4.61
0.15	2,475	850	3,325	4.20

A characteristic pattern of this type of investment is the relatively high proportion of working capital compared to fixed capital. This is due to the necessity of holding stocks of alcohol concentrates during the maturation process and to the importation of a six months' supply of flavour and bottles. However, with the favourable investment climate in the country and the lucrative nature of the project, it is not anticipated that entrepreneurs will have any difficulty in attracting the necessary funds from the financial institutions.

If the factory is organised on a co-operative basis, and members agree to wait for the payments for their concentrates to be made afte the final product has been sold by the society, the required working capital will be much smaller.

The production cost for one bottle of potable spirit is ₦1.77. Taking into account an 11 per cent profit margin for the manufacturer and a further 20 per cent margin for the retail distributer, the fina selling price is anticipated to be ₦2.50. At this price, the product will compete favourably with imported gin, which at present sells for ₦4.40 per bottle, and widen the market to the low-income groups. The manufacturer can expect to make an attractive profit margin on his investment: even if the bottle of spirit is sold at a low price of ₦2.00, he will still have a profit of 11.5 per cent on total sales(1)

There is a wide variation in the prices of palm and raphia wine between the urban and rural communities in the country. This report has assumed a price of ₦0.11 per litre (₦0.50 per gallon) for the palm wine and raphia wine, but if it can be purchased at ₦0.7 per litre (₦0.30 per gallon) - a normal price in the rural communities - the profitability on total sales will correspondingly increase from 29.2 per cent to 45.8 per cent.

CONCLUSION AND RECOMMENDATIONS

In view of the high consumption rate of potable spirit among the low-income group (70 per cent of Nigeria's population), this project offers significantly attractive prospects for the economy. The production technology involves slight modifications to the traditional method, but it is highly efficient and easily adaptable to the rural environment. There are several optimal location sites in the southern parts of the country where palm and raphia trees grow in abundance.

The project is expected to create more employment opportunities; help reduce the large-scale consumption of crude gin which is unhygienically distilled and therefore injurious to health; improve the utilisation of local material resources; and stimulate ancillary industries such as the cork, packaging and glass industries. It is also anticipated that the location of the plants in rural and suburba areas will contribute to reducing the population drift towards the urban centres.

In order to minimise costs, it is recommended that the project be organised in the form of co-operative of rural families who own

1) Of particular interest to the entrepreneur is the fact that the investment in such a plant has a payback of less than one year.

the land on which the palm trees grow. This form of organisation with the usual governmental support would be able to raise the initial working capital required to pay the village distillers while awaiting the sale of the final product. The Nigerian Brewers and Distillers Co-operative Society has already set up a commercially successful system of this type.

The initial product may be gin since it has a minimal maturation period compared to other potable spirits. In this way, large amounts of capital will not be tied up in inventory. Depending on the market acceptability, a further saving in operational costs may be achieved by using plastic rather than glass bottles.

Any private entrepreneur planning to invest in this project is strongly advised to ensure the availability of sufficient working capital before going into production. A total of 20 stills have already been sold to individual entrepreneurs, but a feedback from these operations has yet to come in(1).

Considering the profitability of this project as well as its social benefits to the economy in terms of employment opportunities, technology development and transfer of skills to rural communities, investments in this field should be actively encouraged.

---

1) The Ghana Distillers Co-operative is currently evaluating this small-scale distillation technology with the hope of establishing a pilot production project.

XVI. APPROPRIATE TECHNOLOGY IN ETHIOPIAN FOOTWEAR PRODUCTION

by

Norman S. McBain and James Pickett*

In the light of lively and widespread interest in developing country technology it is not surprising that the search for 'appropriate' technology should be bedevilled by semantic confusion. This is, nevertheless, a nuisance which imposes on those who participate in the search and related discussion a tiresome obligation to try to make more than usually clear the meaning of the terms they use. In this regard, it comes naturally to economists to take as 'appropriate' that technology (set of inputs and processes) which would minimise the cost of production of a given volume of a given good in any particular location. This minimum cost can, of course, be calculated either at market or social prices, and the application of the latter - if they accurately reflected the real opportunity costs to the economy of the various factor inputs - would for many identify the optimum technology.

The economist's view of appropriate technology is not, of course, the only possible one(1). Notwithstanding basic agreement on the need to connote 'appropriate' in the light of local conditions, a variety of meanings can - at least apparently - be attached to the general term. One interesting way of polarising views which has been suggested distinguishes between engineering and economic understanding of local

---

* Mr. McBain is a Senior Research Fellow and Mr. Pickett the Director of the David Livingstone Institute of Overseas Development Studies of the University of Strathclyde in Glasgow, Scotland.
This study is a by-product of a pilot investigation in the choice of techniques in the footwear and sugar industries in Ghana and Ethiopia undertaken under the direction of J. Pickett by the David Livingstone Institute of Development Studies. The investigation was generously financed by the British Ministry of Overseas Development. Thanks are due to the participants of the OECD Seminar on Low-Cost Technology for useful comments on the first draft of the paper and to Professor Gustav Ranis for constructive criticisms which - it is hoped - resulted in the removal of some obscurities in presentation. The authors are, of course, not absolved by the help they have received for the remaining errors or confusion. This study is reprinted by permission from the Journal of Modern African Studies, September, 1975.

1) For one of a number of discussions on meaning, see OECD Development Centre, Choice and Adaptation of Technology in Developing Countries Paris, 1974, particularly Chapter I and pp.161-164.

conditions(1). In the former the need to reduce the scale of production, to make use of local resources, to take account of climate and topology are given prominence, and the emphasis is on the choice, adaptation and design of production (and transport) equipment and systems which would meet these imperatives. At least implicitly, appropriate technology in this usage tends to correlate with the small-scale and the rural, as distinct from the relatively large factories which characterise the modern (largely urban-located) industrial sector. The second usage concentrates on relative factor endowments and differences in these as between developed and developing countries - and as within developing countries - ought, in principle, to be captured in the method of choosing among alternative technologies described briefly in the first paragraph of the present paper.

The distinction between the engineering and the economic approaches to appropriate technology is useful in cataloguing the large, growing and extremely varied volume of work now being done in this area. It should not, however, be overdrawn. In particular it should be recognised that the two categories are not 'pure': the engineering contains important economic elements - including market size and product character; and the economist's most elementary understanding of production is founded on a fusion of economic and technical considerations. This overlap is important because it means that the distinction does not really answer the question of whether, conceptually, there is but one appropriate technology. This question is in turn important because it bears on the sectoral and locational allocation of resources, and thus on the whole problem of urban-rural balance. The answer to the question is difficult to obtain, particularly since it cannot be divorced from the specification of national policy objectives. It is nevertheless worth pursuing, and the identification of the relative economic efficiencies of different forms and locations of industrial production should be an important part of this pursuit. In this regard, rigour requires that like be compared with like and means consequently that definitive views are only likely to emerge after a sufficiently large number of 'controlled' case studies have been made.

Interest in the outcome of such studies is heightened by the persistent belief that small-scale industrialisation has much to contribute to economic development in poor countries, Thus, both the 'regeneration' of the rural areas and the breeding of an indigenous class of entrepreneurs from which future captains of large-scale

---

1) UNDP, Technical Advisory Division, Appropriate Industrial Technologies: A Note for Discussion, (mimeo), New York, 1975.

industry might be drawn have been linked to the establishment of small industrial units. Moreover, the expansion of such units in the countryside and the smaller towns is often seen as a means of increasing attractive employment opportunities outside the main urban centres and thus reducing the employment problem in these centres. That the whole question of urban-rural balance is critically important in development may readily be accepted, as may the importance of small-scale industrialisation in the integration of the rural areas and small towns into an expanding national economy. It is, however, this very importance which makes it more rather than less necessary to approach the development of small-scale, 'low-cost' technologies in a critical spirit. If small-scale industry is to flourish, a judicious choice of products and an awareness of the difficulties confronting would-be entrepreneurs are required.

Against this background, the purpose of this paper is to present and discuss the results of economic appraisal of alternative technologies and factories which could be used in the production of footwear in Ethiopia. The factories considered range from the very large to something like a cottage industry-type of organisation. The quality of the product is, however, invariant, so that it is possible to isolate differences in economic characteristics associated with differences in factory size and form of organisation. The paper implicitly emphasises the importance of economic efficiency to poor countries and sees in carefully defined measures of this some basis for clarifying the meaning and specifying the character of 'appropriate' technology. It is recognised, however, that not everything can be captured in economic criteria, and some attempt is made to set out the conditions under which small-scale options might be preferable even when, at first sight, economic judgement would seem to be against them. It is hoped that the paper has implications for a much larger number of countries than the one to which it explicitly relates.

In comparing technology variants and factory size, the basic criterion used is net present value(1) at a discount rate of 10 per cent per annum. The product in question is men's leather-upper, cement-lasted shoes with cemented-on synthetic soles. It is assumed, on the basis of observation, that these would generally have sold at a retail price of Eth.$10 in 1972, and input prices used in the calculations are those which prevailed in Ethiopia in the same year. Account is also taken of capital allowances, taxes and duties prevailing in 1972, and - unless otherwise stated - the net present value arrived at is a measure of private profitability. The factories

1) It may be noted that using net present value per unit of investment as the criterion would not alter the arguments of the present paper.

compared are 'synthetic' in that the production process they comprise is an amalgam of existing operations in developing and developed countries(1). Project lives are taken as 27 years, except for very small factories when 10 years is chosen, and a one-shift working day is assumed.

The remainder of the paper is in three parts. The first is concerned with economic appraisal of three technology variants for each of three factory sizes (levels of output); the second considers alternative ways of producing a given annual output; and the third draws conclusions from what has gone before.

TECHNOLOGY VARIANTS

Much of the concern with 'low-cost' technology derives from a wish to save capital, which is thought to be relatively scarce, and to provide employment for unskilled labour, which is taken to be relatively plentiful. Given this (and perhaps rashly ignoring the distinction between micro- and aggregate considerations) a pertinent question is: for any given level of output and product specification, what difference does variation in technology make to capital requirement, employment creation and, it is necessary to add, profitability?

The information deployed in Table 1 permits some answers to be returned to this question for each of three levels of footwear output in Ethiopia. These are the production of 50,000, 300,000 and 1.8 million pairs of men's leather-upper, cement-lasted shoes per annum. For each level of output three technology variants are considered - the most machine-intensive, the most labour-intensive and the least-cost(2). Table 1 consequently deals with nine different factories, information concerning which is given in three groups. As already explained, the factories are 'synthetic'. To repeat, they do, however, represent fully articulated, feasible processes. Moreover, in making the calculations which underlie Table 1, care was taken to include every conceivable item of capital and operating expenditure.

In considering the implications of the table, it is convenient to assume that private entrepreneurs and engineers would, left to

---

1) It is worth stressing that the factories are fully designed and their operations fully costed. Their 'construction' is based on visits to shoe factories in a number of developing and developed countries, manufacturers of capital goods, research organisations, government agencies, etc. as well as on careful study of the technical literature.

2) This somewhat confusing terminology is conveniently brief. In principle the least-cost technology could also be either the most machine- or the most labour-intensive technology. In the comparisons reported in Table 1 it never is.

Table 1

SOME CHARACTERISTICS OF MACHINE-INTENSIVE, LABOUR-INTENSIVE AND LEAST-COST VARIANTS OF EACH OF THREE SIZES OF ETHIOPIAN SHOE FACTORIES

Characteristic	Factory size and technology variant								
	A_1	A_2	A_3	B_1	B_2	B_3	C_1	C_2	C_3
No. of unskilled workers	3	6	6	69	92	81	411	556	460
No. of staff and skilled workers	32	39	36	85	126	86	434	683	444
Total number of employees	35	45	42	154	218	167	845	1,239	904
Working capital at full capacity (Eth$'000)	288.9	282.0	281.0	1,590.0	1,605.8	1,590.5	9,397.4	9,504.5	9,410.5
Fixed capital at full capacity (Eth$'000)	327.1	128.4	140.4	793.0	404.7	660.2	4,367.2	2,154.7	3,439.4
Total capital (Eth$'000)	616.0	410.4	421.4	2,383.0	2,010.5	2,250.7	13,764.6	11,659.2	12,840.0
Average annual wage per employee (Eth$)	2,348	2,067	2,136	1,703	1,483	1,614	1,431	1,279	1,384
Fixed capital per employee (Eth$)	9,346	2,853	3,343	5,100	1,856	3,972	5,168	1,739	3,804
Annual net cash flow at full capacity (Eth$'000)	47.4	38.4	40.9	304.6	337.6	393.5	2,533.9	2,275.1	2,480.1
Internal Rate of Return (per cent)	5.8	7.9	8.3	18.8	20.4	20.1	22.6	24.6	24.5
Net Present Value (Eth$'000)	218.8	-77.9	-65.2	1,643.3	1,660.4	1,774.7	13,056.1	12,799.6	14,055.0

Size and Technology Legend:

1 = Most machine-intensive.
2 = Most labour-intensive.
3 = Least-cost.

A = 200 pairs of shoes per shift.
B = 1,200 pairs of shoes per shift.
C = 7,200 pairs of shoes per shift.

themselves, normally tend to install the most machine-intensive technology at any of the three levels of output, rather than search exactingly for the most profitable technology(1). If this is done, then one use to which the results of the table may be put is to answer questions concerning the desirability of technology choice of this kind.

Examination of columns 1, 2 and 3 for output levels A, B and C shows that, in fact, the most machine-intensive technology would not be chosen by profit-maximising entrepreneurs. In each of the three cases, the rational choice would be in column 3. The consequences of making this choice, rather than that described in columns 1 and 2 for each of the three factories, for capital investment and employment are summarised in Table 2. From this it can be seen that the least-cost factory at each level of output lies between the most machine- and the most labour-intensive technology in virtually all respects. For those partisan for labour-intensive technologies the results relating to factory A are the most pleasing. For this factory the least-cost technology comes close to matching the most labour-intensive technology in the extent to which it would save capital and provide additional employment as compared to the most machine-intensive technology. For factories B and C, capital savings and job creation are decidedly less dramatic. They are, however, not insignificant.

With regard to capital it is ultimately savings (investible funds) that are scarce in developing countries, so that it is necessary to consider total and not simply fixed capital invested(2). Even accepting this, savings of 6 and 7 per cent on total capital requirements are not entirely to be sneezed at. In addition overall increases in employment of 6 and 8 per cent, and of 11 and 17 per cent in unskilled employment represents a welcome positive addition

---

1) For an account of the view of the investment decision which underlines this assumption see Pickett, Forsyth and McBain, "The Choice of Technology, Economic Efficiency and Employment in Developing Countries", Journal of World Development, Vol.2, No.3, March, 1974, pp.47-54.

2) Even this should perhaps be qualified. If raw materials, the holding of which accounts for a considerable proportion of working capital, are local, then the relative constancy of working capital across technologies in Table 1 would justify particular focus on the quite marked variations on fixed capital, the financing of which could normally be expected to represent a drain on the balance of payments. It is nevertheless important to accept that this is a qualification to, not a contradiction of, the argument in the text. The force of the argument about savings (investible funds) can quickly be seen by appreciating just how little economic activity could be financed in most developing countries if 'best-practice' techniques were adopted in a handful of industries.

to income earning opportunities. This is particularly so in the light of the fact that footwear manufacturing is, generically, a labour-intensive industry anyway. Even the fairly small decline in the average wage recorded in Table 1 as between technology 1 and technology 3 is perhaps to be welcomed, since it could help to reduce the disparity between urban and rural incomes.

Nothing has yet been said about profitability. Here the hopes of the labour-intensive partisans take a knock. The inescapable evidence of Table 1 is that in circumstances obtaining in Ethiopia in 1972 it would not have been economically justifiable to have established factory A, regardless of the type of technology installed. It would, however, have been defensible to have established both factory B and factory C; and all the technology variants would have been profitable in each factory. Indeed the higher internal rates of return shown in the table indicate that factories B and C would have survived a cost of capital ranging from 18 to almost 25 per cent.

Table 2

CAPITAL AND EMPLOYMENT CHARACTERISTICS OF MACHINE-INTENSIVE LABOUR-INTENSIVE AND LEAST-COST VARIANTS OF EACH OF THREE SIZES OF ETHIOPIAN SHOE FACTORIES

(In index numbers, with machine-intensive = 100)

	Factory size and technology variant								
Characteristic	A_1	A_2	A_3	B_1	B_2	B_3	C_1	C_2	C_3
Total no. employed	100	128	120	100	141	108	100	146	106
No.of unskilled employees	100	200	200	100	133	117	100	135	111
Total Capital Invested	100	66	68	100	84	94	100	84	93
Fixed Capital Invested	100	39	42	100	51	83	100	49	78

Source (and legend): Table 1.

The unprofitability of the smallest of the three factories described in Table 1 suggests that small-scale operations may be at a disadvantage in the footwear industry compared to larger factories. Since, as has already been noted, 'appropriate' technology is often associated with small-scale production - and indeed since small-scale production often connotes much smaller units than any represented in Table 1 - this suggestion is worth pursuing further.

ALTERNATIVE WAYS OF PRODUCING A GIVEN OUTPUT

One interesting way in which a comparison between large and small shoe factories may be made is by answering the following question: if the Ethiopian shoe industry were to be substantially expanded, would this be most efficiently achieved by a single, very large factory, by a limited number of somewhat smaller factories or by a large number of very small factories? Putting the question in this way has a number of advantages. It implicitly places a quality constraint on the comparison, explicitly puts profitability at the centre of it, and although it does not ignore the fact that small shoemakers exist, it does measure their chances of survival as expansion proceeds. Moreover, the comparison can be realistically grounded in an assumed expansion which is sufficiently large to ensure that the biggest, single factory can take advantage of all economies of scale.

In 1972 Ethiopia produced about one million pairs of leather boots and shoes. Virtually all of these were disposed of on the domestic market. There is, however, reason to believe that Ethiopia is well placed substantially to expand its currently miniscule export trade in shoes(1). It is indeed possible that exports could conceivably become at least twice as important as the Ethiopian home market now is. Given this, it is permissible (as an act of creative imagination) to suppose that a decision had been taken to expand the existing capacity to produce men's leather-upper shoes by, say, 1.8 million pairs per annum(2).

Table 3 sets out some characteristics of four different ways in which the extra shoes could be produced. The first comprises a single enterprise which, in producing 7,200 pairs of shoes per day, could alone satisfy the entire additional requirement; the second, 6 enterprises each of which would produce 1,200 pairs of shoes per day; the third, 36 enterprises each of which would produce 200 pairs of shoes per day; and the fourth, 1,200 enterprises each of which would produce 6 pairs per day. The range of factory size thus considered is from that which would employ 904 persons to that which

---

1) See James Pickett and Norman S. McBain, "Developing Country Export Potential and Developed Country Adjustment Policy" in OECD Adjustment for Trade: Studies in Industrial Adjustment Problems and Policies, Paris, 1975.

2) The size of the expansion was arbitrarily but conveniently chosen to accommodate information available from the David Livingstone Institute's pilot investigation. It also amounts to some 8 per cent of British imports of leather shoes in 1972.

would employ no more than 3 persons(1). Again, of course, the main focus of interest is on the consequences of choosing one way rather than another for profitability, capital investment and employment.

The evidence of Table 3 is not, on the face of it, encouraging for those who would put their faith in small-scale organisation. It is clear that if economic efficiency (as measured by the net present value of the various projects) is to be the over-riding criterion then the large single factory (which is the least-cost and, as it happens, not the most machine-intensive variant of its kind) stands out as the obvious choice. Even disaggregating to as few as six enterprises substantially reduces the net present value. On the other two criteria - size of capital investment and of employment - the smaller factories fair better only in job creation. The fourth row of Table 3 makes it clear that the total fixed capital investment rises as the unit size of enterprise declines. Since working capital is relatively invariant with factory size, this inverse correlation holds for total capital investment also. It is true that the smaller units would in aggregate provide greater employment than the large economically most efficient factory. Some measure of the consequent 'displacement' of labour if factory A were to be chosen is given in the third row of the table.

It is worth noting that the smallest factory unit - the three-man enterprise - could be made to appear in a more unfavourable light even with respect to employment if account were taken of the work generated by the establishment of a retail distribution system that the choice of factory A would imply. In the present comparisons it has been assumed that the small-scale enterprises retail directly to the consumer. Moreover even if it were assumed that such enterprises took the form of a cottage industry with a link to a central selling agency or if it were assumed - rather drastically - that the total fixed capital required for the twelve hundred small units (each responsible for its own sales) was to be valued at zero, then the net present value of this form of organisation would still be less than that which could be obtained from factory A. Nor is this all. If it were (realistically) accepted that the very small units could not entirely do without fixed capital, but if it were assumed that they had no need for working capital - because they purchased materials and sold finished shoes instantaneously - then they would still appear economically unattractive. It should also be noted that since all the entries in the fifth row of Table 3 were computed using a discount rate of 10 per cent, no account has been taken of the possibility - which would cost further against the small unit - that very small operators might have to bear relatively high capital costs.

---

1) To write of a (small-scale, clearly not automated) factory employing 3 persons, is clearly straining language somewhat. It is convenient, however, to use the same term for the two basic alternativ forms of organisation.

Table 3

SOME CHARACTERISTICS OF ALTERNATIVE, LEAST-COST WAYS OF PRODUCING 1.8 MILLION PAIRS OF MEN'S SHOES PER ANNUM IN ETHIOPIA

Characteristics	Alternatives (1)			
	A	B	C	D
No. of employees per enterprise	904	167	42	3
No. of employees required to produce 1.8 million pairs	904	1,002	1,512	3,600
No. of employees 'displaced' if Factory A used	0	98	608	2,696
Total fixed capital investment (Eth$'000)	3,439	3,961	5,054	5,940
Net present value	14,055	10,646	-2,340	3,450(2)

1) A = 1 enterprise producing 7,200 pairs per day.
B = 6 enterprises each producing 1,200 pairs per day.
C = 36 enterprises each producing 200 pairs per day.
D = 1,200 enterprises each producing 6 pairs per day.
The project life of A, B and C is 27 years; that of D, 10 years

2) Includes retailing profit and no profits tax.

CONCLUSIONS

Thus far, considerable stress has been placed on economic efficiency and it has been implied that the search for 'appropriate' technology should really be directed towards locating the 'least-cost' technology. This way of looking at the matter is not favourable, on the above evidence, to small-scale units of production. It suggests rather that efforts to save capital and generate more employment than would normally result from the decision-taker's predilection for the highly machine-intensive should be embodied in attempts to find 'appropriate' technology subject to a strong economic efficiency requirement.

It can, of course, always be objected that the implicit assumptions of this paper are too restrictive. Even, for example, on its own terms it could be said that the comparisons made above were 'static' rather than 'dynamic'. In so far as 1972 prices are an inadequate surrogate for real costs throughout the respective project periods there is force to this objection. However, although differential variations in inputs costs - not to mention technical progress - over time would undoubtedly affect the detail of the above findings, extensive sensitivity analysis suggests that the broad conclusions would hold for a wide range of values of the main parameters.

A different objection would claim that the present analysis sees the development process in too narrow a way. This may be so, although the emphasis on the generation of as large a surplus as possible in poor countries is surely proper. Moreover, if wider considerations of income distribution, regional policy and the abolition of poverty are to be entertained - as they should be - then these can, by means of cost-benefit analysis, be accommodated within the framework of economic analysis that underlies the present paper. It is certainly important to ask if the allocative mechanism could not be made to function more effectively in a way that would economise on capital, redistribute income (at least as measured by average income levels in the urban and rural areas) and provide more employment than is presently the case. Even if it were found necessary to adopt radically different criteria in order to obtain the wider objectives, it is surely still important to have some measure of the economic cost of doing this.

Another possible objection to the present line of argument is that in Ethiopia and in other developing countries very small shoemakers survive and are indeed cherished. Their continued presence might be taken as a practical demonstration of viability. It is, however, necessary to ask why the small units survive, and to recognise that to some extent they do so in an economically unexacting way. They undertake a range of functions - repairing as well as making shoes and acting as retail outlets for other manufacturers. These functions need not be efficiently discharged since accounting procedures are very apt to under-value such things as, in effect, managerial time. As national markets expand and become more integrated it is consequently doubtful if the survival of the very small enterprise is likely to continue. In this regard it is worth noting that the largest 'displacement' recorded in Table 3 - that of 2,696 persons - is not much more than could be accommodated within the distribution system associated with the very large enterprise characterised in the same table. It is true that there would be no necessary spontaneous way of spiriting over 2,600 small operators into a relatively large-scale organisation. This is, however, a challenge for policy and not, _per se_, a defence of the small operator.

The four alternatives considered in Table 3 represent two quite distinct forms of organisation. Thus A, B and C are variants of the modern factory system; D is more akin to cottage industry and suggest 'informal' back-street or village-level operations. In some ways, therefore, D might be thought to be closer to the engineering and A, B and C to the economic concepts of appropriate technology discussed above - although each of the four alternatives is capable of making use of Ethiopian rather than imported leather and in conducting the analysis on which Table 3 is based it was assumed that domestic raw material was invariably used. Thus far the very small enterprise has compared unfavourably with the factory system, and -

particularly in the light of the consequences of zero-pricing the capital elements in the cost calculations - it is unlikely that social cost-benefit analysis would alter this result. The sad conclusion is thus that the demise of the very small operator in the shoe industry should be countenanced. This being so, the need to search persistently and imaginatively for relatively labour-intensive but economically acceptable technologies is heightened(1).

In this regard a number of additional points are in order. First, it is worth noting that the appraisals presented in Tables 1 and 3 were conducted in terms of private profitability. This being so it was necessary to take as given the present system of Ethiopian investment allowances and the present pattern of Ethiopian investment duties on imports of capital equipment and raw materials. As it happens, these things are relatively favourable to the machine-intensive modes of production. Consequently the application of social cost-benefit analysis could be expected to make the relatively labour-intensive alternatives considered in Table 1 seem more attractive than they do in the light of private profitability criteria.

The perhaps disappointingly small savings in capital associated with choosing the least-cost rather than the most machine-intensive technology is largely explicable for factories B and C in Table 1 by both the constancy of working capital across technologies and by the relatively high proportion it provides of total capital requirements. Much of the working capital in turn results from what, by developed country standards, is high levels of raw material stocks and finished goods. This identifies a particular point in the production process at which the further search for appropriate technology could be mounted and suggests the general thought that one way of proceeding is to seek to identify 'bottlenecks' of this kind in order that they can be removed.

Finally, it may be seen that the selection of factory A in Table 3 would imply a perhaps unwise degree of centralisation. The establishment of a single large factory in Addis Ababa, say, would certainly concentrate the labour force (and thus reduce the regional distribution of employment opportunities) and would concentrate the capital investment. The success or failure of the plant would depend heavily on the ability of the top managers of the enterprise, whose problems might be compounded by diseconomies of scale. There may, particularly if the desirability of a regional distribution of industry is taken into account, consequently be an argument for spreading

---

1) One possible argument in favour of the very small enterprise is that it might in some circumstances produce more future entrepreneurs than the factory system. This is, however, such a strongly qualitative argument that it cannot be taken either as settled or decisive.

the risk; so that - when fuller consideration is taken of social and economic objectives - alternative B in Table 3 could be preferred to alternative A.

This possibility is worth enlarging on briefly. From Table 3 it can be seen that to locate six factories in different parts of Ethiopia rather than a single factory in Addis Ababa would increase employment by 11 per cent, increase total capital required by 15 per cent and reduce the net present value resulting from the expansion of the shoe industry by 24 per cent. Whether these changes are acceptable depends on the policy objectives of the Ethiopian government and on the various weights attached to these. It is consequently not possible to pass definitive judgement on them here. What can be said, however, is that in a country largely dominated by two urban centres (Addis Ababa and Asmara) the spreading of six shoe factories to other areas might be desirable. It should also be possible, since the skill requirements of a shoe operative are not high.

It is worth recalling that this paper has been concerned with a single product and a single country. This inevitably places limits on sensible generalisation. The country limitation is not, however, a powerful one, and the methodology and the range of questions dealt with - including (at least implicitly) that of the relationship between the economic and non-economic - would be important in most industries in deciding on size, location and form of organisation. It is consequently hoped that the paper throws general light on small- and larger-scale industrialisation, although it has to be recognised that the comparative conclusions of the paper are particular to the footwear industry - so that the finding that economically-acceptable, labour-intensive variants of factory production might be more plausible than very small-scale, 'informal' methods cannot be extended to other products without specific investigation. A main message of the present paper is that such investigation should be conducted over a wide range of possibilities.

XVII. EDUCATION AS A LOW-COST TECHNOLOGY FOR AGRICULTURAL DEVELOPMENT

by

P. Dubin*

The Institut Africain pour le Développement Economique et Social (INADES) working through its "INADES-Formation" department, has been concerned with the training of peasant farmers since 1965. Through its correspondence courses and seminars, its journal "Agripromo" and the field trips it organises, the Institute is involved in nineteen African countries, including all French-speaking Africa, Ethiopia and Ghana. Its headquarters are in Abidjan, Ivory Coast, and offices have been set up in recent years in Bujumbura (Burundi), Douala (Cameroon), Soddo (Ethiopia), Kigali (Rwanda), Dapanga (Togo) and Kinshasa (Zaire). Two other offices, one in Accra (Ghana) and the other in Ouagadougou (Upper Volta), have been opened in 1975.

GENERAL PRINCIPLES

Farmers who are beginning to make progress as a result of the support they receive from different sources - the government, state-owned enterprises, schoolteachers, religious missions, volunteers from a wide range of countries, local movements and so on - want to know the reasons that lie behind the advice they are given. Recently a woman from the Senufo tribe in Northern Ivory Coast said to us: "We've been told we should plant in rows and keep the rows weeded, but we would really like to know why". There are also many instances where the boot has been on the other foot and preposterous mistakes have been made through a failure to appeal to the farmers' good sense. We need only quote two examples: if farmers do not realise that fertiliser is a plant food, it should come as no surprise if they do as one farmer did in the Koudougou region of Upper Volta and apply ten times the proper amount and then completely lose confidence because they do not obtain the results they expected. Similarly, if the role played by water in the spread of disease is not clearly

1) The author is Director General of INADES-Formation in Abidjan, Ivory Coast.

grasped, it ought not to be surprising if, even after a well has been dug, villagers continue to drink the river water which they find tastier.

There can be no long-term increase in agricultural production unless peasant farmers are given proper theoretical instruction and are allowed to assume responsibility for their own development. While this point is particularly important in the present context of food shortages and population growth, it is also closely bound up with the question of the type of society we wish to construct.

The manner in which our agriculture courses are organised has a close bearing on the subject of low-cost technology. We have to tailor the contents of our courses to the technical level of present-day peasant farmers and to what they can be offered in the light of their ability to assimilate agronomic, economic and sociological matters. If a particular form of technology is recommended in a course the assumption is that it is technically feasible, profitable, suitable for widespread application and well adapted to the knowledge and awareness of the people to whom we address ourselves. This is why, when we are drawing up our programme, we have to enlist the aid of agronomists, economists, social anthropologists and educationalists. If we are going to teach mathematics to a student, not only do we have to know the subject but also, and above all, we must know the student.

In this paper, we shall examine the content of the courses developed over the last nine years. Our first-year course has been reprinted four times and the last edition was completely recast. We obtain a considerable amount of feedback from the work done at home by course subscribers as well as from working sessions and tours during which we meet both students and course supervisors.

Correspondence courses, combined with seminars and local tuition, are an example of low-cost technology applied to teaching and training. Although the capital outlay is quite high because a drafting and applications team with high qualifications and considerable practical experience is required, the marginal cost is very low. The number of subscribers to the course can be increased many times over without it being necessary to build teaching premises (course work is usually done in groups at people's homes) or to engage salaried personnel since practical work, which only occupies each group for a few hours a month, is supervised by the existing extension service. The yield on the initial investment is enhanced by the fact that the basic documents in French have been translated into English, Arabic and Indonesian by the Food and Agriculture Organisation (FAO) for projects in India, the Philippines and Latin America; into Amharic and Kirundi by INADES-Formation pending their translation into a number of other African vernacular languages, and into Portuguese by the Caritas Brasileira of the State of Maranhão in Brazil.

The aim of the agriculture courses is to improve the living standards of peasant farmers by teaching them better techniques based on empirical knowledge and the experience acquired over generations. At each stage in the acquisition of new skills, course subscribers take a lead from their experience and learn the technical minimum they need to increase their output and accordingly their income. If the annual growing cycle is followed, farmers who are accustomed to the rhythm of Nature can acquire new knowledge more easily.

THE CHOICE AND PREPARATION OF LAND

In traditional farming, the individual is left little scope for choosing his own land: it is generally allocated to him by the leader of the community. The choice is made on the basis of such empirical criteria as the colour of the earth and the natural vegetation, without any regard for crop rotation or fallow periods. As a result, shifting agriculture is the practice.

In our courses, the peasant farmer learns to know his land. By means of a simple experiment, using the earth from his fields and a box with holes in it, he learns to distinguish the physical components of the soil, such as sand, silt and clay. From observations made on his land, he studies the qualities and defects of his various fields as determined by the degree to which they contain a given component. From the same experiment, the farmer also discovers the presence in the earth of organic matter, such as leaves, roots and stubble, which decompose to form humus. The course goes on to study soil life and the microbes which break down the humus and convert it into mineral salts which provide food for the plants. This piece of technical knowledge is very important because it shows the farmer that manure and chemical fertiliser are necessary to make his fields fertile and that the fields have to be protected from damage by erosion and the devastating bush fires.

In traditional farming, the land is tilled by hand with a hoe, which may take various forms but does not turn over the earth or dig in the grass cover. Tilling does no more than prepare the ground, remove weeds and ventilate the soil. In a very few cases like yam and cassava, the farmer makes mounds or ridges to ensure that as much as possible of the humus in the surface soil layer is concentrated round the plant roots.

We teach simple techniques such as the uprooting of tree stumps with hand tools and then with a winch in cases where groups of farmers are involved. Tree stumps left in the ground form obstacles to ploughs used for land preparation and weeding, and they also compete for food with the roots of planted crops.

By the study of plant roots, knowledge can be gained about the role of manure and fertiliser in soil structure and plant nutrition. Plant feeding is a new technique which our students learn. Manure and compost, which are easy to make and require only the investment of labour, are studied in detail. The study of mineral fertilisers is not omitted but stress is laid more on its profit-yielding features since the purchase of fertilisers implies some familiarity with farm accounting.

SOWING AND TENDING THE CROP

In a traditional situation, the farmer is quite capable of sowing or planting subsistence crops at the proper time and at spacings which take account both of the plant's needs and the fertility of the soil. The technique becomes more difficult when new crops or more productive varieties are introduced. The farmer only makes a rough and ready selection of his seed stock and keeps the best heads of grain, cuttings or tubers for planting in the following season.

With modern cash crops and the improvements made to subsistence crops, we have to study sowing techniques which, although inexpensive, require considerable care when being applied. Sowing has to take place at the right time, especially in the Sahelian region where the growing season is short. Sowing also has to be at the right spacing, so that a larger crop can be obtained from the same area without requiring extra work for weeding, hoeing and plant dressing. The use of more productive seed varieties and treated seeds are two techniques which can be a source of substantial profit for a minimum outlay. By studying the planting-out technique for certain crops, the way is paved for studying market gardening which, with its advanced technology, is particularly attractive as a result of the growing requirements of the urban consumer.

Just as the study of land preparation leads on to the study of animal-drawn cultivation, the logical sequel to an analysis of correct spacings and depths for sowing is the study of simple implements, such as drills and planting wheels, to ensure proper planting. We then go on to the more complex mechanical sower, which is an expensive piece of equipment but without which a really good crop cannot be guaranteed.

In traditional agriculture, the crop is tended merely by weeding with a hand hoe at infrequent intervals - on account of the difficult and tiring nature of the work - at a time of the crop cycle when weeds grow quickly. The farmer is defenceless against the insects and diseases which attack his crops. The only possible means of combatting insects is to destroy them by hand and the only remedy readily available to the farmer against disease is the admittedly clever but scarcely effective technique of spreading wood-ash on the sick plants.

In our courses, the farmer learns why it is important to weed crops at frequent intervals: weeds grow faster than cultivated plants and compete for mineral nourishment and water. Without any extra expenditure, frequent and proper weeding provides the plants with a better food supply and often eliminates the need for costly spraying. But weeding is a tiring business and study of animal-drawn hoes teaches the farmer how to reduce fatigue and increase the efficiency of his work. Weeding with an animal-drawn hoe is a more sophisticated technique than hand ploughing. The seeds have to be properly sown in straight rows and the animals have to be very well trained so that they can pass between the rows without damaging the plants.

By applying the simple and inexpensive technique of harrowing and ridging, the farmer learns to make the most of the water in the soil. Irrigation or drainage are sometimes necessary for some crops and for certain types of soil The basic investment requirement for these two water-utilisation techniques can be quite modest at the outset. Only later do they become more costly when the farmer improves them by building earth dams, irrigation channels and simple drainage systems.

The present-day chemicals industry provides the farmer with a wide and varied range of treatment products for protecting his crops and his animals. In the past, the traditional farmer was completely defenceless but today many resources are at his disposal. However these products are expensive and mistakes in application can prove costly. The study of treatment products and appliances is by no means easy, since it involves a knowledge of chemistry and machines, as well as plant physiology, and yet it is an essential requirement.

Consequently, in our courses, we study the physiology of plant stems, leaves and sap. All this knowledge is useful to the farmer in his day-to-day life since the conversion of raw sap into elaborated sap by the chlorophyl function is at the very heart of agriculture. It is important for the farmer to be aware of all these processes which determine whether harvests will be good or bad.

HARVESTING AND POST-HARVEST TREATMENT

In traditional agriculture the entire crop is harvested by hand. At first, the crop is gathered quickly because difficulty has been experienced in bridging the gap since the previous harvest. Later on, harvesting slackens off because transport conditions are poor, and some of the production is often lost. Once the crop has been gathered it has to be preserved to provide for family needs in the first instance and then for sale at a later stage since it would only fetch a low price at harvest time when the supply is abundant. Storage of the harvest is not an easy matter and the farmer has few means at his disposal for combatting humidity, the damage caused by insects or the rotting of perishable produce.

If better crops result from selected seeds, harvests are more abundant but they are more difficult and trying for the farmer. In our courses, we study simple implements that speed up the farmer's work and make it less back-breaking such as sickles for rice harvesting and scythes and rakes for grass cutting. Animal-drawn implements like the groundnut digger are slightly more expensive but they make the work much easier. Others, like small carts, despite their high price, become increasingly useful to the farmer who has expanded production. A large number of trips have to be made between fields and farm and between farm and market and all too often women have to carry everything on their head. If the farmer has a cart, he and his wife are freed from this irksome constraint. In this way, the draught animals are of better use since they work all the year round. Furthermore, a cart is indispensable if manure is to be used properly.

Problems of harvesting and, in particular, of producing grain and fruit crops lead on to discussions of plant physiology and especially of fertilisation. Although this phenomenon is known to farmers, it is hardly ever explained very clearly and hence it is important to present the subject in as simple and complete a manner as possible in our courses. The cultivation of selected varieties is becoming more widespread, hybrids are beginning to be popular, and the farmer needs the basic knowledge to be able to understand how fertilisation operates.

The storage of agricultural produce in difficult climatic conditions is another problem facing the farmer who has expanded his crops and wants to derive maximum profit from his labour. Our courses go into this problem and attempt to work with the farmers in studying simple and inexpensive means of conservation and storage such as sun-treated seedstocks, hay silage and so on.

STOCKRAISING

Domestic animals are kept all over Africa, from the herds of sheep and cattle of nomadic herdsmen to the chickens, goats and pigs in the villages of sedentary cultivators. However, these animals are virtually unproductive and are often signs of wealth and prestige or a form of stand-by capital. Animals are not slaughtered on any systematic basis but only on special occasions such as religious festivals and family celebrations or in honour of visitors. These animals which are raised in the traditional manner, do not require much care and receive even less. The farmer invests little in them but they also produce little and the mortality rate is high, especially among the young.

Herds are a source of wealth, but wealth that is dormant and our courses endeavour to assist farmers to 'awaken' it. The modern peasant farmer must use his animals to produce more eggs, milk and meat in the same way that he uses his land and crops to produce more food for himself and his family. The first technique taken as a subject of study is that of animal feeding. By improving the feed (grass, grains and tubers) produced on his land, the farmer can obtain more from his cattle without spending much. Even more important he thereby acquires stockbreeding skills and know-how he often lacks. Another valuable low-cost technique studied in our courses is cattle hygiene; this includes the use of vaccinations and first-aid with simple remedies available to everybody, as well as the construction and maintenance of simple but healthy stalls. Once all these techniques have been mastered, the course goes on to study animal physiology and, in particular, nutrition and reproduction. By studying nutritional problems, we teach what animals need by way of upkeep for production purposes and how the farmer can satisfy these needs. In studying reproduction, we tackle the very important problem of selection and the improvement of animal breeds. These are more costly techniques which can only be envisaged when feeding, hygiene and day-to-day care have been improved.

In training draught animals to assist him in his work in the fields, the farmer assimilates a very difficult yet most valuable technique. He really becomes an agriculturalist and acquires a skill that is respected as much as any other. He combines cultivation with stockraising, improves his working and living conditions and diversifies his output. With the manure from his animals, the farmer learns the technique of fertilisation, which is a source of wealth for his land, his crops and himself. Shifting cultivation is no longer necessary.

LOW-COST PRODUCTION TECHNIQUES

Most of the subsistence and cash crops currently produced in Africa are studied in the second year of our courses, as are the different forms of stockraising. The student himself chooses the options he wishes to study, depending on the region he comes from and the crops he produces. The proposed outline is always the same and the course goes over the age-old techniques known to everybody so as to enable the student to identify them, give them careful thought and attempt to explain why they are used.

After studying the day-to-day techniques used in villages, the course suggests an initial series of techniques for improving cultivation. These techniques encourage the student to use his sense of observation and point to improvements that cost little or nothing since the farmer has seldom any cash. These improvements do however entail careful thought and planning on the farmer's part. If seeds are sown at the right time, and if that time is specified for each crop, the chances of success are increased. If there is a delay of only ten days in sowing, half the crop may be lost. This is a technique which is easy to apply and does not require any investment - all it needs is a rational organisation of work. In the same way it may be preferable to weed a millet crop earlier so as to leave more time for sowing cotton on the date recommended.

The number of times a crop is weeded determines whether the harvest will be successful or not. This technique, which is studied in the course, can be easily assimilated and does not require any costly investments. If a cotton crop, for example, is weeded three times instead of two, the production will be markedly higher.

Another example of a simple improvement which can be achieved by everybody at the cost of additional carefully planned labour is pig-breeding. The local breed of pig which can be found in many African countries is left free to find its own food; it reaches a weight of 60 kilos at the end of two years, if it lives that long. On the other hand, an animal of the same breed will only take one year to reach 60 kilos if it is kept in a clean wooden pigsty with a thatch roof and stone floor, if it is fed every day on produce from the farm and on garden and kitchen waste and if it is given plenty of water.

Obviously, these low-cost techniques are no longer sufficient when the farmer wishes to progress further. He will then have to make an investment in cash. But the fact that he will have already been successful in applying simple low-cost techniques will make it all the easier and more profitable for him to assimilate more sophisticated ones. Animal-drawn cultivation is a good example. It requires a high investment and a complete range of skills, including training of oxen, veterinary skills, mechanical know-how, soil fertilisation and accounting. This technique will work to the farmer's advantage only if he has already mastered other techniques requiring physical or mental investment and involving such operations as the uprooting of tree stumps, sowing in rows at the right time with proper spacing and so on. Another example of more advanced techniques requiring the investment of capital is the use of insecticides to treat crops. Complex appliances have to be purchased and the handling of expensive products that are toxic for both man and animals must be careful. This technique, which calls for a knowledge of chemistry, zoology and machines, can give high yields only if it is applied to a crop on which all the other work (ploughing, sowing, weeding, pruning, and so on) has been carried out properly.

THE PEASANT FARMER AS AN 'INDUSTRIALIST'

The peasant farmer is no longer merely a producer of crops and cattle and his occupation can be likened to that of the head of an industrial firm who has to forecast production, organise sales, calculate expenditure and administer his assets. This implies a very different role for a cultivator who was accustomed to relying on other people such as the authorities, technical advisors, and traders to think out and solve all these managerial problems.

In the traditional system, the choice of crops and animals was governed by the habits of the village community, and the individual did not have to decide for himself. The amount of food that had to be produced was determined by the consumption requirements of the village, which lived in a subsistence economy. The peasant farmer has now entered a market economy. He still produces food for his own use but he now also produces food crops for the urban consumer, raw materials for industry and cash crops for export. All this involves new skills. In our courses, concrete examples of potentially attractive crops are given to encourage students to reflect on the matter, to envisage what work and expenditure will be needed if a particular crop is chosen, to calculate the yield from the land and to estimate the volume of the harvest. The combination of cultivation and stock-raising obliges the farmer to organise his production and make provision for feeding his cattle, and they both require forethought, method and information.

THE MARKETING AND FOOD DISTRIBUTION SYSTEM

In traditional agriculture, the peasant farmer does not have to give much thought to the problems of marketing. As soon as the harvest is gathered he sells part of it to a trader because he needs money quickly. The trader offers a price which the farmer can scarcely question, because he has no other means of disposing of his crop. Frequently, the need for money is so acute that the farmer sells more than he ought and no longer has enough food left to tide over his family until the next harvest. He often has to buy back food at high prices, usually from the same trader to whom he sold his crop at harvest time.

The title of one of our courses is "How to sell crops at a higher price". This is a technique which is learnt in a market economy and one that the farmer must acquire if the profit of his labour is not to go entirely to other people. He has to learn the law of supply and demand, he must find out about the tastes of customers and consumers, assess the risks and hazards of transportation and be aware of the prices prevailing on urban and village markets. Most of the problems involved can be solved if he picks up a number of simple techniques, such as harvesting the crop when it is at the proper degree of ripeness, drying it perfectly, sorting it out into different qualities and presenting and packaging the product to the consumer's taste. Storage of the crop in whole or in part, so that it can be sold at the best time, is a more advanced technique which requires substantial investment and considerable technical knowledge if, for example, a watertight silo has to be built or if the foodstuffs which are stored have to be treated to protect them from insects. Transporting the produce to larger markets where prices are more attractive can often be envisaged only if an association is formed with other farmers: the normal transport capacity of a truck is too large for the produce of a single farmer. This technique which is studied on the basis of specific examples and precise figures, enables the farmer to consider and understand the need for group marketing.

THE ROLE OF MONEY AND CREDIT

In a traditional barter economy, money is of little importance. When cash crops are introduced, money begins to circulate in the villages but the farmer is not familiar with it and has difficulty in mastering it. The largest amount of money he ever sees is when the sale of his crop is negotiated. This money soon usually passes through his fingers: creditors claim their entitlements, taxes have to be paid at once, repayments on loans are demanded, and a host of touts often make sure that what little money is left to the farmer is spent on the very same day.

Money is a necessity in a market economy and people have to learn how to handle it. It is one of the techniques studied in our courses. First we teach some arithmetic since the farmer must know how to count if he is not to be caught out by traders who are more skilled than he is at handling figures. Our peasant farmer's arithmetic book provides the student with an opportunity to revise his knowledge of the four basic operations. With simple concrete examples it shows him how to calculate yields and percentages and how to check the amounts shown on weighing machines.

Another part of the course covers the farmer's accounts by means of a simple system which enables him in a first stage to ascertain his income and expenditure. It teaches him how to make provision for his expenses, investments and repayments on loans; it then shows him how to draw up a technical budget to keep a close watch on his money. In order to convey this accounting technique more clearly, the course goes step-by-step through the story of a village family which owns various fields and animals, and has to solve its day-to-day problems of spending and buying. By following the story of this family, whose sons go to school and can write and calculate the cost of everything, the student learns the technique. The course concludes with concrete examples of data charts for crops and stockraising which teach the student to calculate the prospective yield of a particular crop, and understand why one crop brings in more money than another. By calculating the amortisation of the family's agricultural holding, its production costs and expenditure on equipment, he can work out a balance sheet, understand what it means and find out if money has been made or lost.

The farmer has always used credit, even if he does not call it by that name. In order to bridge the gap between one harvest and the next he borrows a certain amount of food from a neighbour which he returns, together with an additional amount, after the harvest. If he needs money outside the market time, he borrows from a trader and pays back in kind at harvest time. The interest rates on such loans are usually very high: if the farmer has borrowed 20 francs, three months later when the crop is harvested, he will hand over a sack of coffee worth say 30 francs. Moreover, he often contracts loans for consumption expenditures but seldom for production purposes; this makes repayment very difficult. Another widespread form of credit is the 'tontine' or 'kitty' - a group credit arrangement to which each member contributes the same amount each month; the total amount is allocated to each member in turn. The farmer who wishes to make progress has to turn to modern credit facilities: however useful traditional forms of credit may be, their volume is not very great and they are a heavy burden on the farmer's budget. Money from the common purse of the 'tontine', although less costly, is often used for non-productive expenses such as traditional celebrations, marriages or consumer goods.

Modern agricultural credit is much more complex than the traditional forms of credit. Information has to be obtained about the different types of loans, the interest rates, the collateral required, mutual guarantees and personal contributions. Cash flows have to be worked out and annual repayments calculated. In an endeavour to provide the farmer with guidance in acquiring all this knowledge, our course on credit uses the story of a farmer who wishes to progress, and who is in need of money. With the assistance of an extension worker, he weighs up the problem and tries to find out how he can manage. The story goes step-by-step through all the representations the farmer has to make to apply for and obtain the loan. It then shows him how he organises his work to produce results from the money obtained so as to be able to meet the annual repayments. This simple but comprehensive story enables the farmer to have an all-round picture of the problem. He acquires a very thorough knowledge of the credit system, of ways and means of obtaining a loan and of the terms and conditions of repayment.

UPGRADING THE TRADITIONAL VILLAGE GROUPINGS

There have always been groupings of peasant farmers in villages, including work groups to do difficult jobs in each member's fields in turn, mutual assistance groups to care for the fields of the sick or the absent, or celebration groups whose members work together and save the proceeds of their labour to have a celebration.

A strong spirit of village solidarity and an ingrained habit of working together have grown out of these groups. However, owing to their relative lack of structure, these groups no longer meet the present-day requirements of agriculture where the energies of all must be harnessed if there is to be some progress. The old spirit can and should be upheld and adapted to more firmly structured and more efficient types of association. Our course on associations attempts to meet this need. It uses concrete stories to explain why it is difficult for one farmer on his own to make progress without arousing the jealousy of others, and to show that a united and determined group can produce much more for each of its members. The course then goes on to explain the conditions that have to be fulfilled if a group is to be successful. So many attempts at co-operatives have failed that it is very important for the farmer to have a clear understanding of the factor governing the success of an association. The farmers themselves have to choose the members of the group and then the group selects an immediate aim that can be easily achieved so as not to dissipate its efforts. The selection of group leaders is important since these people will be responsible for organising the work, representing the group on outside matters and ensuring compliance with the rules.

In narrating the story of a village which has set up an association, the course explains the reasons behind the decision, the motivations of its members and the pitfalls which have to be avoided if the initial enthusiasm is not to turn sour when difficulties crop up. As the story progresses, the course brings the student to reflect on the actions he has just read about, to understand the reactions of the characters in the story and to draw conclusions that will guide his own personal actions.

The techniques taught in our courses are not theoretical. They are based on the observations which our student subscribers have made in their villages and their fields. They are illustrated by concrete examples and they are dynamic in that they incite the students to take action to improve their own lives and the life of their community.

XVIII. THE ROLE OF NON-GOVERNMENTAL INSTITUTIONS IN THE INNOVATION PROCESS

by

John W. Pilgrim*

Maize was introduced to African peasants in Kenya by European farmer settlers at the turn of this century, and was actively encouraged as a crop in African farming areas during the First World War. The concomitant introduction of new ploughing, milling, storage and transport technologies in these areas led to the development in the 1920's of a market economy based on maize. The new crop progressively came to replace or supplement the traditional cultivation of finger millet, and the technical innovations which accompanied it paved the way for the growth of Kenya's present agricultural market economy.

Three aspects of this innovation process are of particular relevance here. The first is the appropriateness of these new technologies. The second is the adaptive character of the organisations and institutions which either fostered the innovations or were directly affected by them. And the third, which bears directly upon a country's development policies, is the effect of technological innovation on social and agrarian structures. The main theme underlying the present paper is that non-governmental organisations (churches, educational institutions, research centres, private groups, etc.) can play, and have played in the past, a major role in the development process by fostering technical and social innovation, and in particular appropriate low-cost innovations.

THE INTRODUCTION OF MAIZE-GROWING IN KENYA

One of the main reasons for the successful adoption of maize in the Kericho District of Kenya(1) was the existence of cattle herding

---

* The author was the Director of, and is currently adviser to, the Centre for Applied Research of the Panafrican Institute for Development in Douala, Cameroon.

1) Although this section focuses on one particular district, it must be emphasised that the processes of innovation in maize cultivation and processing in other parts of Kenya were not fundamentally different.

as the main male economic activity which made it possible to use oxen as draft animals. Production of maize for the market thus became mainly a man's activity, while women continued to produce finger-millet for home consumption. The basic innovation in production technology was the iron plough, manufactured in England and introduced to Kenya at the same time as maize. This innovation had an immediate and profound effect on the agrarian structure and induced far-reaching changes in the use of land and labour and in the property system.

The tradional method of milling finger-millet, by rubbing it between flat stones, was not applicable to maize. Morever maize was unsuitable to traditional methods of grain storage, being more subjec to infestation by weevils, and for these reasons was not immediately acceptable for domestic use.

Its gradual adoption as a dual crop, for home consumption and cash sale, was therefore dependent on the introduction of appropriate milling and storage methods. The first of these technologies in the form of imported iron hand mills and locally-made water mills, was learned from missionaries; the second, in the form of open crib stores, from both missionaries and European farmers.

Both technologies had been developed over a period of hundreds of years in Europe, and were still used in Europe and the United States at the beginning of this century when they were introduced into Kenya. The missionaries who played the most active part in this transfer of technology were mainly American. The water mill which employs the principle of gravity drive by water was not a scaled-down modern technology, but a traditional medieval technology. It continues to be the main means of milling for home consumption in Kenya today. Such mills are generally owned and managed either by individuals or by small groups of three or four men.

THE MAIN FEATURES OF THE INNOVATION PROCESS

Research into the overall agrarian and institutional processes which were associated with the development of maize cultivation and post-harvest technology has brought to light a number of important elements in the diffusion of these innovations(1).

One of these is the critical role of environmental circumstances in the initial adoption of maize cultivation. Soils and climate were suitable and maize presented an attractive alternative to traditional forms of capital accumulation like livestock ownership. This was in itself a differential and differentiating process.

---

1) See J.W. Pilgrim, Agrarian Development and Social Change in an African Society, Centre of Applied Research, Panafrican Institute for Development, Douala, 1975.

The innovators were entrepreneurial in character, and when starting to grow maize for the market, were prepared to engage in conflict with fellow-villagers. The difficulties this created are clearly spelled out in two government reports which can be quoted here.

"A situation is arising in South Lumbwa District which may well prove a matter for anxiety in a year or two. A number of the more advanced Kipsigis are plough owners, and as the tribe has no system of land tenure, other than as a community, these plough owners tend to cultivate very large areas indeed, thus reducing the available amount of grazing....".(1)

"A growing clash of interests between the Kipsigis agriculturist and the Kipsigis pastoralist is noticeable. The one uses a plough and brings large areas under cultivation and the other allows his herds and flocks to trample and consume the ensuing crops. Fencing of small gardens is the general practice but large acreages cannot be fenced because in most areas there is an almost complete lack of fencing material."(2)

These problems were also noted by an anthropologist from Oxford University:

"In every community one or two wealthy people possess a plough, and they hire them out to the others ... People who have sold a few cows to buy a plough have made large profits by this transaction, as they hire the plough at a very high rate.... Sometimes money is not demanded, but the man to whom the plough is lent is asked to assist the owner in ploughing or weeding his fields ...."(3)

THE DIFFERENTIAL NATURE OF THE INNOVATION PROCESS

These differential, capital accumulative responses were reinforced by institutional and cognitive support to the individuals and groups involved. This support was provided mainly by mission membership and schooling, by employment experience with European farmers and by recruitment to administrative posts within the local colonial government. The 'agriculturalists' ploughing up large areas and building water mills were most often mission-educated and held positions as headmen and chiefs, often after spending several years working on local European farms.

---

1) District Commissioner's Report, Kericho District, 1934.
2) District Commissioner's Report, Kericho District, 1936.
3) J.G. Peristiany, The Social Institutions of the Kipsigis, London, 1939.

Access to new technology therefore tended to be accompanied by exceptional access to initial capital inputs, to administrative and judicial protection and to educational resources.

During a critical period of ten to twenty years, plough agriculture and innovations in economic and domestic behaviour were generally closely associated with mission membership. The first water mills belonged to notable mission adherents, who had been helped to construct them by the missionaries. The mills were sited at 'Christian villages' - groups of adherents who had clustered together, usually around a school.

Land enclosure (i.e. the dividing up of the community's land into individual or nuclear family holdings and fencing or hedging) was an eventual direct result of these innovative land-use practices, and again tended to reinforce the technical and economic advantage of the innovators. In 1960 this advantage was still discernible as a correlation between mission adherence and land ownership.

In general, however, differential adaptation to innovation and to a cash economy was related more directly to the size of land ownership. Research at one village in Kericho District showed for instance that in 1960, 66 of the 112 land owners had 16 acres or more and that 57 owned 15 acres or less(1). The affluent farmers employed a much larger number of hired labourers: Twenty-eight (42.4 per cent) of the large land owners employed one labourer or more, as against only one out of the 57 small holders. Twenty-four of them (36.4 per cent) had planted tea as against nine (15.8 per cent) of the farmers with 15 acres or less. Twenty-two (33.3 per cent) of the affluent farmers had sons who had passed through some secondary education and were now in some form of government or white collar employment, as against two (3.5 per cent) of the farmers with 15 acres or less.

THE ROLE OF NON-GOVERNMENTAL INSTITUTIONS IN THE INNOVATION PROCESS

The story of maize-growing in Kenya shows not only that technological innovation was accompanied by the emergence and formation of innovative, entrepreneurial men, but also that the institutional support which was a necessary corollary of this process led inevitably to internal institutional change. One effect of this has been the creation of different classes within the rural population. The relatively wealthy and privileged in terms of material resources had

---

1) These figures do not reveal the extreme cases of innovative individuals, such as the man owning 168 acres, who had benefited from the fact of playing dominating roles in the early processes of agrarian change. For further details, see Pilgrim, <u>op.cit</u>. pp.109-111.

access to capital, technology and influence, while the relatively poor and underprivileged did not.

This is a very general phenomenon. In most developing countries, the agricultural extension system,credit organisations and other support services tend to continue this distinction by operating with the 'progressive' and relatively wealthy farmers rather than with the 'backward' and relatively poor(1).

But in more general terms the lessons from this experience of technological innovation at the level of a poor farming population are relevant to the more consciously designed 'development project' (unheard of in 1920). One of them in particular is the role of non-governmental institutions, of which the mission is a prime example.

The mission's function as a successful agent of technical change rests upon a number of factors. The first is its stability which allows it to act as an ongoing institutional base for innovation. The second is its conscious effort in training and education; this work incidentally matches man to technology and technology to man. The third is the fact that it operates on a low-cost, local material basis. The fourth is that it does not consciously set time limits, and is thus prepared to continue its efforts over an indefinitely long period. Finally, it does not set out to provide any superior external expertise, nor does it base its authority on technical know-how or on privileged links with a national or international techno-cracy.

By and large these assets continue to be those of non-governmental development agencies. It is worth while asking why in particular organisations with missionary origins should operate in this way. One answer is that their main purpose is to integrate themselves into the local structures and local culture, and that it is not any specifically religious need that induces them to behave in this way. A research organisation, or an educational institution, has similar motivations and similar capacities for endurance and adaptation.

Historically the same role was played in the United States by the land-grant institutions with their commitment to a tripartite function of training research and extension. In West Africa, the Institut Africain pour le Développement Economique et Social, the Panafrican Institute for Development and such universities at Ife,

---

1) See Niels Röling, "Forgotten Farmers in Kenya", paper prepared for the OECD Seminar on Development Projects to Reach the Lowest Income Groups. Paris, June 1974, and Joseph Ascroft, Niels Röling, Joseph Karinki and Fred Chege, Extension and the Forgotten Farmer, Landbouwhogeschool, Wageningen, 1973.
Röling and his colleagues demonstrate that, in a neighbouring area of Kenya, credit and agricultural extension services are similarly channelled to the wealthier farmers, and that in consequence the diffusion of hybrid maize is impeded over a major part of the area.

Ibadan and Ahmadu Bello in Nigeria, Njala in Sierra Leone, and Kumasi Institute of Science and Technology in Ghana are playing similar roles of technological extension(1).

THE SHORTCOMINGS OF GOVERNMENTAL INSTITUTIONS

These non-governmental institutions are not primarily those which governments and external development agencies think of or seek to create in order to accelerate development or to achieve a specific technological change. Generally speaking, the major institutional machinery used for these purposes in rural areas is that of routine extension or other rural development services. Universities and other research and educational institutions may contribute to these services indirectly or in a limited context, but they are in themselves usually completely different in structure from the centrally-based bureaucracies of the government sector.

If research and educational institutions do not as a rule provide an adequate machinery for the diffusion of new technology, it is equally true that the track record of agricultural extension services has been marked by a constant tendency to fictionalise achievement. Thus in Ghana, while the agricultural services saw no difficulty in evaluating increases in cocoa production in terms of the technical services they offered to cocoa farmers, subsequent research has shown that the cocoa farmers in fact succeeded in increasing production by virtue of their own economic understanding of land and labour utilisation. They had early learned the lesson that the economics of cocoa farming required that the farmer ignore the technical advice offered year-in year-out over a quarter of a century by the agricultural services(2).

In Kenya the government's agrarian policy has been based on the understanding that individual holdings were the only path to the development of agriculture and in Kericho District in the 1950's the enclosure programme conducted under the guidance of the agricultural service was claimed to have been the basis of a totally changed economy based on commercially-oriented agriculture(3). In fact, there had been virtually no overall change in agricultural production over

---

1) See the papers by P. Dubin on the INADES, and by B.A. Ntim and J. Powell on Kumasi University in this book.

2) See Polly Hill, Migrant Cocoa Farmers of Southern Ghana - A Study in Rural Capitalism, Cambridge University Press, Cambridge, 1963.

3) See C.W. Barwell, "A Note on Some Changes in the Economy of the Kipsigis Tribe", Journal of African Administration, Vol.8, No.2, April 1956, pp.95-101.

a thirty year period, and enclosure had in any case taken place spontaneously through internal processes to which the government had previously been for many years opposed. Recent prosperity in the district has largely come about through the effects of migrant wage labour. Where technical innovations and a new commercially-oriented agriculture were developing, this was basically, as we have noted above, the work of men with entrepreneurial characteristics who learned from the experiences of a European-based agricultural economy and from the missionaries. These innovations, which date from the introduction of maize as a cash crop in the 1920's, were a cause rather than a result of land enclosure, and government agencies played little if any part in the process.

One reason for the general weakness of agricultural extension services is the absence of an R and D capability of the type which universities or other research institutions could provide. But the importance of R and D in low-cost technology should not be over-emphasised. Account must be taken of the fact that all innovative technologies are experimental and that the social processes of adoption are often very long.

In this context the R and D function may best be carried out by an institutional structure with rugged and primitive capabilities of learning and perseverance rather than by one with sophisticated technical means at its disposal. Success will come from an ability to communicate in the language of the people, to use local resources and to operate at the cost level of a local poor uneducated society, rather than in the language and with the resources of external technological or administrative institutions. These are the qualities of the missions rather than those of the research station.

Any attempt to emulate these qualities must however be conducted in a way which will avoid the defects of the missionary approach. In fact, the examples of effective action on the part of non-governmental organisations are greatly outnumbered by the cases of failure due to irrelevance, poor learning capacity, lack of means for wide-scale extension, poor links with the major social and economic development efforts provided by public agencies, and lack of awareness of the broad changes in land and labour use. There is no merit in rugged and irrelevant endurance.

There is, equally, no virtue in slowness for its own sake in the processes of learning and response. It could be argued that where a missionary type involvement has led to technological innovation, as in the Kenyan case cited above, this is usually because the learning and testing process is in fact taken over by sharp-minded entrepreneurs. Entrepreneurship will, however, by definition, always lead to a differential process of adoption of new technology. Development objectives (as against the objectives of simply introducing technological change) may therefore need specifically to be framed in terms of control over these differential processes which discriminate betwee individual and individual, group and group, region and region. In this context, the R and D which is required may be essentially of the overall planning type.

THE EXPERIENCE OF THE PANAFRICAN INSTITUTE FOR DEVELOPMENT

Apart from the missions, another type of non-governmental organisation which can play an important part in rural development and the diffusion of low-cost technology is the training and educational institution. One such case is the Panafrican Institute for Development (PAID). It is an international non-governmental and private organisation, affiliated in various ways to the United Nations and other international organisations and supported financially by some twenty-five governments and independent aid agencies. It has a small office in Geneva and four institutions in Africa, all located for the moment in the United Republic of Cameroon.

PAID started in 1965 as an international college, whose function was primarily to train middle level development staff from francophone African territories, whether from government or from non-governmental agencies. Its main concern is with rural development (and especially with agricultural and rural training generally), community development ('animation rurale'), sub-regional planning, co-operative organisation and project management. Five years later a similar college was set up for English-speaking African trainees in Buea, in the anglophone part of Cameroon. In 1972 and 1973 respectively, the Centre of Applied Research and the Centre for Project Management were created

A major aim of PAID has been to maintain the link between training, research and management and at the same time to strengthen what has always been a primary concept of the Institute, namely, direct involvement in the field in collaboration with the local population and the administration. Until now this has been achieved only to the extent that national agencies use PAID trainees in functions which correspond with their training. More recently with the creation of capability to conduct research on a sustained basis and to provide assistance with planning and management, PAID has begun to find itself involved directly with government and other agencies in the formulation of development programmes, and has had to recast its own structure and aims to maintain and improve control of its operations.

The pattern which has emerged is one in which PAID can operate over a relatively long period at the sub-regional level without relinquishing its traditional training role but linking it to research, planning and management assistance for the integrated development of a zone (which might be a province or district of the assisted country). This pattern is illustrated by programmes currently undertaken in Cameroon.

One question which arises in this context concerns the nature of integrated rural development as a planned operation. In the South West Province of Cameroon, for example, feasible low-cost technological innovations can be introduced in the village-level rubber and oil palm plantations, in the storage of maize and root crops and in the raising, processing and marketing of pigs. Much of PAID's present research is concerned with the simultaneous introduction of these innovations at the farm level, in the curricula of primary schools, in out-of-school vocational training centres and through the training of development agents.

This has, in our mind, particular relevance to the working out of methodologies for integrated rural development planning within which flows of capital and technology can occur. In the absence of comprehensive data for the zone, planning may be effectively provided by the linking of specific technological innovation to educational and other informational processes on the one hand and to market structures on the other. The other essential prerequisite for such a process of sub-optimal planning is that of an on-going institutional capacity to evaluate and feed back the experience of innovative action. This in our present operations may be provided by PAID initially but ultimately it poses the need to provide for an indigenous institutional capacity for R and D, which we believe has to operate at the level of a geographical and administrative zone.

XIX. LOW-COST TECHNOLOGY, COST OF LABOUR MANAGEMENT AND INDUSTRIALISATION

by

A.J. Bhalla*

INTRODUCTION

What is low-cost technology? While terminology is not always what matters very much, it is useful to define the context in which the practitioners of low-cost technology operate. Low-cost technology may be defined as technology which is low in terms of capital (physical plus working), foreign exchange, labour skills and any other inputs which are often scarce in the rural areas of developing countries in particular, and in the economy in general. On the other hand, it is high in terms of abundant local resources (e.g. indigenous materials and unskilled labour). In addition, if one looks at the technology from the output point of view, it may be low in terms of scale of output, and sometimes even in terms of the quality of the product.

In this short study, the cost of labour management will be related to the choice of technology in the process of rural industrialisation. It is often argued that low-cost technology might often be labour-saving relative to indigenous technology, yet labour-intensive enough to raise questions of high costs of supervision and management. We shall examine this hypothesis in the light of empirical evidence about industries in developing countries. It will be maintained that management costs arising from such factors as rates of labour discontinuities, increase in supervisory wages, and social resistance to labour dismissal, etc. are also influenced by the nature of industrialisation strategies that are adopted, and the modes of production (wage-based or family-based) under which different technologies operate. It is for this reason that a brief section below

---

* The author is Chief of the Technology and Employment Branch, World Employment Programme, of the International Labour Office in Geneva. This paper is written in his personal capacity and does not necessarily reflect the views of the organisation with which he is associated. He has received helpful suggestions from Prof. Subbiah Kannappan.

is devoted to technological transformation processes in industrialisation before analysing the relationship between labour management costs and low-cost technology.

TECHNOLOGICAL TRANSFORMATION STRATEGIES

It is often argued that the rural areas and the urban 'informal' sector are the main sources of low-cost technologies. Even if the technologies currently prevailing in these sectors may not necessarily be economically efficient, it is possible to achieve lower costs of production by raising the technological level of traditional technologies. This process of modernisation could either be gradual or too rapid and discontinuous(1). However, the latter 'crash modernisation' strategy under which the rural and urban 'informal' sectors play only a passive role in the development process, and under which the bulk of the limited resources are concentrated on the small segment of large-scale modern industry is likely to militate against rural industrialisation and low-cost technologies. In fact, a policy of rapid technological transformation is only likely to displace the rural industry and artisan sectors unless the latter are protected through some sort of subsidy. On the other hand, the gradual transformation policies are likely to ensure a complementary development of rural and small-scale industry and modern large-scale industry.

The experience of rural industrialisation in China and of its 'walking on two legs' policy clearly demonstrates that leg-one (larger-scale and capital-intensive), and leg-two (small-scale and labour-intensive) projects were linked to ensure productivity improvements in the latter. Since the Great Leap Forward, increasing attention has been paid to narrowing the gaps between the two legs: emphasising fuller utilisation of existing capital stock in large modern industry to make resources available for ensuring technical improvements in the traditional technologies, small-scale and rural industries.

The linkages between agriculture, rural industry and modern industry in China are illustrated in the diagram below which is borrowed from Sigurdson(2). It is shown that rural industry is

---

1) For a comparison and contrast between 'progressive transformation' and 'crash modernisation' programmes in a quantitative framework, see Keith Marsden, "Integrated Regional Development: A Quantitative Approach", International Labour Review, June 1969.

2) Jon Sigurdson, The Role of Small-Scale and Rural Industry and its Interaction with Agriculture and Large-Scale Industry in China. The Economic Research Institute at the Stockholm School of Economics (mimeo), Stockholm, 1974.

expected to be complementary to both agriculture and modern industry. Rural industrialisation is introduced in gradual stages. The first stage represents rural industry as an intermediary between agriculture and modern industry: producing industrial inputs for agriculture and processing agricultural output, and receiving equipment, technology and training from modern industry. At the second stage, rural industry and agriculture become partly integrated (shown by intersecting circles), with agriculture supplying more and more raw materials for light industrial products and with an emerging two-way relationship in modern industry. In the third stage, rural industry is complementary to both agriculture and modern industry, and integrated with rather than eliminated by the latter. Quality standards of products of rural industry and improvements in technology take on additional importance at this stage.

Although the schema below may be a rather simplified version of real-world situations in many developing countries, it does show the working of a gradual transformation of rural industry in the context of which the cost of labour management issue needs to be examined.

DEVELOPMENT OF RURAL INDUSTRY:
INTERACTION WITH AGRICULTURE AND MODERN INDUSTRY

STAGE 1

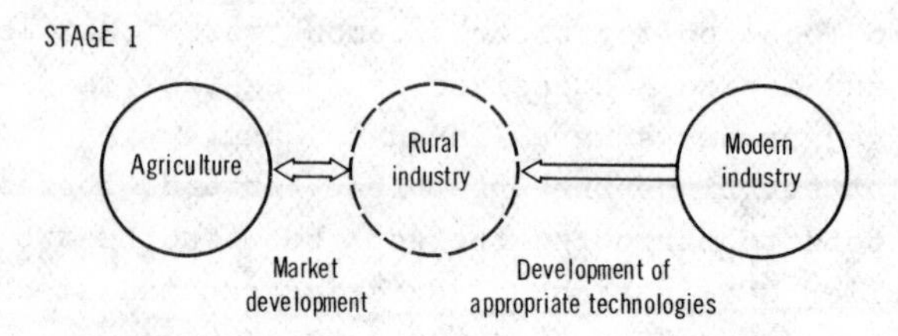

STAGE 2

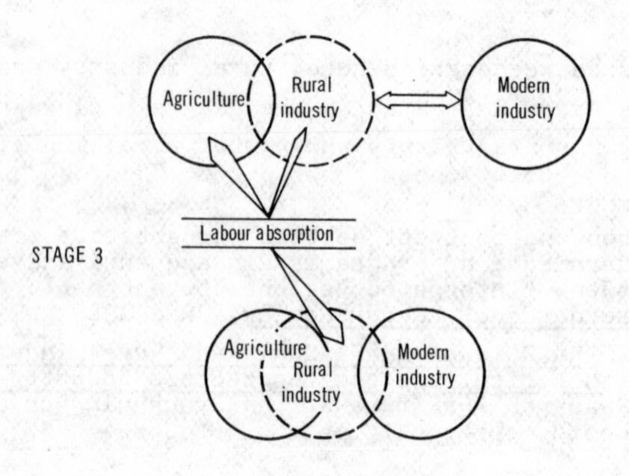

TECHNOLOGY, MANAGEMENT AND ORGANISATION

Hirschman can be considered as one of the early and controversial writers who pointed out that in developing countries the requirements of labour are proportionately much larger when less capital-intensive methods are used, due to lower labour efficiency and lower managerial expertise(1). His prescription is that a developing economy can make more efficient use of capital-intensive methods, especially in machine-paced (as against operator-paced) industries, since the labour force is unstable, inexperienced and less disciplined, and managerial talent very scarce. If these technologies are used, the differential in industrial productivity between developed and developing countries will tend to be narrower than might otherwise be expected.

Implicit in the above prescription is also the choice of a particular industry-mix in the developing countries. In support of his hypothesis, Hirschman considered the following types of industries: (a) process-centred machine-paced industries, e.g. oil-refining, iron and steel smelting, and cement manufacture, and (b) product-centred operator-paced industries, e.g. construction and fabrication industries, carpentry, brick-laying, metal-working, where tools and machines are largely accessory to workers. By the nature of these industries, it would seem that those in category (b) lend themselves to rural areas and small-scale operations much more than the latter.

Equating process-centred industries with machine-paced capital-intensive ones somewhat arbitrarily, Hirschman argues that the latter are more efficient. This is so because "machines set a rhythm of work, a steady pace and standard norms of efficiency difficult to ensure when the work-force is undisciplined". Secondly, management costs may be lower with machine-intensive methods because the mechanical or chemical processes define precisely for the manager "what is to be done where and at what point of time ...". Thirdly, another argument in favour of machine-paced integrated continuous flow processes is that they require adequate maintenance. However, since machine operations are also dependent on adequate maintenance skills which may be in short supply in developing countries, it does not necessarily follow that capital-intensive operations have as high a productivity as in the developed countries. Even within a developing country, given the capital-intensive technology in two different plants, large differences in productivity may be observed. This is due not to differences in capital per worker, but mainly to differences in managerial efficiency (caused perhaps by motivational

---

1) Albert O. Hirschman, The Strategy of Economic Development, Yale University Press, New Haven and London, 1958.

factors). Higher productivity may also result from a simple reorganisation involving plant lay-out, materials handling, waste controls, etc.(1)

EMPIRICAL TESTS OF HIRSCHMAN'S HYPOTHESIS

Hirschman's general hypothesis that the greater the degree of capital-intensity in industry, the smaller the labour productivity differential between developed and developing countries, has been empirically tested on the basis of rather aggregate data published in national censuses of manufacturing. We shall briefly examine the conclusions of three tests, the last of which (by Gouverneur) is based on the use of plant data.

Carlos Alejandro has examined the comparative productivity of industry in the US and Argentina(2). By modifying the Hirschman hypothesis he has tried to find out if, when the labour-intensity of Argentinian industry increases, the productivity differential with respect to American industry also tends to rise. His results provide "only modest support for the assumption that industries with low labour-intensity tend to be associated with import-substitution, while those with high labour-intensity fall more into the category of home or non-tradeable goods". To the extent that rural industrialisation through a "gradual transformation strategy" requires wage-goods manufactured for limited local markets, the greater labour-intensity of industries would prevail at least initially. It may also be, as we shall examine in the following section, that home industries in rural areas or small towns are more labour-intensive or use low-cost technologies because the international companies with advanced management and production techniques are much more likely to invest capital in import-substituting large-scale industries than in the small-scale domestic ones.

The second empirical test is based on a comparison of 110 broadly similar industries in the UK and India for the years 1958-59 and 1960(3). There is no conclusive or definite evidence to support or refute Hirschman's hypothesis. In some extreme cases, labour-intensive

---

1) See Peter Kilby, "Organisation and Productivity in Backward Economies", Quarterly Journal of Economics, May 1962; and Harvey Leibenstein, "Allocative Efficiency versus "X-Efficiency", American Economic Review, June 1966.

2) Carlos F. Diaz Alejandro, "Industrialisation and Labour Productivity Differential", Review of Economics and Statistics, May 1965.

3) J.M. Healey, "Industrialisation, Capital-Intensity and Efficiency", Bulletin of the Oxford University Institute of Economics and Statistics, November 1968.

industries may be particularly inefficient; this does not necessarily mean that capital-intensive industries are especially efficient in an industrially less developed country. From the observed association between capital-intensity and efficiency in some industries, Healey concludes that this "may be more plausibly attributed to a 'technological gap' hypothesis, rather than the Hirschman type sociological/ psychological thesis".

Technological gaps between the developed and developing countries are likely to be narrowest in the case of the most capital-intensive rather than in the case of labour-intensive industries. The reason for this is the greater international mobility of physical equipment as compared to labour skills. This factor is relevant to the consideration of supervisory costs associated with alternative technologies (see the following section).

The third and perhaps the most recent empirical test relates to individual plants in the shipbuilding and flour-milling industries in Zaire (formerly Congo-Kinshasa) and Belgium(1). The second of these two industries is process-centred and therefore relevant to rural industrialisation. The plants analysed by Gouverneur were all run by experienced Belgian entrepreneurs. This means that productivity differentials were mainly due to the nature of the industries and to their type of operation. The empirical analysis shows that the productivity differential between Zaire and Belgium in 1968 was 48 per cent in shipbuilding and 30 per cent in flour-milling. This may be seen as broadly supporting the Hirschman hypothesis, namely that productivity differentials would be higher in product-centred (operator-paced) industries than in process-centred machine-paced industries _even when similar technologies are used in the developed and developing countries_ (emphasis added).

However, the Gouverneur study has a merit over the earlier two. It brings to the forefront an assumption implicit in Hirschman's arguments, namely that productivity differentials between developed and developing countries will tend to be small for _all_ operations since the central mechanical processes indirectly set the pace of all other ancillary operator-paced operations. This hypothesis however is contradicted by the facts, as can be seen from the data presented in the table below.

The productivity differential between the Zairian and Belgian flour mills is 30 per cent for _all_ production workers; it is only 4 per cent for those involved in the 'technical flow-sheet' (machine-paced) operations.

---

1) J. Gouverneur, "Hirschman on Labour Productivity Differentials: An Empirical Analysis", _Bulletin of the Oxford University Institute of Economics and Statistics_, August 1970.

LABOUR PRODUCTIVITY DIFFERENTIALS BETWEEN ZAIRE AND BELGIUM
(Flour-Milling, 1964)

	Belgium	Zaire (3)
A. Production workers		
All production workers	100	70 (60)
Workers in 'technical flow sheet' (1)	100	96 (71)
Workers in 'general handling'(2)	100	46 (46)
B. General maintenance workers	100	66 (54)
C. All production and maintenance workers	100	69 (58)

Notes:(1) Machine-paced operations.
(2) Man-paced operations.
(3) The figures in brackets are not adjusted to the rate of capacity (90 per cent) of the Belgian plant.

A significant conclusion is that labour productivity differentials should be explained not simply as due to process- or product-centred industries but the extent to which individual operations are machine-paced and operator-paced, and the degree of capital-intensity.

The Hirschman assumption that all machine-paced operations are necessarily capital-intensive is a little over-simplified. Although the control and correction of mechanical processes is treated as machine-paced operations, they could well be done by manual low-cost technology. Similarly, operator-paced operations like internal transportation also lend themselves to varying degrees of labour/capital intensity and to different technological choices.

SUPERVISORY COSTS

It is often assumed that although labour-intensive technologies may be low-cost in the sense of economising a scarce factor such as physical equipment, they can be highly skill-intensive. Since some types of skills tend to be extremely scarce in most developing countries, it is worth examining whether these technologies are also low-cost in terms of skills required. In this context, a distinction between different kinds of labour skills (managerial, supervisory, and other types) is quite useful(1). This distinction is important because the use of low-cost technologies in the process of industrialisation need not necessarily imply high requirements for all these types of skills. In fact, in many cases, the small-scale of operations within non-wage modes of production may imply an economy

1) See also A.S. Bhalla, "Small Industry, Technology Transfer and Labour Absorption", in OECD Development Centre, Transfer of Technology for Small Industries, Paris, 1974.

of supervisory, operative, and managerial skills. The scale as well as the technologies employed may be conducive to family style non-specialised management which can dispense with complicated control procedures and hierarchical authority, required by sophisticated technology in modern industry(1).

Case Study Illustrations

a) In Tanzania, attempts to introduce mechanisation in rural areas required constant recourse to government staff (e.g. agricultural engineers) with sophisticated management to ensure regular supplies of fuel and spare parts, and the proper use of machinery. When government workers were withdrawn, much of the machinery fell into disuse for lack of the above "external" inputs. It was therefore decided that the work undertaken by two ILO/UNDP experts in Arusha on the design and construction of equipment prototypes should conform to the needs and capacity of the villagers themselves in the use of intermediate 'village' technologies(2).

b) In a pilot survey of 37 plants in three tropical African countries done by the ILO(3), questions were asked as to whether technology in African industry was influenced by the quality of labour, the training possibilities, and the ease or difficulty of changing the skill-mix. The responses of managers varied, depending on the nature of the industry, the size of operations and the skills required. These responses led to the following conclusions: small-scale saw-milling enterprises, engineering plants, and some wood-working enterprises felt no shortages of skilled labour; but bigger plants (cement, steel and engineering) using capital-intensive methods reported skill shortages at both higher and lower levels.

c) In the Nigerian manufacturing sector, the use of a greater number of older type of machines accompanied by the use of labour-intensive handling techniques is reported to have lowered the number of senior (still largely expatriate) technicians required to maintain the more mechanised processes. However, this choice would involve

---

1) For a discussion of technologies under different modes of production and employment, see Amartya Sen, Employment, Technology and Development, Oxford University Press (on behalf of ILO), London, 1975; and "Employment, Institutions and Technology - Some Policy Issues", International Labour Review, February 1975.

2) For an account of the Tanzanian experience, see George Macpherson and Dudley Jackson, "Village Technology for Rural Development", International Labour Review, February 1975.

3) See ILO, Some Factors Affecting Employment and Choice of Technology in African Industry - A Pilot Survey (mimeo), Geneva 1972; see also A.S. Bhalla, "Implications of Technological Choice in African Countries", Afrika Spectrum (Hamburg), No.1, 1973.

higher operative skills, greater reliance on Nigerian supervisors and greater vulnerability to interruptions as a result of human failure(1).

d) In Kenya, in connection with the ILO-sponsored employment mission, a case study was undertaken on the semi-automatic and automatic processes in the manufacture of tin cans(2). There was some suggestion that supervisory costs as a proportion of total unit costs rise in line with the degree of labour-intensity. The main problem about semi-automated processes seemed to be the requirement of supervisory labour which acted essentially as a substitute for worker skills(3). In the case of round open-top cans, supervision costs were 37 per cent of total costs on the very labour-intensive lap-seaming line; they were proportionally lower on the more capital-intensive lines (16 per cent and 13 per cent respectively). However, the supervisory costs rose sharply with the most capital-intensive process, rising to 21 per cent of total unit costs. It is interesting to note that supervisory costs are lowest with intermediate technologies, and rise with greater labour-intensity and greater capital-intensity.

A related case study of can manufacturing in Thailand yielded plant data similar to that for East Africa, but showed no clearcut association between labour-intensity and supervisory costs: the latter were generally in the region of 15-20 per cent. However, the Thai case study "suggested an interesting qualitative difference between supervision on capital-intensive lines and the labour-intensive ones". The supervisory wage rates are much higher on the automated than on the 'normal' processes, and this is only partly due to the former being in the high-wage modern than the low-wage 'informal' sector. An important factor explaining this wage differential is the type of supervision needed. On the automated processes, the supervisors are required to possess a high level of technical ability to manage machines, whereas on the labour-intensive lines, they are mainly asked to organise large numbers of workers. The ILO case study argues that in developing countries in general, the skills needed for organising workers may be more readily available than those needed

---

1) Peter Kilby, Industrialisation in an Open Economy: Nigeria 1945-66, Cambridge University Press, Cambridge, 1969.

2) ILO, Employment, Incomes and Equality - A Strategy for Increasing Productive Employment in Kenya, Technical Paper No.7, Geneva, 1972.

3) This situation in can-sealing needs to be distinguished from that of labour-intensive small works where supervision is a necessary complementary organisational input for managing large numbers of men. In the case of manufacturing, a more labour-intensive process may not necessarily require supervisory skills. Much depends upon the precise pattern of skill inputs required for various techniques

for supervising machines on automated processes. This leads the authors of the study to suggest that "the supervision may not be quite as general a problem in the manufacturing sector as it is sometimes thought to be"(1).

e) A report on Indian industry showed that good management was more important for task simplification than for mechanisation. Assuming one shift, poor management and poor incentive systems would lower production in semi-automatic operations by one-fourth and in hand-fed operations by one-third, implying that good management is more important in hand-fed operations(2).

The foregoing illustrations give no clear-cut support to the hypothesis that the cost of supervision and management are higher with the low-cost labour-intensive technologies. Of course, one difficulty with most of these case studies is that they did not assess directly the effect of scale or volume on the managerial and supervisory costs. However, to the extent that low-cost technologies are more suited to small-scale operations, it would seem _a priori_ that management and supervisory problems could be reduced through the scaling-down of plants. Besides, in the context of rural industrialisation, small firms run by owner-operators and unpaid family workers do not require intermediaries, since organisation is much simpler. With each production unit producing for its own consumption, or with each individual working in producing a single commodity, the problem of co-ordination of simultaneous operations, supervision, and marketing management (typical of intra-commodity division of labour) are either avoided or considerably reduced.

One could also argue that given the scale of production units, more disciplined cultures would perhaps reduce the need for tight supervision. In support of this hypothesis, the Japanese system of management may be put forward as a case in point. Unlike many Western capitalist countries, in Japan traditional culture has not been replaced but rather has been assimilated by industrial development. In an industrial enterprise, or in any other type of organisation, it is the group rather than the individual which assumes responsibility for decisions. And as Harbison puts it "from the individual the ethical code demands unqualified loyalty to his group, subordination to his superiors, respect for his elders and complete identification with the goals of the house"(3). These non-economic factors by which

---

1) See A.S. Bhalla (ed.), _Technology and Employment in Industry_, ILO, Geneva, 1975, Chapter 4 on "Choice of Techniques for Can-Making in Kenya, Tanzania and Thailand" by Charles Cooper _et al_., p.112.

2) United Nations, _Choice of Capital-Intensity in Operational Planning for Underdeveloped Countries_, paper presented by the Centre for Industrial Development at the São Paulo Seminar on Industrial Programming, São Paulo, March 1963.

3) Frederick Harbison, "Management in Japan", in Frederick Harbison and Charles Myers, _Management in the Industrial World - An International Analysis_, McGraw Hill, New York, 1969, p.254.

employment within an enterprise is regarded as a life-time commitment may tend to generate a sense of discipline which makes a high degree of supervision of workers unnecessary. The question that arises is whether the Japanese-style 'paternalistic' system, under which the relation between the head of an enterprise and the rest of its members creates an atmosphere of co-operation and discipline can be found in other cultures. If this situation is highly peculiar to Japan, what kinds of institutions are required in developing countries to relax the constraint of supervision in the application of intermediate technologies? This is an area which deserves much more attention and research than it has received so far.

Supervisory costs may also be lowered if clerical or other less expensive staff could be substituted to do the job of supervisors. This brings up the question of substitution possibilities among different skills. The Chinese experience in road construction in Nepal where they used local village headmen to supervise local labour(1) demonstrates that such substitution possibilities can at times be successfully exploited.

LABOUR DISCONTINUITIES

Hirschman's hypothesis reviewed above implies that the structuration and stabilisation of the labour force in developing countries would raise the chances of efficient use of operator-based technologies in product- and process-centred industries.

One of the parameters indicating discontinuities of labour is the rate of turnover and absenteeism which may both account for low labour productivity. An attempt to reduce labour turnover is designed to raise labour productivity by stabilising the labour force and creating work discipline. Before considering the implications of these labour discontinuities for the use of low-cost technologies, some concrete examples will be examined.

Case Study Illustrations

a) It has been reported that in about three-quarters of Puerto Rican manufacturing plants, absenteeism was higher than 4 per cent, although even an occasional 20-30 per cent did not strike some managers as unusual. Strassmann notes that an "electrical manufacturer with over 400 employees found that absenteeism remained at 6 to 7 per cent in spite of new employees' contracts with threats of

---

1) See ILO, Comparative Evaluation of Road Construction Techniques in Nepal, Geneva (forthcoming).

dismissal after the second occurrence......"(1).

b) In a sample of 67 plants, Gregory and Reynolds pointed out that 63 per cent had more than 4 per cent absenteeism, and mostly over 7 per cent(2).

c) In the Indian textile industry, Myers and Kannappan note however that turnover rates are relatively low, and absenteeism rates higher. In Sudanese textiles, even turnover rates appear high.

Strassmann distinguishes between 'proficiency' and 'consistency' aspects of productivity and argues that absenteeism is not so important in lowering consistency as in avoiding breakdowns in production. One may hypothesise conditions of a simplified technology when even high turnover and absenteeism may not be very troublesome to management since replacements can be easily found or trained. On the other hand, in the early stages of industrialisation, the costs of structuring an undisciplined and inexperienced labour force are likely to be higher.

One element of these costs will be the increase in wage rates of skilled and unskilled labour offered either as an incentive to attract a stable work-force in the first place, or to induce greater stability among those who are already hired. Higher wages can be paid more easily by larger firms using modern technology than by the smaller ones. Although the latter tend to have higher turnover rates they may be less able to afford these extra wage costs. Another indirect element in the cost is the "auxiliary management costs which are directly related to the use of the factor labour and which pertain to such management functions like hiring, induction, placement and supervision"(3). For any given managerial capability, these costs will vary, depending on the quality of the hired labour and the general labour market environment.

The above costs of stabilising the work-force may be more relevant to complex operations which also involve considerable skill accumulation in the course of work. This would be less true of labour-intensive innovations. The mode of production in a given industry is relevant in determining whether these costs are justified in relation to potential benefits, and whether they deter innovations. Where production is organised partly or mainly outside the wage system,

---

1) W. Paul Strassman, Technological Change and Economic Development: The Manufacturing Experience of Mexico and Puerto Rico, Cornell University Press, Ithaca, 1968, p.76.

2) Lloyd G. Reynolds and Peter Gregory, Wages, Productivity and Industrialisation in Puerto Rico, Homewood, Illinois, 1965.

3) Subbiah Kannappan, "The Economics of Structuring an Industrial Labour Force: Some Reflections on the Commitment Problem", British Journal of Industrial Relations, November 1966.

the varying needs of production can be met by using a less stabilised work-force. The casual labour market is a source of cheap labour where wages fluctuate more freely in response to variations in supply and demand. The incentives to stabilise labour force may be slight in these circumstances.

RESISTANCE TO LABOUR DISMISSAL

The costs of management may also be higher when there are institutional pressures against laying off workers. When regulations either prohibit the dismissal of workers or enforce additional employment in industry (as in Kenya under the Tripartite Agreements), labour becomes a quasi-fixed cost to management rather than a variable cost. The result is that there may not really be an incentive on the part of the managers to introduce labour-intensive techniques or innovations.

Case Study Illustrations

a) In the Thai case study of can manufacturing, management's attitude to labour was considered as one of the factors influencing the choice of technology. One of the firms in the sample, which used high-speed automatic machines, ran into a number of labour-management problems. Given inadequate supervisory skills, line-operating labour for example had to be trained into supervisory positions. There was some apparent resentment on the part of the labour force that promotion possibilities to supervisory positions within the plant were limited. Besides, the difficulties in running a high-speed line to produce a wide range of products resulted in large machine down-time and a high propensity for machinery to "strike". The Thai case study arrives at the rather unexpected conclusion that labour management problems would have been reduced if more labour-intensive techniques had been used.

The following reasons are put forward in support of this hypothesis:

- a larger line-operating force would have meant a greater scope for potential supervision;
- a larger number of supervisory positions might have ensured greater promotion possibilities;
- and a simpler operating routine of very low change-over frequency in a more labour-intensive technique would have provided greater scope for training and utilising supervisory skills.

b) In the case of sugar processing in India, it is noted that the mechanisation of such operations as cane unloading has been avoided because of trade-union and social pressures for maintaining the present level of employment(1). There is another reason for not substituting machines for men: wages constitute only a small fraction of total costs in these operations.

c) According to a Tripartite Agreement signed in 1970 by the employers' federations, the trade-unions and the Kenyan Government, the employers were to expand employment by 10 per cent of their regular establishment(2). This was a short-term measure rather than a long-term one. Among the firms interviewed, one can distinguish between those which intended to expand in any case and those which were required to hire additional employees although they did not need them. No firm reported that its output had gone up as a result of extra employment. Many employers complained that additional workers were a pure cost since they contributed nothing to output. One firm even claimed that the administrative and supervisory cost involved in managing the additional men had a sizeable negative effect on output. None of the firms interviewed reported "any major change in method of production, organisation, or capital intensiveness of technique in order to accommodate itself to the Agreement".

These case studies do not suggest any definitive conclusion as to the effect of legislation against dismissal on management behaviour with respect to innovations, investment and production decisions. In the case of Kenya, the absence of changes in organisation or technical methods may reflect only the temporary nature of the Agreement. Even so, firms which are investing for expansion are unlikely to find additional employment costly or burdensome. At any rate, the case study on Thai can manufacturing also casts doubts on the oft-quoted hypothesis that increased labour-intensity of production (often implied by "low-cost" technology) raises labour management problems. It may actually be a positive influence.

CONCLUDING REMARKS

In this study we have attempted to show how the cost of labour management, labour discontinuities, and social resistance to labour dismissal vary according to different technological levels. Emphasis has been laid particularly on the nature of modes of production and employment, and on the particular types of industrialisation strategies in the light of which different technologies and their application need to be considered.

---

1) See C.G. Baron, "Sugar Processing Techniques in India", in A.S. Bhalla (ed.), Technology and Employment in Industry, op.cit.

2) See ILO, Employment, Incomes and Equality, op.cit., Technical Paper No.26.

Our tentative conclusion is that in the present state of knowledge there is no definitive evidence in favour of the argument that low-cost technologies are necessarily management- and supervision-intensive. The influence of economic factors is not yet clear-cut: much less is the influence of cultural and other non-economic factors. Here is an area where further research is badly needed.

OECD SALES AGENTS
DEPOSITAIRES DES PUBLICATIONS DE L'OCDE

ARGENTINA – ARGENTINE
Carlos Hirsch S.R.L.,
Florida 165, BUENOS-AIRES.
☎ 33-1787-2391 Y 30-7122

AUSTRALIA – AUSTRALIE
International B.C.N. Library Suppliers Pty Ltd.,
161 Sturt St., South MELBOURNE, Vic. 3205.
☎ 69.7601
658 Pittwater Road, BROOKVALE NSW 2100.
☎ 938 2267

AUSTRIA – AUTRICHE
Gerold and Co., Graben 31, WIEN 1.
☎ 52.22.35

BELGIUM – BELGIQUE
Librairie des Sciences
Coudenberg 76-78, B 1000 BRUXELLES 1.
☎ 512-05-60

BRAZIL – BRESIL
Mestre Jou S.A., Rua Guaipá 518,
Caixa Postal 24090, 05089 SAO PAULO 10.
☎ 256-2746/262-1609
Rua Senador Dantas 19 s/205-6, RIO DE JANEIRO GB. ☎ 232-07. 32

CANADA
Information Canada
171 Slater, OTTAWA. KIA 0S9.
☎ (613) 992-9738

DENMARK – DANEMARK
Munksgaards Boghandel
Nørregade 6, 1165 KØBENHAVN K.
☎ (01) 12 69 70

FINLAND – FINLANDE
Akateeminen Kirjakauppa
Keskuskatu 1, 00100 HELSINKI 10. ☎ 625.901

FRANCE
Bureau des Publications de l'OCDE
2 rue André-Pascal, 75775 PARIS CEDEX 16.
☎ 524.81.67
Principaux correspondants :
13602 AIX-EN-PROVENCE : Librairie de l'Université. ☎ 26.18.08
38000 GRENOBLE : B. Arthaud. ☎ 87.25.11
31000 TOULOUSE : Privat. ☎ 21.09.26

GERMANY – ALLEMAGNE
Verlag Weltarchiv G.m.b.H.
D 2000 HAMBURG 36, Neuer Jungfernstieg 21
☎ 040-35-62-500

GREECE – GRECE
Librairie Kauffmann, 28 rue du Stade,
ATHENES 132. ☎ 322.21.60

HONG-KONG
Government Information Services,
Sales of Publications Office,
1A Garden Road,
☎ H-252281-4

ICELAND – ISLANDE
Snaebjörn Jónsson and Co., h.f.,
Hafnarstræti 4 and 9, P.O.B. 1131,
REYKJAVIK. ☎ 13133/14281/11936

INDIA – INDE
Oxford Book and Stationery Co. :
NEW DELHI, Scindia House. ☎ 47388
CALCUTTA, 17 Park Street. ☎ 24083

IRELAND – IRLANDE
Eason and Son, 40 Lower O'Connell Street,
P.O.B. 42, DUBLIN 1. ☎ 74 39 35

ISRAEL
Emanuel Brown :
35 Allenby Road, TEL AVIV. ☎ 51049/54082
also at :
9, Shlomzion Hamalka Street, JERUSALEM.
☎ 234807
48 Nahlath Benjamin Street, TEL AVIV.
☎ 53276

ITALY – ITALIE
Libreria Commissionaria Sansoni :
Via Lamarmora 45, 50121 FIRENZE. ☎ 579751
Via Bartolini 29, 20155 MILANO. ☎ 365083
Sous-dépositaires :
Editrice e Libreria Herder,
Piazza Montecitorio 120, 00186 ROMA.
☎ 674628
Libreria Hoepli, Via Hoepli 5, 20121 MILANO.
☎ 865446
Libreria Lattes, Via Garibaldi 3, 10122 TORINO.
☎ 519274
La diffusione delle edizioni OCDE è inoltre assicurata dalle migliori librerie nelle città più importanti.

JAPAN – JAPON
OECD Publications Centre,
Akasaka Park Building,
2-3-4 Akasaka,
Minato-ku
TOKYO 107. ☎ 586-2016
Maruzen Company Ltd.,
6 Tori-Nichome Nihonbashi, TOKYO 103,
P.O.B. 5050, Tokyo International 100-31.
☎ 272-7211

LEBANON – LIBAN
Documenta Scientifica/Redico
Edison Building, Bliss Street,
P.O.Box 5641, BEIRUT. ☎ 354429 – 344425

THE NETHERLANDS – PAYS-BAS
W.P. Van Stockum
Buitenhof 36, DEN HAAG. ☎ 070-65.68.08

NEW ZEALAND – NOUVELLE-ZELANDE
The Publications Manager
Government Printing Office
Mulgrave Street (Private Bag)
WELLINGTON, ☎ 737-320
and Government Bookshops at
AUCKLAND (P.O.B. 5344). ☎ 32.919
CHRISTCHURCH (P.O.B. 1721). ☎ 50.331
HAMILTON (P.O.B. 857). ☎ 80.103
DUNEDIN (P.O.B. 1104). ☎ 78.294

NORWAY – NORVEGE
Johan Grundt Tanums Bokhandel,
Karl Johansgate 41/43, OSLO 1. ☎ 02-332980

PAKISTAN
Mirza Book Agency, 65 Shahrah Quaid-E-Azam,
LAHORE 3. ☎ 66839

PHILIPPINES
R.M. Garcia Publishing House,
903 Quezon Blvd. Ext., QUEZON CITY,
P.O. Box 1860 – MANILA. ☎ 99.98.47

PORTUGAL
Livraria Portugal,
Rua do Carmo 70-74. LISBOA 2. ☎ 360582/3

SPAIN – ESPAGNE
Libreria Mundi Prensa
Castelló 37, MADRID-1. ☎ 275.46.55
Libreria Bastinos
Pelayo, 52, BARCELONA 1. ☎ 222.06.00

SWEDEN – SUEDE
Fritzes Kungl. Hovbokhandel,
Fredsgatan 2, 11152 STOCKHOLM 16.
☎ 08/23 89 00

SWITZERLAND – SUISSE
Librairie Payot, 6 rue Grenus, 1211 GENEVE 11.
☎ 022-31.89.50

TAIWAN
Books and Scientific Supplies Services, Ltd.
P.O.B. 83, TAIPEI.

TURKEY – TURQUIE
Librairie Hachette,
469 Istiklal Caddesi,
Beyoglu, ISTANBUL, ☎ 44.94.70
et 14 E Ziya Gökalp Caddesi
ANKARA. ☎ 12.10.80

UNITED KINGDOM – ROYAUME-UNI
H.M. Stationery Office, P.O.B. 569, LONDON
SE1 9 NH, ☎ 01-928-6977, Ext. 410
or
49 High Holborn
LONDON WC1V 6HB (personal callers)
Branches at: EDINBURGH, BIRMINGHAM, BRISTOL, MANCHESTER, CARDIFF, BELFAST.

UNITED STATES OF AMERICA
OECD Publications Center, Suite 1207,
1750 Pennsylvania Ave, N.W.
WASHINGTON, D.C. 20006. ☎ (202)298-8755

VENEZUELA
Libreria del Este, Avda. F. Miranda 52,
Edificio Galipán, Aptdo. 60 337, CARACAS 106.
☎ 32 23 01/33 26 04/33 24 73

YUGOSLAVIA – YOUGOSLAVIE
Jugoslovenska Knjiga, Terazije 27, P.O.B. 36,
BEOGRAD. ☎ 621-992

Les commandes provenant de pays où l'OCDE n'a pas encore désigné de dépositaire peuvent être adressées à :
OCDE, Bureau des Publications, 2 rue André-Pascal, 75775 Paris CEDEX 16
Orders and inquiries from countries where sales agents have not yet been appointed may be sent to
OECD, Publications Office, 2 rue André-Pascal, 75775 Paris CEDEX 16

OECD PUBLICATIONS, 2, rue André-Pascal, 75775 Paris Cedex 16 - No. 35729 1976

PRINTED IN FRANCE